T0250379

Wireless Ad Hoc and Sensor Networks

Management, Performance, and Applications

OTHER COMMUNICATIONS BOOKS FROM AUERBACH

Wireless Ad Hoc and Sensor Networks

Management, Performance, and Applications

Jing (Selena) He
Shouling Ji
Yi Pan
Yingshu Li

CRC Press
Taylor & Francis Group
Boca Raton London New York

CRC Press is an imprint of the
Taylor & Francis Group, an **Informa** business

CRC Press
Taylor & Francis Group
6000 Broken Sound Parkway NW, Suite 300
Boca Raton, FL 33487-2742

© 2014 by Taylor & Francis Group, LLC
CRC Press is an imprint of Taylor & Francis Group, an Informa business

No claim to original U.S. Government works

Version Date: 20130624

International Standard Book Number-13: 978-1-4665-5694-2 (Hardback)

Visit the Taylor & Francis Web site at
http://www.taylorandfrancis.com

and the CRC Press Web site at
http://www.crcpress.com

Dedicated to our beloved.

Contents

SECTION III: APPLICATIONS   175

8   Reliable and Energy Efficient Target Coverage for Wireless Sensor Networks   177

9   CDS-based Multi-regional Query Processing in Wireless Sensor Networks   195

List of Figures

List of Tables

Preface

Wireless sensor networks are being employed in a variety of applications ranging from medical to military, and from home to industry. There is no recent book emphasizing the algorithm description, performance analysis, and applications of the network management techniques in wireless sensor networks. The principal aim of this book is to provide a reference tool for the increasing number of scientists who depend upon sensor networks in some way. The book summarizes not only the traditional and classical network management techniques, but also the state-of-the-art techniques in this area. The book is organized into three sections, each including chapters exploring a specific topic.

The articles presented in the book are expository, but of a scholarly nature, including the appropriate history background, a review of the state-of-the-art thinking relative to the topic, as well as a discussion of unsolved problems that are of special interest.

The target readers of this book include the researchers in computer science, computer engineering, and applied mathematics, as well as students in these subjects. Specialists as well as general readers will find the articles stimulating and helpful.

Book Organization

The book is organized into three sections. Section I introduces the basic concepts of wireless sensor networks (WSNs) and their applications, followed by the summarization of the network management techniques used in WSNs.

Section II reviews the literature of the virtual backbone-based network management techniques. Because there are some drawbacks to the classical and existing methods, we propose several new network management techniques for WSNs and analyze their performance subsequently. Each new proposed network management technique is shown in one chapter with the following structure: introduction, literature review, network model, algorithm description, theoretical analysis, and conclusion.

Section III applies proposed new techniques to some important applications in WSNs, such as, routing, data collection, data aggregation, query processing, etc. We also conduct simulations to verify the performance of the proposed techniques. Each application is shown in one chapter with the following structure: brief application overview, application design and implementation, performance analysis, simulation settings, and comments for different test cases/scenario configurations. These approaches are much more intuitive than a lecture giving the readers almost hands-on experiences of the various network management issues.

Acknowledgments

We would like to acknowledge all the contributing authors, who are experts in sensor networks and have done extensive work in this area. We also appreciate the help of the referees who have devoted much time in reviewing the submissions. We are also grateful to the publisher who made this book possible.

Authors

Dr. Jing (Selena) He is currently an assistant professor in Department of Computer Science at Kennesaw State University. She received her BS of Electric Engineering from Wuhan Institute of Technology and MS of Computer Science from Utah State University, respectively. Her research interests include wireless ad hoc networks, wireless sensor networks, cyber-physical systems, social networks, and cloud computing. She is now an IEEE member and an IEEE COMSOC member.

Shouling Ji is currently a PhD student in the Department of Computer Science at Georgia State University. He received his BS and MS in computer science from the School of Computer Science and Technology at Heilongjiang University, China, in 2007 and 2010, respectively. His research interests include wireless sensor networks, data management in wireless networks, cognitive radio networks, and cyber-physical systems. He is now an ACM student member, an IEEE student member, and an IEEE COMSOC student member.

Dr. Yi Pan is the chair and a professor in the Department of Computer Science and a professor in the Department of Computer Information Systems at Georgia State University. Dr. Pan received his BEng and MEng in computer engineering from Tsinghua University, China, in 1982 and 1984, respectively, and his PhD in computer science from the University of Pittsburgh, in 1991. Dr. Pan's research interests include parallel and distributed computing, optical networks, wireless networks, and bioinformatics. Dr. Pan has published more than 100 journal papers with about 50 papers published in various IEEE/ACM journals. He is a co-inventor of three US patents (pending) and 5 provisional patents, and has received many awards from agencies such as NSF, AFOSR, JSPS, IISF and the Mellon Foundation. Dr. Pan has served as an editor-in-chief or editorial board member for 15 journals including 6 IEEE *Transactions* and a guest editor for 10 special issues for 9 journals including 2 IEEE *Transactions*.

Dr. Yingshu Li received her PhD and MS from the Department of Computer Science and Engineering at University of Minnesota-Twin Cities. She received her BS from the Department of Computer Science and Engineering at Beijing Institute of Technology, China. Dr. Li is currently an associate professor in the Department of Computer Science at Georgia State University. Her research interests include wireless networking, cyber-physical systems, and phylogenetic analyses. Her research has been supported by the National Science Foundation (NSF) of the US, the National Science Foundation of China (NSFC), the Electronics and Telecommunications Research Institute (ETRI) of South Korea, and GSU internal grants. Dr. Li is the recipient of an NSF CAREER Award.

Contributors

Raheem R. Beyah
Georgia Institute of Technology
Atlanta, Georgia

Zhiepeng Cai
Georgia State University
Atlanta, Georgia

Mingyuan Yan
Georgia State University
Atlanta, Georgia

BACKGROUND

I

Chapter 1

Introduction

CONTENTS

1.1 Wireless Sensor Networks

1.1.1 Basic Idea

Wireless sensor networks (WSNs), consisting of small nodes with sensing, computation, and wireless communications capabilities, are now widely used in many applications, including environment and habitat monitoring, traffic control, etc. Although sensor networks share many common aspects with generic ad hoc networks, several important constraints in WSNs introduce a number of research challenges. First, due to the relatively large number of sensor nodes, it is impossible to build a global addressing scheme for the deployment of a large number of sensor nodes as the overhead of ID maintenance is high. Thus, traditional IP-based protocols may not be applied to WSNs. Second, sensor nodes are tightly constrained in terms of energy, processing, and storage capacities. Thus, they require careful resource management. Thirdly, the requirements regarding dependability and quality of services (QoS) are quite different. In ad hoc networks, each individual node should be fairly reliable, while in WSNs, an individual node is next to irrelevant. The QoS issues in an ad hoc network are dictated by traditional applications, while for WSNs, entirely new QoS concepts are required, which also take energy explicitly into account. Fourth, redundant deployment will make data-centric protocols attractive in WSNs. Finally, although position awareness of sensor nodes is important, it is not feasible to use global positioning system (GPS) hardware for this purpose. GPS can only be used outdoors and without the presence of any obstruction. Moreover, GPS receivers are expensive and unsuitable for the construction of small cheap sensor nodes. In summary, there are commonalities, but the fact that WSNs have to support very different applications, that they have to interact with the physical environment, and that they have to carefully adjudicate various trade-offs justifies considering WSNs as a system concept distinct from ad hoc networks.

1.1.2 *Deterministic Wireless Sensor Networks and Probabilistic Wireless Sensor Networks*

WSNs are usually modeled using the deterministic network model (DNM) in recent literature [1]. Under this model, there is a transmission radius of each node. According to this radius, specific pairs of nodes are always connected to be neighbors if their physical distance is less than this radius, while the rest of the pairs are always disconnected. The unit disk graph (UDG) model is a special case of the DNM model if all nodes have the same transmission radius. When all nodes are connected to each other, via a single-hop or multi-hop path, the WSN is said to have full connectivity. In most real applications, however, the DNM model cannot fully characterize the behavior of wireless links. This is mainly due to the transitional region phenomenon which has been revealed by many empirical studies [2][3][4][5]. Beyond the "always connected" region, there is a transitional region where a pair of nodes is probabilistically connected. Such pairs of nodes are not fully connected but reachable via the so-called lossy links [5]. As reported in [5], there are often many more lossy links than fully connected links in a WSN. Additionally, in a specific setup [6], more than 90% of the network links are lossy links. Therefore, their impact can hardly be neglected.

The employment of lossy links in WSNs is not straightforward, since when the lossy links are employed, the WSN may have no guarantee of full network connectivity. When data transmissions are conducted over such topologies, they may degrade the node-to-node delivery ratio. Usually a WSN has large node density and high data redundancy, thus this certain degraded performance may be acceptable for many WSN applications. Therefore, as long as an expected percentage of the nodes can be reached, that is the node-to-node delivery ratio satisfies some preset requirement, lossy links are tolerable in a WSN. In other words, full network connectivity is not always a necessity. Some applications can trade full network connectivity for a higher energy efficiency and larger network capacity [6].

1.2 Topology Control in Wireless Sensor Networks

1.2.1 Motivation

One perhaps typical characteristic of wireless sensor networks is the possibility of deploying many nodes in a small area, for example, to ensure sufficient coverage of an area or to have redundancy present in the network to protect against node failures. These are clear advantages of a dense network deployment; however there are also disadvantages. In a relatively crowded network, many typical wireless networking problems are aggravated by the large number of neighbors: many nodes interfere with each other, there are a lot of possible routes, nodes might needlessly use large transmission power to talk to distant nodes directly (also limiting the reuse of wireless bandwidth), and routing protocols might have to recompute routes even if only small node movements have happened.

Some of these problems can be overcome by topology-control techniques. Instead of using the possible connectivity of a network to its maximum possible extent, a deliberate choice is made to restrict the topology of the network. The topology of a network is determined by the subset of active nodes and the set of active links along which direct communication can occur. Formally speaking, a topology-control algorithm takes a graph $G = (V, E)$ representing the network, where V is the set of all nodes in the network and there is an edge $(v_1, v_2) \in E \subseteq V^2$ if and only if nodes v_1 and v_2 can directly communicate with each other. Hence all active nodes form an induced graph $T = (V_T, E_T)$ such that $V_T \subseteq V$ and $E_T \subseteq E$.

1.2.2 Options for Topology Control

To compute an induced graph T out of a graph G representing the original network G, a topology control algorithm has a few options:

- The set of active nodes can be reduced ($V_T \subset V$), for example, by periodically switching off nodes with low energy reserves and activating other nodes instead, exploiting redundant deployment in doing so.

- The set of active links/the set of neighbors for a node can be controlled. Instead of using all links in the network, some links can be disregarded and communication is restricted to crucial links. When a flat network topology (all nodes are considered equal) is desired, the set of neighbors of a node can be reduced by simply not communicating with some neighbors. There are several possible approaches to choose neighbors, but one that is obviously promising for a WSN is to limit the reach of a node's transmissions typically by power control, but also by using adaptive modulations (using faster modulations is only possible over shorter distances) and using the improved energy efficiency when communicating only with nearby neighbors. In essence, power control attempts to optimize the trade-off between the higher likelihood of finding a (useful) receiver at higher power values on the one hand and the increased chance of collisions/interference/reduced spatial reuse on the other hand.

■ Active links/neighbors can also be rearranged in a hierarchical network topology where some nodes assume special roles. One example, illustrated in Figure 1.1, is to select some nodes as a Virtual Backbone (VB) for the network and to only use the links within this backbone and direct links from other nodes to the backbone. To do so, the backbone has to form a **dominating set**(DS): a subset $D \subset V$ such that all nodes in V are either in D itself or are one-hop neighbors of some node $d \in D$ ($\forall v \in V : v \in D \lor \exists d \in D : (v,d) \in E$). Then, only the links between nodes of the dominating set or between other nodes and a member of the active set are maintained. For a backbone to be useful, it should be connected. A related, but slightly different, idea is to partition the network into clusters, illustrated in Figure 1.2. Clusters are subsets of nodes that together include all nodes of the original graph such that, for each cluster, certain conditions hold (details vary). The most typical problem formulation is to find clusters with cluster heads, which is a representative of a cluster such that each node is only one hop away from its cluster head. When the (average) number of nodes in a cluster should be minimized, this is equivalent to finding a maximum (dominating) independent set (a subset $C \subset V$ such that $\forall v \in V - C : \exists c \in C : (v,c) \in E$ and no two nodes in C are joined by an edge in E, $\forall c_1, c_2 \in C : (c_1, c_2) \notin E$). In such a clustered network, only links within a cluster are maintained (typically only those involving the cluster head) as also selected links between clusters to ensure connectivity of the whole network. Both problems are intrinsically hard and various approximations and relaxations have been studied.

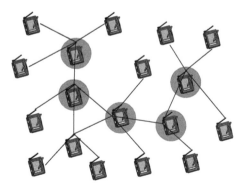

Figure 1.1: Restricting the topology by using dominating sets.

In conclusion, there are three main options for topology control: flat networks with a special attention to power control on the one hand, hierarchical networks with backbones or clusters on the other hand.

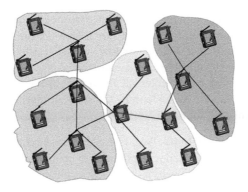

Figure 1.2: Using clusters to partition a graph.

1.2.3 Measurements of Topology Control Algorithms

There are a few basic metrics to judge the efficacy and quality of a topology-control algorithm [7]:

■ **Connectivity** Topology control should not disconnect a connected graph *G*. In other words, if there is a multihop path in *G* between two nodes *u* and *v*, there should also be some such path in *T* (clearly, it does not have to be the same path).

■ **Stretch factors** Removing links from a graph will likely increase the length of a path between any two nodes *u* and *v*. The hop stretch factor is defined as the worst increase in path length for any pair of nodes *u* and *v* between the original graph G and the topology-controlled path *T* . Formally,

$$hop\ stretch\ factor = \max_{u,v \in V} \frac{|(u,v)_T|}{|(u,v)_G|} \qquad (1.1)$$

where $(u,v)_G$ is the shortest path in graph *G* and $|(u,v)|$ is its length.

Similarly, the energy stretch factor can be defined:

$$hop\ stretch\ factor = \max_{u,v \in V} \frac{E_T(u,v)}{E_G(u,v)} \qquad (1.2)$$

where $E_G(u,v)$ is the energy consumed along the most energy-efficient path in graph *G*. Clearly, topology-control algorithms with small stretch factors are desirable. It particular, stretch factors in $O(1)$ can be advantageous.

■ **Graph metrics** The intuitive examples above already indicated the importance of a small number of edges in *T* and a low maximum degree (number of neighbors) for each node.

■ **Throughput** The reduced network topology should be able to sustain a comparable amount of traffic as the original network (this can be important even in wireless sensor networks with low average traffic, in particular, in case of event showers). One metric to capture this aspect is throughput competitiveness (the largest $\varphi \leq 1$ such that, given a set of flows from node s_i to node d_i with rate r_i that are routable in G, the set with rates φr_i can be routed in T), see reference [7] for details.

■ **Robustness to mobility** When neighborhood relationships change in the original graph other nodes might have to change their topology information (for example, to reactivate links). Clearly, a robust topology should only require a small number of such adaptations and avoid having the effects of a reorganization of a local node movement ripple through the entire network.

■ **Algorithm overhead** It almost goes without saying that the overhead imposed by the algorithm itself should be small (low number of additional messages, low computational overhead). Also, distributed implementation is practically a condition.

In the present context of WSNs, connectivity and stretch factors are perhaps the most important characteristics of a topology-control algorithm, apart from the indispensable distributed nature and low overhead.

MANAGEMENT AND PERFORMANCE

Chapter 2

Greedy-based Construction of Load-balanced Virtual Backbones in Wireless Sensor Networks

CONTENTS

2.1 Introduction

Wireless sensor networks (WSNs) are deployed for monitoring and controlling of systems where human intervention is not desirable or feasible. One typical characteristic of WSNs is the possibility of deploying many nodes in an area to ensure sufficient coverage of an area or/and to have redundancy against node failures. However, in a relatively crowded network, many problems are aggravated: 1) Many nodes interfere with each other. 2) There are a lot of possible routes. 3) Nodes might needlessly use large transmission power to talk to distant nodes directly. 4) Routing protocols might have to recompute routes even if only a small number of nodes changed their locations. These problems can be overcomed by selecting some nodes as a *virtual backbone* (VB) for a network, in which only the links within this backbone and direct links from other nodes to the backbone nodes are mainly used in the WSN. Usually, we use a *dominating set* (DS) to serve as a backbone for a WSN, which is a subset of nodes in the network where every node is either in the subset or a neighbor of at least one node in the subset. For a backbone to be useful, it should be connected, namely by a, *connected dominating set* (CDS). The nodes in a CDS are called *dominators*, otherwise, *dominatees*. In a WSN with a CDS as its VB, dominatees only forward their data to their connected dominators. Moreover, the CDS with the smallest size (the number of nodes in the CDS) is called a *minimum-sized connected dominating Set* (MCDS).

Since only dominators need to maintain the routing information, CDS-based routing becomes more easier and more qualified to accommodate topology change. To be specific, the search space for the available routes is reduced within a CDS. Moreover, if there is no topology change in the subgraph induced by the CDS, there is no need to update the routing information at all. In addition to routing [8][9], a CDS has many other applications in WSNs, such as data collection [10][11], broadcasting [12][13], topology control [14], coverage [15][16], data aggregation [17], and query scheduling [18]. Clearly, the benefits of a CDS can be magnified by making its size smaller. Therefore, it is desirable to build an MCDS to reduce the number of nodes and links involved in communication. As a matter of fact, constructing a CDS, especially an MCDS for WSNs is one way to extend network lifetime.

Ever since the idea of employing a CDS for WSNs was introduced in [19], a huge amount of effort has been made to find CDSs with variety of features for different applications, especially the MCDS. In the seminal work [20], Guha and Kuller first modeled the problem of computing the smallest CDS as the MCDS problem in a general graph, which is a well-know, NP-hard problem [21]. After that, to make a CDS more resilient in mobile WSNs, the fault tolerance of a VB is considered. In [22][23][24], k-connected and m-dominated sets are introduced as a generalized abstraction of a fault-tolerance VB. In [25], the authors proposed a minimum routing cost connected dominating set (MOC-CDS), which aims to find a minimum CDS while assuring that any routing path through this CDS is the shortest in WSNs. Additionally, the authors investigate the problem of constructing a quality CDS in terms of size, diameter, and average backbone path length (ABPL) in [26][27].

Unfortunately, to the best of our knowledge, all the aforementioned works did not consider the *load-balance* factor when they construct a CDS. If the workload on each dominator in a CDS is not balanced, some heavy-duty dominators deplete their energy quickly. Then, the whole network might be disconnected. Hence, intuitively, we not only have to consider to construct an MCDS, but also need to consider to construct a load-balanced CDS (LBCDS). An illustration of an LBCDS is depicted in Figure 2.1, in which dominators are marked as black nodes, while white nodes represent dominatees. Moreover, the number besides each node represents the node's degree. In Figure 2.1(b) and Figure 2.1(c), solid lines represent that the dominatees are allocated to the connected dominators, while the dashed lines represent the communication links in the original graph shown in Figure 2.1(a). According to the traditional MCDS construction algorithms, a CDS $\{s_4, s_7\}$ with size 2 is obtained for the network shown in Figure 2.1(a). However, There are two severe drawbacks of the CDS shown in Figure 2.1(a). For convenience, the set of neighboring dominatees of a dominator s_i is denoted by $ND(s_i)$. First, $ND(s_4) = \{s_1, s_2, s_3, s_5, s_6\}$, which represents that dominator s_4 connects to five different dominatees, and $ND(s_7) = \{s_6, s_8\}$. If every dominatee has the same amount of data to be transferred through the connected dominator at a fixed data rate, dominator s_4 must deplete its energy much faster than dominator s_7, since dominator s_4 has to forward the data collected from five connected dominatees. Second, dominatee s_6 connects to both dominators. If s_6 chooses dominator s_4 as its data forwarder, obviously, only one dominatee s_8 can forward its data to dominator s_7. In this situation, the workload imbalance in the CDS is further amplified. Consequently, the entire network lifetime is shortened. We show a counterexample in Figure 2.1(b), where the constructed CDS is $\{s_3, s_6, s_7\}$. According to the topology shown in Figure 2.1(b), we can get the dominatee sets of each dominator: $ND(s_3) = \{s_1, s_2, s_4\}$, $ND(s_6) = \{s_4, s_5\}$, and $ND(s_7) = \{s_4, s_8\}$. Compared with the MCDS constructed in Figure 2.1(a), the numbers of dominatees of all the dominators in Figure 2.1(b) are very similar. For convenience, we use $A(s_i) = \{s_j | s_j$ is a dominatee and s_j forward its data to $s_i\}$ to represent the dominatees allocated to a dominator

s_i. Thus, we can have two different dominatee allocation schemes shown in Figure 2.1(b) and Figure 2.1(c) respectively. One is: $A(s_3) = \{s_1, s_2, s_4\}$, $A(s_6) = \{s_5\}$, and $A(s_7) = \{s_8\}$. The other one is: $A(s_3) = \{s_1, s_2\}$, $A(s_6) = \{s_4, s_5\}$, and $A(s_7) = \{s_8\}$. Apparently, the workload on each dominator is almost evenly distributed in the CDS constructed in Figure 2.1(c). Intuitively, the construction algorithm and dominatee allocation scheme shown in Figure 2.1(c) can extend network lifetime notably, since the traffic loads on each dominator shown in Figure 2.1(c) are almost evenly distributed.

To benefit from the CDS-based VB in WSNs and also take the load-balance factor into consideration, few attempts have been carried out to construct a VB in this manner [28]. In our previous work [28], we proposed a genetic algorithm-based method to build a load-balanced CDS (LBCDS) in WSNs. However, there is no performance ratio analysis in that work. In this research, we first investigate how to construct an LBCDS. When we build the LBCDS, we consider the degree of each dominator as the indicator of potential future workload. Taking the degree of each dominator into consideration, we use the *p-norm* to measure how balanced the LBCDS can make a workload. The details are introduced in Section 2.4. After constructing an LBCDS, we explore how to load-balancedly allocate dominatees to dominators (LBAD). We propose a novel probability-based algorithm to solve this problem. The detailed design, algorithm description and theoretical analysis are presented in Section 2.5.

Particularly, our contributions mainly include four aspects as follows:

(i) We point out the disadvantages and advantages of the existing MCDS construction algorithms. Based on the observation of workload imbalance in MCDS, we propose to take the load-balance factor into consideration when building a CDS and allocating dominatees.

(ii) For the LBCDS problem, we design a greedy algorithm based on dominators' degree values. Furthermore, we analyze that our proposed algorithm can find a more load-balanced CDS than the existing MCDS algorithms.

(iii) For the LBAD problem, rather than allocating a dominatee to a dominator in a naive way [17], we introduce a new term, *expected allocation probability* (EAP), which represents, for any connected dominatee and dominator pair, the expected probability that the dominatee is allocated to the dominator. Based on the *EAP* value, we formulate the LBAD problem into a constrained non-linear programming optimization problem, by solving the non-linear programming equations, an optimal solution can be found. Subsequently, we propose a probability-based distributed algorithm to dynamically allocate dominatees to dominators. Furthermore, we theoretically analyze the upper bound and lower bound performance ratios of the distributed algorithm.

(iv) We also conduct extensive simulations to validate our proposed algorithms. The simulation results show that the constructed LBCDS and the load-balanced dominatee allocation scheme extend network lifetime significantly compared with the existing algorithms. Particularly, when the node number changes from 200 to 450, our proposed methods extend network lifetime by

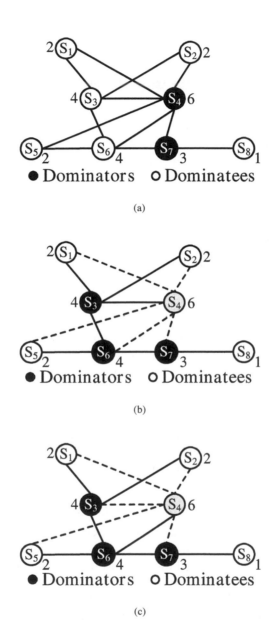

(a)

(b)

(c)

Figure 2.1: Illustration of regular CDS and LBCDS.

80% compared with the work in [17], which is the most recently published CDS construction algorithm.

The rest of this chapter is organized as follows: in Section 2.2.4, we review some related works on CDSs. In Section 2.3.1, we introduce the network model and formally define the LBCDS problem and the LBAD problem respectively. In Section 2.4, how to build an LBCDS is presented through an example and a greedy algorithm is shown. In Section 2.5, we first illustrate the LBAD problem using an example, and then propose a centralized constrained non-linear programming algorithm and a probability-based distributed algorithm to load-balancedly allocate dominatees. Subsequently, we give some theoretical analysis of our proposed algorithm. The simulation results are presented in Section 2.6 to validate our proposed algorithms. Finally, the chapter is concluded in Section 2.7.

2.2 Related Work

2.2.1 Centralized Algorithms for CDS

Constructing an MCDS for a WSN is a well-known NP-hard problem even under the *unit disk graph* (UDG) [29] model. Many research works have been devoted to achieve a better performance ratio. For centralized CDS algorithms, they generally can be categorized into two types: 1-stage algorithms and 2-stage algorithms. 1-stage algorithms [30] aim to construct a CDS directly. In contrast, 2-stage algorithms construct a CDS in two stages. The first stage is to select a minimum DS and the second stage is to construct a CDS using the technique of Steiner tree [31]. In [30], two centralized greedy algorithms were proposed. The first algorithm is a 1-stage strategy with an approximation ratio of $2H(\Delta) + 2$ where Δ is the maximum node degree in a network and $H(\cdot)$ is a harmonic function. The second strategy is a 2-stage strategy with an approximation ratio of $H(\Delta) + 2$. Later, based on the main idea of [30], Ruan et al. [32] proposed a 1-stage algorithm with an approximation ratio of $3 + \ln(\Delta)$.

2.2.2 Distributed Algorithms for CDS

Due to the instability of network topology in WSNs, it is necessary to update topology information periodically. However, to deal with the network scalability and dynamical changes in a centralized way, the cost is extremely high. Therefore, many distributed algorithms are proposed.

These distributed algorithms can be classified into two categories: subtraction-based [33][34], and addition-based [17][35][36]. The subtraction-based algorithms begin with the set of all the nodes in a network, then remove some nodes according to pre-defined rules to obtain a CDS. The best known algorithms in this category include Wu [33] and Dai's [34] algorithms.

The addition-based CDS algorithms start from a subset of nodes (usually disconnected), then include additional nodes to form a CDS. Depending on the type of the initial subset, the addition-based CDS algorithms can be further divided into MIS-based [17] and tree-based [35][36]. The MIS-based CDS algorithm [17] obtains a CDS by selecting a maximal independent set (MIS) first (which is also a DS actually), and then finding connectors between the MIS nodes. For convenience, we use *opt* to denote the size of any optimal MCDS. For UDGs, Wan et al.'s [37] approach guarantees that the approximation factor on the size of a CDS is at most $4\ opt + 1$. Later, many attempts tried to improve this approximation factor on the size of a CDS based on this idea. Wu et al. reported an approximation factor of $3.8\ opt + 1.2$ [38]. Yao et al. improved it to $3.67\ opt + 1.33$ [39]. The factor is further improved to $3.478\ opt + 4.874$ [40] and $3.4306\ opt + 4.8185$ [41], subsequently. The current best is that the size of an MIS is at most $3.399\ opt + 4.874$ [42].

The tree-based CDS algorithms, for instance, single-initiator (SI) version in [35] and multi-initiator (MI) version in [36], start from a subset of nodes called initiators, and grow a dominator tree from each of the initiators. Particularly, there are three phases in the tree-based algorithms. In phase one, a number of initiators are elected. In phase two, each initiator utilizes a timer to grow a tree so that the nodes with more neighbors can be added to the tree. In phase three, additional bridge nodes are added to connect neighboring trees. It has been shown in [36] that the addition-based algorithms generally produce smaller CDSs than the subtraction-based algorithms. Moreover, the tree-based algorithms incur less communication overhead.

2.2.3   Other Algorithms for CDSs

Because a CDS can benefit a lot to WSNs, a variety of other factors are considered when constructing a CDS. There are more than one CDS can be found for each WSN. To conserve energy, all CDSs are constructed and each CDS serves as the virtual backbone duty cycled in [43]. For the sake of fault tolerance, k-connect m-dominating sets [44] are constructed, where k-connectivity means between any pair of backbone nodes there exists at least k independent paths, and m-dominating represents that every dominatee has at least m adjacent dominator neighbors. To minimize delivery delay, a special CDS problem — minimum routing cost CDS (MOC-CDS) [25] is proposed, where each pair of nodes in MOC-CDS has the shortest path. The work [26] considers more than one factor — size, diameter, and average backbone path length (ABPL) [26] in order to construct a CDS with better quality.

2.2.4   Other Load-balancing Related Work

In this subsection, we summarize the recent work to improve the load balancing of other applications in WSNs.

The authors in [45] proposed an even energy dissipation protocol (EEDP) for efficient cluster-based data gathering in wireless sensor networks. In EEDP, sensor data are forwarded to the base station (BS) via multiple chains of cluster heads. Each chain uses a rotation scheme to balance energy consumption among cluster heads and avoid the formation of a hot spot. Achieving efficient bandwidth utilization in multichannel sensor networks is a challenging research problem. In [46], the authors presented a cognitive load balance algorithm for single-hop multi-channel sensor networks. Based on the load distribution of all base stations, the proposed algorithm dynamically alternated the communication channels. As a result, the extra load from over-loaded channels is directed to under-loaded channels with a computed switch probability. Moreover, the authors also proved that a high throughput can be achieved if the load is balanced. In order to balance power usage in heterogeneous sensor networks, the load balancing group clustering (LBGC) strategy is proposed in [47] based on the clustering model. The LBGC protocol periodically selects cluster heads and implements dynamic route calculation according to the condition of energy distributing of network, which could make full use of the heterogeneous energy to realize load balance. The authors in [48] proposed a novel approach for load balance using compressive sensing. The proposed approach offered accurate recovery of sampled data from a small number compressed data. Simulation results show that the proposed method achieves a good performance in terms of energy balancing among all nodes in a network, which outperforms the tree-based collection protocols. In [49], a three-layer framework was proposed for mobile data collection in wireless sensor networks, which includes the sensor layer, cluster head layer, and mobile collector. The framework employs distributed load balanced clustering and MIMO uploading techniques. The objective is to achieve good scalability, long network lifetime and low data collection latency.

2.2.5   Remarks

All the above mentioned existing works consider to construct an MCDS, a k-connect m-dominating CDS, a minimum routing cost CDS or a bounded-diameter CDS. Unfortunately, they do not consider the load-balance factor at all when constructing a CDS. Dissimilarly, in this chapter, we first show an example to illustrate that a traditional MCDS cannot prolong network lifetime by reducing the communication cost. Instead, MCDS increases the workload imbalance among dominators, which leads to the reduction of network lifetime. Based on this observation, we then built an LBCDS. After constructing it, we investigated how to load-balancedly allocate dominatees to dominators. The probability (EAP)-based centralized and distributed algorithms are proposed to obtain an optimal allocation scheme. The upper bound and lower bound of the performance ratios of the proposed algorithms are analyzed in Section 2.5.

2.3   Problem Statement

In this section, we introduce the network model and define the LBCDS problem and the LBAD problem formally.

2.3.1   Network Model

We assume a static connected WSN is deployed in a square with area size $A = cn$, where c is a constant and the WSN is consisting of n sensors, denoted by $s_1, s_2, \ldots, s_n$ respectively. All sensors are independent identically distributed (i.i.d.) over the whole network. We also assume all nodes have the same transmission range. We modeled the WSN as a connected undirected general graph $G = (V, E)$, in which V represents node set and E represents the link set. $\forall u, v \in V$, there exists an edge (u, v) in G if and only if u and v are in each other's transmission range. In this chapter, we assume edges are undirected (bidirectional), which means two linked nodes are able to transmit and receive information from each other.

2.3.2   Preliminary

The *load-balance* factor is our major concern in this work. Thus, finding an appropriate measurement to evaluate load-balance is the key to solve the LBCDS and LBAD problems. We use *p-norm* to measure load-balance in this chapter. The definition of *p-norm* is given as follows:

Definition 2.1   p-norm. *The p-norm of an $n \times 1$ vector* $\mathbb{X} = (x_1, x_2, \cdots, x_n)$ *is:*

$$|\mathbb{X}|_p = \left(\sum_{i=1}^{n} |x_i|^p \right)^{\frac{1}{p}} \tag{2.1}$$

The authors in [50] stated that *p*-norm shows interesting properties for different values of p. If p is close to 1, the information routes resemble the geometric shortest paths from the sources to the sinks. For $p = 2$, the information flow shows an analogy to an electrostatics field, which can be used to measure the load-balance among x_i. More importantly, the smaller the *p-norm* value, the more load-balanced the interested feature vector $\mathbb{X}$. For simplicity, we use $p = 2$ in this chapter.

In this chapter, we use node degree (*Definition 6.10*) and the number of dominatees connected to a dominator (*Definition 2.3*) of the interested node set as the information vector $\mathbb{X}$, since the degree of each node and the number of the dominatees connected to a dominator is only a potential indicator of traffic load.

We still use the WSN shown in Figure 2.1 to illustrate how to use *p-norm* to measure the load-balance of CDSs. Two different CDSs for the same network are identified in Figure 2.1. The degree of the node s_i is denoted by d_i. Thus, $|d_i - \bar{d}|$ are used as the information vector $\mathbb{X}$, where $\bar{d}$ is the mean degree of the graph in Figure 2.1. Therefore, the *p-norm* value of the CDS shown in Figure 2.1(a) is $\sqrt{9}$. Similarly, in Figure 2.1(b), the *p-norm* value is $\sqrt{2}$. Clearly, $\sqrt{2} < \sqrt{9}$, which implies that the CDS in Figure 2.1(b) is more load-balanced than the CDS in Figure 2.1(a).

After we construct an LBCDS, the next step is to allocate dominatees to each dominator in the LBCDS. The *p-norm* can again be used to measure the load-balance of different allocation schemes, in which the number of dominatees connected to a dominator of the interested node set is used as the information vector $\mathbb{X}$. An illustration example is shown in Section 2.5.

2.3.3 Problem Definition

Now we give the formal definition of the problems we investigate in this chapter.

Definition 2.2 Load-balanced CDS (LBCDS). *For a WSN represented by graph $G = (V, E)$, the LBCDS problem is to find a node set $D \subseteq V$, $D = \{s_1, s_2, \cdots, s_M\}$ such that:*

1. $G[D] = (D, E')$, where $E' = \{e|\ e = (u, v), u \in D, v \in D, (u, v) \in E)\}$, is connected.

2. $\forall\, u \in V$ and $u \notin D$, $\exists\, v \in D$, such that $(u, v) \in E$.

3. $min|D|_p = (\sum_{i=1}^{M} |d_i - \bar{d}|^2)^{\frac{1}{2}}$.

Definition 2.3 Load-balanced allocation of dominatees (LBAD). *For a WSN represented by graph $G = (V, E)$, and a CDS $D = \{s_1, s_2, \cdots, s_M\}$. The number of the dominatees connecting to each dominator s_i $(1 \le i \le M)$ is denoted by $|A(s_i)|$, and the expected allocated dominatees of each dominator is denoted by $|\bar{A}|$. The LBAD problem is to find M disjoint sets on V, i.e. $A(s_1), A(s_2), \cdots, A(s_M)$, such that:*

1. *Each set $A(s_i)$ $(1 \le i \le M)$ contains exactly one dominator s_i.*

2. $\bigcup_{i=1}^{M} A(s_i) = V, A(s_i) \bigcap A(s_j) = \emptyset$ $(1 \le i \ne j \le M)$.

3. $\forall\, u \in A(s_i)$ $(1 \le i \le M)$ and $u \ne s_i$, such that $(u, s_i) \in E$.

4. $min|D|_p = (\sum_{i=1}^{M} ||A(s_i)| - |\bar{A}||^2)^{\frac{1}{2}}$.

2.4 Load-Balanced CDS

2.4.1 Algorithm Description

In essence, we design a greedy algorithm to solve the LBCDS problem. The algorithm starts from an empty dominator set (DS). Each time, it adds the node into the DS set that has the smallest $|d_i - \bar{d}|$ value (where $1 \le i \le n$). If there exists a tie on $|d_i - \bar{d}|$ value, we use greater d_i value to break the tie, since the nodes with higher degree can make the algorithm converge faster. The algorithm terminates when the nodes in the DS set form a CDS.

The pseudocode of the greedy algorithm is shown in Algorithm 1.

LBCDS-approximate algorithm as shown in Algorithm 1 is a centralized one-phase greedy algorithm. Initially, all the nodes are white. All black nodes form an LBCDS finally. We use the following terms in the algorithm: d_i: degree of node s_i, and $\bar{d}$: mean degree of G.

Algorithm 1: : LBCDS-Approximate

Require: A WSN represented by graph $G = (V, E)$; Node degree d_i; Mean degree of the graph $\bar{d}$.

  1: Make all nodes white {dominatee nodes}.

  2: Sort the n sensors based on their $|d_i - \bar{d}|$ values. If there exists a tie, use greater d_i value to break the tie, where $1 \leq i \leq n$. The node IDs are stored in the sorted array denoted by A.

  3: **for** $i = 1$ **to** n **do**

  4:    **if** the node represented by A_i and all its 1-hop neighbors are not dominated **then**

  5:      Mark the node represented by A_i black {dominator node}.

  6:      **if** All black nodes form a connected dominating set (CDS) **then**

  7:        **return** all black nodes.

  8:      **end if**

  9:    **end if**

10: **end for**

Initially, all nodes in the WSN are marked as dominatee nodes. Then, sorting the n sensors based on their $|d_i - \bar{d}|$ values, the sorted node ID are stored in an array called A (shown in Line 2). Starting from the lowest subscript i of the sorted array A, if the nodes represented by $A[i]$ and all its 1-hop neighbors are not dominated, then mark the node represented by $A[i]$ as the dominator node (shown in Line 5). Keep continuing the process till all the black nodes form a CDS, then an LBCDS is constructed (from Line 3 to 10).

The time complexity of the sorting process shown in Line 2 is $O(n\log(n))$, whereas the time complexity of the process to form an LBCDS (shown from Line 3 to 10) is $O(n)$. Hence, the time complexity of Algorithm 1 is $O(n\log(n))$. Since Algorithm 1 is a centralized algorithm, no extra storage memory is required for each sensor in the WSN.

2.4.2 Example Illustration

We use the WSN shown in Figure 2.2 to illustrate how to build an LBCDS. Based on each node's degree, we can calculate $\bar{d} = 3$. According to the aforementioned LBCDS construction algorithm, in the first round, all the nodes with degree 3 are added into the DS set (shown in Figure 2.2(a)). Thus, node s_7 is added into the DS set, since $d_7 = \bar{d} = 3$. In the next round, nodes s_3 and s_6 with degree 4 are added

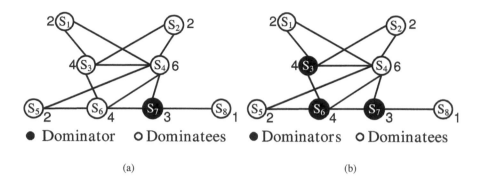

Figure 2.2: Construction of an LBCDS.

into the DS set (shown in Figure 2.2(b)), since s_3 and its 1-hop neighbors, s_6 and its 1-hop neighbors are not dominated. Nodes s_1 and s_2 are not added into the DS set, because they have smaller degree values than nodes s_3 and s_6. So far, there are three nodes in the DS set, which forms a CDS, therefore the algorithm terminates. Finally, we get an LBCDS which is $\{s_3, s_6, s_7\}$.

2.4.3 Remarks

According to Definition 6.10 on Page 128, *p-norm* is the measurement of the load-balance for CDSs. The smaller the *p-norm* value, the more load-balanced the CDS is. Moreover, $|d_i - \bar{d}|$ is the information vector $\mathbb{X}$ in Equation (2.1). The LBCDS construction algorithm greedily searches the dominators with the smallest $|d_i - \bar{d}|$ values. Based on this greedy criterion, the algorithm can output a CDS with a small *p-norm* value.

2.5 Load-balanced Allocation of Dominatees

Constructing an LBCDS is the foundation to solve the LBAD problem. In this section, we introduce how to use an existing LBCDS for load-balanced allocation of dominatees.

2.5.1 Terminologies

In a traditional/naive way, such as the work in [17], each dominatee sets its data forwarder to be the connected dominator with the smallest ID. Thus, the load-balance factor is not taken into account. In some environment, the dominator with the smallest ID, which is chosen by majority dominatees, probably has a heavier workload than the other dominators with a smaller number of dominatees. Therefore, the node degree cannot imply the potential workload precisely. In a WSN with a CDS as the VB, only the dominator and dominatee links contribute to the workload. Based on this observation, we define the following:

Definition 2.4 Valid degree (VD). *For each dominatee s_i, VD_i is the number of its connected dominators. For each dominator s_j, VD_j is the number of its allocated dominatees.*

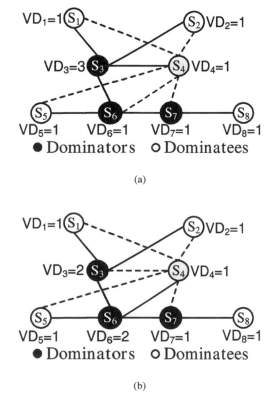

(a)

(b)

Figure 2.3: Allocation examples.

Figure 2.3(a) and Figure 2.3(b) illustrate an imbalanced and a balanced allocations of dominatees. The number beside each node s_i is its VD_i value. Using $|VD_i - \bar{d}|$ as the information vector $\mathbb{X}$, we still can use *p-norm* to measure the load-balance factor of the dominatee allocation scheme. Therefore, the *p-norm* value of the allocation scheme shown in Figure 2.3(a) is $\sqrt{8}$. Similarly, in Figure 2.3(b), the *p-norm* value is

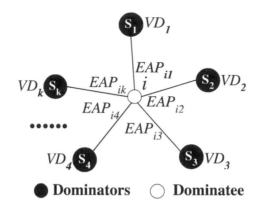

Figure 2.4: Property of expected allocation probability (EAP).

$\sqrt{6}$. Clearly, $\sqrt{6} < \sqrt{8}$, which implies the allocation scheme shown in Figure 2.3(b) is more load-balanced than the scheme shown in Figure 2.3(a).

Due to the instability of network topology, it is not practical to always allocate one dominatee to one dominator. In order to adapt to network topology change, the terminology *expected allocation probability (EAP)* is proposed as follows:

Definition 2.5   Expected allocation probability (EAP). *For each dominatee and dominator pair, there is an* EAP, *which represents the expected probability that the dominatee is allocated to the dominator.*

The *EAP* value associated on each dominatee and dominator pair directly determines the load-balance factor of each allocation scheme. We conclude the properties of the *EAP* values as follows:

1) For each dominatee s_i (as shown in Figure 2.4),

$$\sum_{j=1}^{|NE(s_i)|} EAP_{ij} = 1 \tag{2.2}$$

where $NE(s_i)$ is the set of neighboring dominators of dominatee s_i, and $|NE(s_i)|$ is the cardinality of set $NE(s_i)$;

2) The most load-balanced allocation scheme, which is obtained when the expected numbers of allocated dominatees of all the dominators are the same, can be formulated as follows:

$$EAP_{i1} \times VD_1 = \cdots = EAP_{i|NE(s_i)|} \times VD_{|NE(s_i)|} \tag{2.3}$$

An example about how to calculate *EAP* values is shown in Figure 2.5. The gray nodes, i.e., s_6 in Figure 2.5(a) and s_4 in Figure 2.5(b) are dominatees connected to

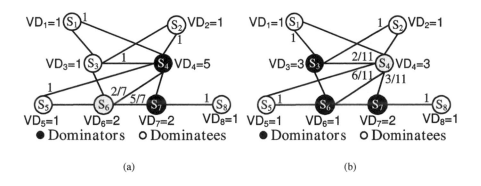

(a)   (b)

Figure 2.5: An illustration of calculating expected allocation probability (EAP).

more than one dominator. The numbers shown on the links are the *EAP* values of each dominatee and dominator pair. If a dominatee only connects to one dominator, the *EAP* value associated with the pair is equal to 1. Otherwise, Equation (2.3) can be used to calculate the *EAP* values of all connected dominator links.

As mentioned in Definition 2.4, the degree of each dominator is not a good indicator of workload. Hence, after allocating dominatees through the *EAP* scheme, the information vector $\mathbb{X}$ in Equation (2.1) is the summation of the *EAP* values of dominator s_j minus the expected number of allocated dominatees of each dominator, which is formulated by $\sum_{i=1}^{ND(s_j)} |EAP_{ij} - \bar{p}|$, where $\bar{p} = \frac{n-M}{M}$ representing the expected number of allocated dominatees of each dominator. In Figure 2.3, $\bar{p} = \frac{5}{3}$. Therefore, the *p-norm* value of the the allocation scheme shown in Figure 2.3(a) is $\sqrt{1.16}$. Similarly, the *p-norm* value of the allocation scheme shown in in Figure 2.3(b) is $\sqrt{0.57}$. Apparently, $\sqrt{0.57}$ is smaller than $\sqrt{1.16}$, which means the allocation scheme in Figure 2.3(b) is more load-balanced than the scheme in Figure 2.3(a). There are two reasons to have a very small *p-norm* value in Figure 2.3(b). First, an LBCDS is used. Second and more important, we adopt the probability-based dominatee allocation scheme. The allocation criterion is based on making the expected number of allocated dominatees of each dominator the same. The criterion implies, on average, the expected number of allocated dominatees of all the dominators are the same. If every dominatee has the same amount of data to be transferred through the allocated dominator at a fixed data rate, then the probability-based allocation scheme can achieve the maximized load-balance on the expected workload among dominators.

2.5.2   Algorithm Description

The allocation system starts from finding an LBCDS using the aforementioned LBCDS construction algorithm. Then the *EAP* value is calculated for each dominatee and dominator pair. *EAP* only indicates the probability the dominatee will be assigned to the dominator for each dominator and dominatee pair. Thus the final step is allocate the dominatees to the dominators. We use the stochastic allocation, which is that a dominatee is randomly assigned to an adjacent dominator based on the *EAP* value.

Figure 2.5(b) shows an example about how to perform the stochastic dominatee allocation. In Figure 2.5(b), only dominatee s_4 connects to more than one dominator and its associated *EAP* values are: $EAP_{43} = \frac{2}{11} = 0.18$; $EAP_{46} = \frac{6}{11} = 0.55$; and $EAP_{47} = \frac{3}{11} = 0.27$. Dominatee s_4 generates a random number $\delta = 0.358$. If $\delta \in [0, 0.18]$, s_4 chooses dominator s_3, else s_4 chooses dominator s_6 if $\delta \in (0.18, 0.73]$, otherwise s_4 chooses dominator s_7 if $\delta \in (0.73, 1]$. Since $\delta = 0.358$, dominatee s_4 is assigned to dominator s_6.

Each time a dominatee which is connected to more than one dominator wants to send data, it must redo the last step to pick a proper dominator based on the *EAP* probability and then forward its data.

In Section 2.4, the detailed description of how to construct an LBCDS is introduced. Additionally, the third step, allocating the dominatees to the dominators, is a trivial process as we just explained. In the rest of this section, we design two algorithms to implement the second step, namely how to calculate the *EAP* value for each dominator and dominatee pair. We introduce the centralized algorithm first:

2.5.2.1 Centralized Algorithm

We propose a constrained non-linear programming scheme to solve the LBAD problem. The essence of allocating dominatees is to achieve maximum load-balance among dominators. We use the *p-norm* value to measure the load-balance factor. Consequently, the objective of the optimization problem is to minimize the *p-norm* value of the dominatee allocation scheme. In addition, the constraint is to guarantee Property 1 of *EAP* values (Equation. 2.2). To conclude, the optimization problem is formulated as follows:

$$
\text{Minimize: } |EAP|_p = \sum_{j=1}^{M} \left(\sum_{i=1}^{|ND(s_j)|} |EAP_{ij} - \bar{p}| \right)^2
$$

$$
\text{Subject to: for dominatee } s_i, \sum_{j=1}^{|ND(s_i)|} EAP_{ij} = 1 \tag{2.4}
$$

Where: $0 \leq EAP_{ij} \leq 1$.

The centralized algorithm is shown in Algorithm 2.

The objective function in Equation (2.4) is the *p-norm* value on *EAP* values. The constraint states that the sum of the *EAP* values on each dominatee is equal to 1, which is the first property of *EAP* (Equation 2.2). The centralized algorithm can guarantee to find the optimal solution. However, solving the non-linear programming is too time and energy consuming. If precision is the major concern, we can solve the non-linear programming formulas at the base station. Nevertheless, if the energy and time are the primary concern, a distributed algorithm to find a near-optimal solution is preferred. We therefore propose the distributed algorithm as follows:

Algorithm 2: : LBAD-Centralized

Require: A WSN represented by a graph $G = (V, E)$; an LBCDS: $G[D] = (D, E')$
1: Solve the constrained non-linear programming formulated in Equation 2.4. Let EAP_{ij} be the optimal solution of the non-linear programming.
2: **for** each dominatee s_i **do**
3:     Generate a number δ between 0 and 1
4:     **if** $\delta \in [\sum_{j=0}^{k-1} EAP_{ij}, \sum_{j=0}^{k} EAP_{ij}]$, where $0 < k \leq NE(s_i)$ **then**
5:         mark the link between dominatee s_i and dominator s_k black
6:     **end if**
7: **end for**
8: **return** all black links.

2.5.2.2 Distributed Algorithm

The objective of the LBAD problem is to find a load-balanced dominatee allocation scheme. The most load-balanced allocation scheme is that the expected numbers of allocated dominatees of all the dominators are the same, which is formulated in Equation (2.3). Additionally, we guarantee Property 1 of *EAP* values (Equation. 2.2). By listing all the equations, we can solve them to get EAP_{ij} of each connected dominatee s_i and dominator s_j, which is formulated as follows:

$$
\begin{aligned}
EAP_{i1} : EAP_{i2} : \cdots : EAP_{i|ND(s_i)|} = \\
VD_2 \times VD_3 \times \cdots \times VD_{|ND(s_i)|} : \\
\cdots : \prod_{j=1, i \neq j}^{|ND(s_i)|} VD_j : \cdots \\
: VD_1 \times VD_2 \times \cdots \times VD_{|ND(s_i)|-1}
\end{aligned}
\tag{2.5}
$$

Therefore, the distributed LBAD problem can be transformed to calculate the *EAP* value of each dominatee locally. The dominatee stochastic allocation step is the same as the centralized algorithm.

The distributed algorithm is a localized two-phase algorithm where each node only needs to know the connectivity information within its 1-hop-away neighborhood. All the nodes get the *VD* values by broadcasting messages to all its neighbor nodes, and then store the values locally. Each dominatee calculates the *EAP* values using Equation (2.5).

The pseudocode is given in Algorithm 3. We call it *LBAD-distributed* algorithm. We use the following terms in Algorithm 3,
VD_k: The VD value of each node s_k.
$ND(s_k)$: The set of neighboring dominatees of dominator s_k.
$|ND(s_k)|$: The number of the nodes in set $ND(s_k)$.
$NE(s_k)$: The set of the neighboring dominators of dominatee s_k.
$|NE(s_k)|$: The number of the nodes in set $NE(s_k)$.

EAP_{ij}: The EAP value of each connected dominatee s_i and dominator s_j pair.

Each node s_i maintains the following data structures:

(i) s_i's *ID*, initialized to 0.

(ii) The dominator/dominatee flag f: 1 means dominator; 0 means dominatee. It is initialized to 0.

(iii) $|ND(s_i)|$, if s_i is a dominator; $|NE(s_i)|$, if s_i is a dominatee, initialized to 0.

(iv) Neighboring dominator/dominatee lists. A list contains: a dominator/dominatee's *ID*, its VD value, and EAP_{ij}, initialized to $\emptyset$.

Initially, each node initializes its data structures and broadcasts a *hello* message containing its *ID*, VD, and f to its 1-hop neighbors to exchange neighbors' information. All the nodes run the following:

• For any dominator s_i, upon receiving a *hello* message from node s_j: if s_j is a dominator, ignore the message. If s_j is a dominatee, update $|ND(s_i)|$ and dominatee s_j's *ID* and VD value in the neighboring dominatee list of the dominator s_i.

• For any dominatee s_i, upon receiving a *hello* message from node s_j: if s_j is a dominatee, ignore the message. If s_j is a dominator, update $|NE(s_i)|$ and dominator s_j's *ID* and VD value in the neighboring dominator list of the dominatee s_i. Calculate and store EAP_{ij} based on the VD values stored in the neighboring dominator list using Equation 2.5.

Algorithm 3: : LBAD-Distributed

1: *Initialization Phase:*

2: For each dominatee s_i, get the number of neighboring dominators (denoted by $|NE(s_i)|$) and store locally.

3: For each dominator s_j, get the number of neighboring dominatees (denoted by $|ND(s_j)|$) and store locally.

4: *Allocation Phase:*

5: For each dominatee s_i, calculate its neighboring dominators' EAP_{ij} by the following formula:

6: $EAP_{i1} : EAP_{i2} : \cdots : EAP_{i|ND(s_i)|} = VD_2 \times VD_3 \times \cdots \times VD_{|ND(s_i)|} : \cdots :$
$$VD_1 \times VD_2 \times \cdots \times VD_{|ND(s_i)|-1} = \prod_{j=1, i\neq j}^{|ND(s_i)|} VD_j$$

7: A dominatee is assigned to an adjacent dominator based on the above calculated EAP_{ij} value.

The distributed algorithm is a two-phase algorithm. The first phase is the initialization phase, where all the nodes get their neighborhood information and update their own data structure locally. The message complexity of this phase is $O(n)$, since each node only needs to communicate with its 1-hop neighbors. In practice, it is hard to decide when the initialization phase completes. Hence we set a timer. If the timer expires, the second phase, allocation phase, starts to work. In the allocation phase, every dominatee calculates the *EAP* values of its connected dominators using Equation (2.5). We only use 1-hop-away neighborhood information to calculate the *EAP* values locally. Therefore, it is an easy and efficient algorithm. In other words, the time complexity of the allocation phase is $O(1)$. Nevertheless, only using the 1-hop-away neighborhood information to calculate the *EAP* values may lead us to find a local optimal solution instead of a global optimal solution. Note that we sacrifice the storage space to save the time complexity in the distributed algorithm. Every node must store the neighbor's dominatee/dominator's list locally.

2.5.3 Analysis

Based on the assumptions mentioned in Section 2.3.1, n sensors are i.i.d. in a square with area size $A = cn$. The communication range of each sensor is 1. Thus, we denote the unit circle associated with each sensor s_i by c_i. According to the network model, the following lemma can be proved:

Lemma 2.1
For any unit circle c_i, let the random variable Z_i denote the number of the sensors within it. Then, the probability that c_i contains more than $\ln n$ sensors is no greater than $\frac{\exp((\exp(\gamma)-1)\times(\frac{\pi}{c}))}{\exp(\gamma\times\ln n)}$, i.e., $\Pr[Z_i > \ln n] \leq \frac{\exp((\exp(\gamma)-1)\times(\frac{\pi}{c}))}{\exp(\gamma\times\ln n)}$, for any $\gamma > 0$.

Proof: Since all the sensors are i.i.d., the number of the sensors in c_i satisfies the binomial distributions with parameters $(n, \frac{\pi}{A})$ [51]. Applying the Chernoff bound and for any $\gamma > 0$, we have

$$
\begin{aligned}
\Pr[Z_i > \ln n] &\leq \frac{\mathrm{E}[\exp(\gamma Z_i)]}{\exp(\gamma\ln n)} = \frac{[1+(\exp(\gamma)-1)\frac{\pi}{A}]^n}{\exp(\gamma\ln n)} \\
&\leq \frac{\exp((\exp(\gamma)-1)\frac{\pi}{A}\times n)}{\exp(\gamma\ln n)} \quad (by\ 1+x\leq e^x) \\
&= \frac{\exp((\exp(\gamma)-1)\times(\frac{\pi}{c}))}{\exp(\gamma\times\ln n)} \quad (by\ A=cn)
\end{aligned}
$$

□

From Lemma 12.1, the probability that a unit circle contains more than $\ln n$ sensors is zero when $n \to \infty$. Hence, we can use $\ln n$ as the upper bound of the number of the sensors in an unit circle in our analysis. Then, we can get the following theorem which states the upper and lower bounds of the *p-norm* of the distributed algorithm.

Theorem 2.1
The upper bound of the p-norm value in the distributed algorithm is $M(\ln n - 1 - \frac{2\pi}{3\times c} - \frac{\sqrt{3}}{2\times c} - \frac{n-M}{M})^2$; The lower bound of the p-norm value in the distributed algorithm is $M(\frac{\pi}{c} - \frac{n-M}{M})^2$.

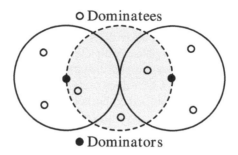

Figure 2.6: Connectivity guarantee.

Proof: According to Definition 2.3, the *p-norm* value of the allocated dominatees with *EAP* is formulated as: $|EAP|_p = \sum_{j=1}^{M}(|\sum_{i=1}^{|ND(s_j)|} EAP_{ij} - \bar{p}|)^2$. The *p-norm* value depends on how many dominatees are adjacent to each dominator, namely $ND(s_j)$ in the formula. So the upper bound and lower bound of the number of neighboring dominatees are the key challenges to analyze the performance ratio. The upper bound of the number of the sensors in a unit circle occurs when there is only one dominator in a unit circle and all the other dominatees connect to the dominator, then we can get the upper bound of the *p-norm* value. However, we are considering a CDS. In order to maintain the connectivity, at least two dominators must be within each other in the transmission range, namely in one unit circle. Figure 2.6 illustrates the situation. There are some overlapped areas shown by the gray in the figure. The gray area is a sector with 120 degrees and its size is: $\frac{2\pi}{3} - \frac{\sqrt{3}}{2}$. Because all the sensors are i.i.d., the expected number of the sensors is: $\frac{\frac{2\pi}{3} - \frac{\sqrt{3}}{2}}{c \times n} \times n = \frac{2\pi}{3 \times c} - \frac{\sqrt{3}}{2 \times c}$. Therefore, the upper bound of the *p-norm* value is:

$$|EAP|_p \le \sum_{j=1}^{M}(\ln n - 1 - \frac{2\pi}{3 \times c} - \frac{\sqrt{3}}{2 \times c} - \frac{n-M}{M})^2$$
$$= (\ln n - 1 - \frac{2\pi}{3 \times c} - \frac{\sqrt{3}}{2 \times c} - \frac{n-M}{M})^2$$

The lower bound of the number of the sensors in each unit circle can be estimated by $\frac{\pi}{c \times n} \times n = \frac{\pi}{c}$. Hence the lower bound of the *p-norm* value is:

$$|EAP|_p \ge \sum_{j=1}^{M}(\frac{\pi}{c} - \frac{n-M}{M})^2 = M(\frac{\pi}{c} - \frac{n-M}{M})^2$$

$\square$

2.6 Performance Evaluation

In this section, we validate our proposed algorithms through implementing the CDS-based data aggregation protocol and CDS-based data collection protocol.

2.6.1 Scenario 1 - Data Aggregation Communication Mode

In this subsection, we evaluate our proposed algorithms by comparing our work with the work in [17], in which each dominatee chooses the neighboring dominator of the smallest ID as its parent. Four different schemes are implemented:

■ LBCDSs with LBAD, noted by LB-A.

■ LBCDSs with the smallest ID dominator selection scheme, noted by LB-ID.

■ MIS-based CDSs with LBAD, noted by MIS-A.

■ MIS-based CDSs with the smallest ID dominator selection scheme, which is the work in [17], noted by MIS-ID.

We compare them in terms of the *p-norm* value, network lifetime, which is defined as the time duration till the first dominator's energy is depleted, and the standard derivation of the residual energy of all the nodes.

2.6.1.1 Simulation Environment

We build our own simulator where all the nodes have the same transmission range (10 *m*). *n* nodes are randomly deployed in a fixed area of 100 *m* × 100 *m*. *n* is incremented from 200 to 450 by 50. For a certain *n*, 100 instances are generated. The results are averaged among 100 instances. Moreover, we use the CDS-based data aggregation as the communication mode.

2.6.1.2 Simulation Results

Figure 2.7 shows the *p-norm* values of the four schemes. The *X*-axis represents the number of the nodes. The *Y*-axis represents the *p-norm* values of the four schemes. With the increase of the number of the sensor nodes, the *p-norm* values increase correspondingly. This is because when the number of the nodes increases, we need more nodes to build an LBCDS. According to Definition 6.4, more nodes imply more sum subitems, so the *p-norm* values increase. As mentioned in Section 2.3.1, the smaller the *p-norm* value is, the more load-balanced the scheme is. From Figure 2.8(a), we know that the MIS-ID scheme has the largest *p-norm* values while the LB-A scheme has the smallest *p-norm* values. This is because the MIS-ID scheme did not consider the load-balance factor when building a CDS and allocating dominatees to dominators. For clearly to see the *p-norm* values of the LB-A scheme, we redraw the curve using smaller scale in Figure 2.7(b) for LB-A. Additionally, Figure 2.7 demonstrates that the LBAD algorithm fits for any type of CDSs. The MIS-A

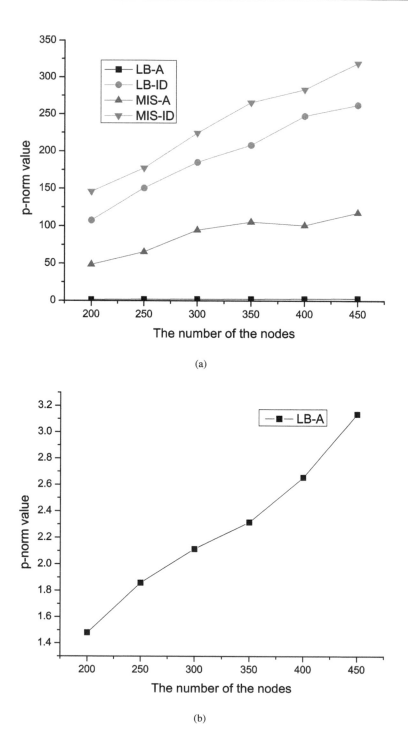

(a)

(b)

Figure 2.7: *p-Norm* value.

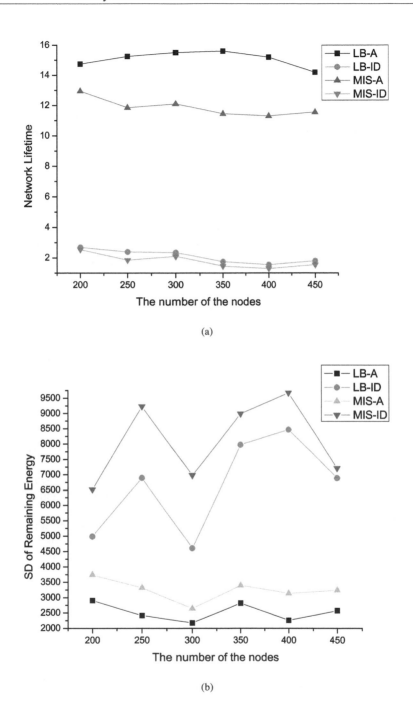

(a)

(b)

Figure 2.8: Simulation results for a square area of *100 m × 100 m*, the node transmission range is *10 m*, and the number of nodes changes from *200* to *450*: (a) network lifetime; (b) SD of residual energy.

scheme still has smaller *p-norm* values than the other two schemes using smallest ID allocation scheme, namely LB-ID and MIS-ID.

Figure 2.8(a) shows the network lifetime of the four schemes. The X-axis represents the number of the nodes and the Y-axis represents the network lifetime of the four schemes. The simulated energy consumption model is that every node has the same initial 100 units of energy. Receiving a packet consumes 1 unit of energy, while transmitting a packet consumes 2 units of energy. From Figure 2.8(a), we can see the load-balanced schemes (LB-A and MIS-A) prolong network lifetime by 80% compared to non-balanced schemes (LB-ID and MIS-ID). With the number of the node increases, there is no obvious increase or decrease trend of network lifetime, since the locality of the network topology mainly decides the network lifetime. A network topology is generated randomly, so we cannot control the locality of the network. From Figure 2.8(a), we also find the network lifetimes of imbalanced schemes (LB-ID and MIS-ID) are close to 1, 2 or 3. This is because some critical smaller ID dominators are connected to many dominatees. They deplete energy very quickly, then the whole network is disconnected.

Figure 2.8(b) shows the standard derivation of the residual energy of the four schemes. The X-axis represents the number of the nodes. The Y-axis represents the standard derivation of the average residual energy of all the nodes. We use the standard derivation here to observe whether the residual energy is balanced or not. From Figure 2.8(b), we know the balanced schemes (LB-A and MIS-A) have more balanced residual energy than imbalanced schemes (LB-ID and MIS-ID). This is because we consider the load-balance factor when building a CDS and allocating dominatees to dominators.

The simulation results can be summarized as follows:

■ The LB-A scheme always has the best performance according to the *p-norm* value, network lifetime, and the standard derivation of residual energy. The results demonstrate building a load-balanced CDS and then load-balancedly allocating dominatees can increase network lifetime significantly.

■ The load-balanced dominatee allocation algorithm can be applied to not only load-balanced CDSs but also imbalanced CDSs to achieve good performances. The LB-A and MIS-A schemes have better performances over the LB-ID and MIS-ID schemes among all measures, namely the *p-norm* value, network lifetime, and the standard derivation of residual energy.

■ The balanced schemes (LB-A and MIS-A) have better scalability than the imbalanced schemes (LB-ID and MIS-ID).

2.6.2   Scenario 2 - Data Collection Communication Mode

In this subsection, we evaluate our proposed algorithms by comparing our work denoted by LBCDS with the tree-based CDS construction work [17] and the subtraction-based CDS construction work [34]. We compare them in terms of the number of dominators, network lifetime which is defined as the time duration till the first dominator's energy is depleted, and the average residual energy among all the dominators.

2.6.2.1   Simulation Environment

We build our own simulator where all the nodes have the same transmission range (10 m). n nodes are randomly deployed in a fixed area of 100 m × 100 m. n is incremented from 200 to 700 by 100. For a certain n, 100 instances are generated. The results are averaged among 100 instances. Moreover, we use the CDS-based data collection as the communication mode. The simulated energy consumption model is that every node has the same initial 2000 units of energy. Receiving and transmitting a packet both consume 1 unit of energy.

2.6.2.2   Simulation Results

From Figure 2.9(a), we can see that, with the increase of the number of sensor nodes, the number of dominators almost keeps stable for all the three algorithms (tree-based, LBCDS, and subtraction). This is because the area of the network deployed region keeps fixed. The result implies that, if the network deployed area keeps unchanged, the density of the WSN has little effect on the size of the constructed CDS.

Despite few changes in the number of dominators, different dominatee allocation schemes do affect network lifetime as shown in Figure 2.9(b). From Figure 2.9(b), we know that the network lifetime decreases for all algorithms with the number of nodes increasing. This is because there are more neighbors of each dominator in a more and more crowded network. As shown in Figure 2.9(a), the number of dominators almost remains unchanged. It implies that with the number of neighbors increasing, the collected data on each dominator becomes heavier. Hence, the network lifetime decreases for all the three algorithms. Additionally, we can see LBCDS outperforms tree-based and subtraction. Furthermore, LBCDS prolongs network lifetime by 31% on average compared with tree-based, and by 26% on average compared with subtraction. The results demonstrate that load-balancedly allocating dominatees to dominators can improve network lifetime notably. On the other hand, subtraction outperforms tree-based, since the size of the constructed CDS by subtraction is larger than the size of the constructed CDS by tree-based.

Figure 2.9(c) shows the average residual energy among all the dominators of the three algorithms. With the increase of the number of nodes, the average residual energy decreases for all algorithms. As the WSN becomes denser and denser, the dominators collect more and more data in the WSN. From Figure 2.9(c), we know that LBCDS has less average residual energy than tree-based and subtraction. This is because LBCDS considers the load-balance factor when building a CDS and allocating dominatees to dominators. Thus, the lifetime of the whole network is extended, which means the average residual energy of the dominators is less than tree-based and subtraction. Additionally, subtraction has less average residual energy than tree-based. Since subtraction has longer network lifetime than tree-based as shown in Figure 2.9(b), subtraction has less residual energy than tree-based. In summary, Fig-

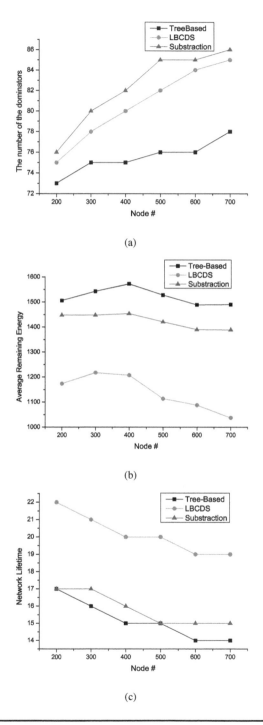

Figure 2.9: Simulation results for a square area of *100 m* × *100 m*, the node transmission range is *10 m*, and the number of nodes changes from *200* to *700*: (a) the number of dominators; (b) average residual energy of dominators; (c) network lifetime.

ure 2.9 indicates that constructing an LBCDS can balance the energy consumption on each backbone node, and make the lifetime of the whole network prolonged considerably.

2.7 Conclusion

In this chapter, we propose a new LBCDS concept, which is a CDS with the minimum *p-norm* value in order to assure that the workload among each dominator is balanced. We also propose an LBAD problem. It aims to load-balancedly allocate each dominatee to a dominator. We use *EAP* value to represent the expected probability of the allocation between each dominatee and dominator pair. An optimal centralized algorithm and an efficient distributed algorithm for the LBAD problem are proposed in the chapter. The lower bound and upper bound of the approximation ratio are proved in the chapter. The extensive simulation results demonstrate that compared to Wan's work [17], using an LBCDS as a virtual backbone and *EAP* values to load-balancedly allocate dominatees can prolong network lifetime significantly.

Chapter 3

Load-balanced CDS Construction in Wireless Sensor Networks via Genetic Algorithm

CONTENTS

3.1   Introduction

Wireless sensor networks (WSNs) are deployed for monitoring and controlling systems where human intervention is not desirable or feasible. Therefore, WSNs are widely used in many military and civilian applications such as battlefield surveillance, health care applications, environment and habitat monitoring, and traffic control [52]. Due to the lack of physical backbone infrastructure in WSNs, most routing protocols in WSNs, such as flooding, usually cause a serious broadcasting storm [53]. A connected dominating set (CDS) has become a well known approach for constructing a virtual backbone (VB) to alleviate the broadcasting storm, thus improving the performance and increasing the efficiency of routing protocols in WSNs. A dominating set (DS) is defined as a subset of nodes in a WSN such that each node in the network is either in the set or adjacent to some node in the set. If the induced graph by the nodes in a DS is connected, then this DS is called a CDS. The nodes in a CDS are called *dominators* or otherwise *dominatees*. The set of dominators is denoted by $\mathbb{B}$, whereas, the set of dominatees is denoted by $\mathbb{W}$. In a WSN with a CDS as its VB, dominatees may forward their data only to their neighboring dominators. With the help of a CDS, the average message burden of a WSN could be reduced so that routing becomes much easier and can adapt quickly to network topology changes [54]. Furthermore, using dominators as forwarding nodes can efficiently reduce energy consumption, which is also a critical concern in WSNs. In addition to routing protocols, a CDS has many other applications in WSNs, such as topology control [14], coverage [16], data collection [10], and data aggregation [17]. Clearly, the benefits of a CDS can be magnified by making its size (the number of the nodes in the CDS) smaller. In general, the smaller the CDS is, the less communication and storage overhead are incurred. Hence, it is desirable to build a minimum-sized CDS (MCDS).

Ever since the idea of employing a CDS for WSNs was introduced [19], a huge amount of effort has been made to construct different CDSs for different applications, especially MCDSs. In the seminal work [20], Guha and Kuller first modeled the problem of constructing the smallest CDS as the MCDS problem in a general graph, which is a well-known NP-hard problem [21]. Subsequently, many polynomial time approximation algorithms for constructing an MCDS have been proposed in recent literatures such as abstraction-based algorithms [33][34], and addition-based algorithms [17]. After that, to make a CDS more resilient in mobile WSNs, the fault tolerance of a VB is considered. In [23], k-connected and m-dominated sets are introduced as a generalized abstraction of a fault-tolerance VB. In [25], the authors proposed a minimum routing cost connected dominating set (MOC-CDS), which aims to find a minimum CDS while assuring that any routing path through this CDS is the shortest in WSNs. Additionally, the authors investigated the problem of constructing a quality CDS in terms of size, diameter, and average backbone path length (ABPL) in [26].

Unfortunately, to the best of our knowledge, all the related works did not consider the **load-balance** factor when they construct a CDS. If the workloads on each dominator in a CDS are not balanced, some heavy-duty dominators deplete their energy quickly. Then, the whole network might be disconnected. Hence, intuitively, we not only have to consider to construct an MCDS, but also need to consider to construct a load-balanced CDS (LBCDS). An illustration of an LBCDS is depicted in Figure 3.1, in which dominators are marked as black nodes, while white nodes represent dominatees; solid lines represent that the dominatees are allocated to the neighboring dominators, while the dashed lines represent the communication links in the original network topological graph. For convenience, the set of neighboring dominatees of a dominator $v_i \in \mathbb{B}$ is denoted by $\mathbb{U}(v_i)$. The set of dominatees allocated to a dominator v_i is denoted by $\mathbb{A}(v_i) = \{v_j | v_j \in \mathbb{W}, v_j$ forwards its data only to $v_i\}$. According to the traditional MCDS construction algorithms, a CDS $\{v_4, v_7\}$ with size 2 is obtained for the network shown in Figure 3.1(a). However, There are two severe drawbacks of the CDS shown in Figure 3.1(a). First, $\mathbb{U}(v_4) = \{v_1, v_2, v_3, v_5, v_6\}$, which represents that dominator v_4 connects to five different dominatees, and $\mathbb{U}(v_7) = \{v_6, v_8\}$. If every dominatee has the same amount of data to be transferred through the neighboring dominator at a fixed data rate, dominator v_4 must deplete its energy much faster than dominator v_7, since dominator v_4 has to forward the data collected from five neighboring dominatees. Second, dominatee v_6 connects to both dominators. If v_6 is allocated to dominator v_4, shown in Figure 3.1(a), obviously, only one dominatee v_8 forwards its data to dominator v_7. In this situation, the workload imbalance in the CDS is further amplified. Consequently, the entire network lifetime is shortened. We show a counter-example in Figure 3.1(b), where the constructed CDS is $\{v_3, v_6, v_7\}$. According to the topology shown in Figure 3.1(b), we can get the dominatee set of each dominator: $\mathbb{U}(v_3) = \{v_1, v_2, v_4\}$, $\mathbb{U}(v_6) = \{v_4, v_5\}$, and $\mathbb{U}(v_7) = \{v_4, v_8\}$. Compared with the MCDS constructed in Figure 3.1(a), the numbers of neighboring dominatees of all the dominators in Figure 3.1(b) are very similar. On the other hand, we have two different dominatee allocation schemes shown in Figure 3.1(b) and Figure 3.1(c) respectively. One is: $\mathbb{A}(v_3) = \{v_1, v_2, v_4\}$, $\mathbb{A}(v_6) = \{v_5\}$, and $\mathbb{A}(v_7) = \{v_8\}$. The other one is: $\mathbb{A}(v_3) = \{v_1, v_2\}$, $\mathbb{A}(v_6) = \{v_4, v_5\}$, and $\mathbb{A}(v_7) = \{v_8\}$. Apparently, the workload on each dominator is almost evenly distributed in the CDS constructed in Figure 3.1(c). Intuitively, the construction algorithm and dominatee allocation scheme shown in Figure 3.1(c) can extend network lifetime notably. Obviously, constructing an LBCDS and then load-balancedly allocate dominatees to dominators are equally important when considering the load-balance factor to construct a CDS. Neither of these two aspects can be ignored.

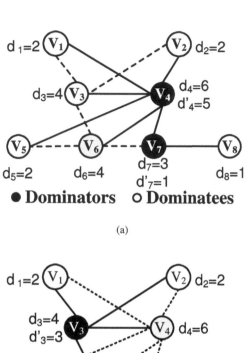

(a)

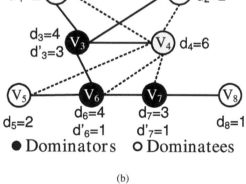

(b)

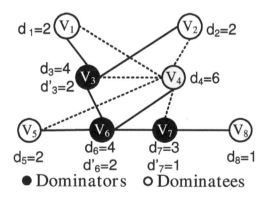

(c)

Figure 3.1: Illustration of a regular CDS and an LBCDS.

To solve the workload imbalance problem of an MCDS, in this chapter, we investigate how to construct an LBCDS and how to load-balancedly allocate dominatees to dominators simultaneously. To address this problem, we explore the genetic algorithm (GA) optimization approach. GAs are numerical search tools which operate according to the procedures that resemble the principles of nature selection and genetics [55]. Because of their flexibility and widespread applicability, GAs have been successfully used in a wide variety of problems in several areas of WSNs [56] [57].

Particularly, our contributions mainly include four aspects as follows:

1) To the best of our knowledge, this is the first work that studies the LBCDS problem. We point out the disadvantages and advantages of the existing MCDS construction algorithms. Based on the observation of workload imbalance in MCDS, we propose to take the load-balance factor into consideration when building a CDS and allocating dominatees to each dominator.

2) In order to measure the load-balance factor of a CDS and a dominatee allocation scheme, we define two new metrics *CDS p-norm* and *allocation scheme p-norm* based on the degrees of dominators and the number of allocated dominatees to each dominator, respectively.

3) We claim that the LBCDS problem is an NP-hard problem, thus we propose an effective GA to solve the LBCDS problem, called LBCDS-GA. Based on the characteristics of GAs and the searching heuristic, lots of possible solutions are searched and the best one is selected as the final result. Additionally, GAs have been proven to be effective at escaping local optima and discovering the global optimum in even a very complex searching space [58].

4) We also conduct extensive simulations to validate our proposed algorithm. The simulation results show that the constructed LBCDS can extend network lifetime significantly compared with the existing algorithms. Particularly, when the node number changes from 100 to 1000, our proposed method extends network lifetime by 65% on average compared with the work in [17], which is the latest MCDS construction algorithm.

The rest of this chapter is organized as follows: in Section 3.2, we review some related works on CDSs. In Section 3.3, we introduce the network model and formally define the LBCDS problem. The design of a GA for the LBCDS problem is presented in Section 3.4. The simulation results are presented in Section 3.5 to validate our proposed algorithm. Finally, the chapter is concluded in Section 3.6.

3.2   Related Work

The idea of using a CDS as a virtual backbone was first proposed by Ephremides et al. in 1987 [19]. Since then, many algorithms that construct CDSs have been reported and can be classified into the following four categories based on the network information they used:

■   Centralized algorithms;

■   Subtraction-based distributed algorithms;

- ■ Distributed algorithms using single leader;

- ■ Distributed algorithms using multiple leaders.

We use n to denote the number of sensors in a WSN, Δ to denote the maximum degree of nodes in the graph representing a WSN, and *opt* to denote the size of any optimal MCDS.

3.2.1   Centralized Algorithms for CDSs

Guha et al. [20] first proposed two centralized greedy algorithms with performance ratios of $2(H(\Delta)+1)$ and $H(\Delta)+2$ respectively, where H is a harmonic function. The greedy function is based on the number of white neighbors of each node. At each step, the one with the largest such number becomes a dominator.

Due to the instability of network topology in WSNs, it is necessary to update topology information periodically. Therefore, many distributed algorithms are proposed. These distributed algorithms can be classified into two categories: subtraction-based and addition-based algorithms. The subtraction-based algorithms begin with the set of all the nodes in a network, then remove some nodes according to predefined rules to obtain a CDS. The addition-based CDS algorithms start from a subset of nodes (usually disconnected), then include additional nodes to form a CDS. Depending on the type of the initial subset, the addition-based CDS algorithms can be further divided into single-leader and multiple-leader algorithms.

3.2.2   Subtraction-based Distributed Algorithms for CDSs

Wu and Li first proposed a completely distributed algorithm in [33] to obtain a CDS. The CDS construction procedure consists of two stages. In the first stages, each node collects its neighboring information by exchanging message with the one-hop neighbors. If a node finds that there is a direct link between any pair of its one-hop neighbors, it removes itself from the CDS. In the second stage, additional heuristic rules are applied to further reduce the size of the CDS. Wu's algorithm [33] uses Rule 1 and Rule 2, where a node is removed from the CDS, if all its neighbors are covered by its one or two direct neighbors. Later, Dai's [34] work generalizes this as Rule k, in which coverage is defined by an arbitrary number of connected neighbors.

3.2.3   Addition-based Distributed Algorithms for CDSs

Single-leader distributed algorithms for CDSs use one initiator to initialize the distributed algorithms. Usually, a base station could be the initiator for constructing CDSs in WSNs. In these distributed algorithms, a spanning tree rooted at the initiator is first constructed, and then maximal independent sets (MISs) are identified layer by layer, finally a set of connectors to connect the MISs is ascertained to form a CDS. Wan et al. [37] presented an ID-based distributed algorithm to construct a CDS using a single initiator. For UDGs, Wan et al.'s [37] approach guarantees that the approximation factor on the size of a CDS is at most 8 $opt + 1$. The algorithm has $O(n)$ time complexity and $O(n\log n)$ message complexity. Subsequently, the approximation factor on the size of a CDS was improved in another work reported by Cardei et al. [59], in which the authors used the degree-based heuristic and degree-aware optimization to identify Steiner nodes as the connectors in the CDS construction. The approximation factor on the size of a CDS is improved to 8 opt, while this distributed algorithm has $O(n)$ message complexity, and $O(\Delta n)$ time complexity. Later, Li et al. [60] reported a better approximation factor of $5.8 + \ln 4$ by constructing a Steiner tree when connecting all the nodes in the MISs. Later, many attempts tried to improve this approximation factor on the size of a CDS based on this idea. Wu et al. reported a approximation factor of 3.8 $opt + 1.2$ [38]. Yao et al. improved it to 3.67 $opt + 1.33$ [39]. The factor is further improved to 3.478 $opt + 4.874$ [40] and 3.4306 $opt + 4.8185$ [41], subsequently. Currently, the best result is that the size of an MIS is at most 3.399 $opt + 4.874$ [42].

Distributed algorithms with multiple leaders do not require an initiator to construct a CDS. Alzoubi et al.'s technique [61] first constructs an MIS using a distributed approach without a leader or tree construction and then interconnects these MIS nodes to get a CDS. Li et al. proposed a distributed algorithm r-CDS in [62], whose performance ratio is 172. r-CDS is a completely distributed one-phase algorithm where each node only needs to know the connectivity information within its 2-hop-away neighborhood. An MIS is constructed based on each nodes r value which is defined as the number of this nodes 2-hop-away neighbors minus the degree of this node. The nodes with smaller r values are preferred to serve as MIS nodes. Adjih et al. [63] presented an approach for constructing an MCDS based on multi-point relays (MPR), but there is no approximation analysis of the algorithm yet. Recently, in [64], another distributed algorithm was proposed whose performance ratio is 147. This algorithm contains three steps. Step 1 constructs a forest in which each tree is rooted at a node with the minimum ID among its 1-hop away neighbors. Step 2 collects the neighborhood information, which is used in Step 3 to connect neighboring trees.

3.2.4 Other Algorithms

Because CDSs can benefit WSNs, a variety of other factors are considered when constructing CDSs. More than one CDS can be found for each WSN. To conserve energy, all CDSs are constructed and each CDS serves as the virtual backbone duty cycled in [43]. For the sake of fault tolerance, k-connect m-dominating sets [23] are constructed, where k-connectivity means between any pair of backbone nodes there exist at least k independent paths, and m-dominating represents that every dominatee has at least m adjacent dominator neighbors. To minimize delivery delay, a special CDS problem — minimum routing cost CDS (MOC-CDS) [25] is proposed, where each pair of nodes in MOC-CDS has the shortest path. The work [26] considers size, diameter, and average backbone path length (ABPL) of a CDS in order to construct a CDS with better quality.

3.2.5 Remarks

All the above mentioned existing works consider a k-connect m-dominating CDS, a minimum routing cost CDS or a bounded-diameter CDS. Unfortunately, they do not consider the load-balance factor when constructing a CDS. In contrast, in this chapter, we first show an example to illustrate that a traditional MCDS cannot prolong network lifetime by reducing the communication cost. Instead, it actually increases the workload imbalance among dominators, which leads to the reduction of network lifetime. Based on this observation, we then study an LBCDS and load-balancedly allocate dominatees to dominators. An effective GA named LBCDS-GA is proposed to solve the problem in Section 3.4.

3.3 Problem Definition

In this section, we give an overview of the LBCDS problem. We first present the assumptions and introduce the network model. Then, we give the problem definition. Finally, we point out the key issues and main challenges we are facing when solving the problem.

3.3.1 Network Model

We assume a connected static WSN and all the nodes in the WSN have the same transmission range. Hence, we model a WSN as an undirected graph $\mathbb{G}(\mathbb{V}, \mathbb{E})$, where $\mathbb{V}$ is the set of n sensor nodes, denoted by $v_1, v_2, \ldots, v_n$; $\mathbb{E}$ represents the link set, $\forall\, u, v \in \mathbb{V}$, there exists an link (u, v) in $\mathbb{E}$ if and only if u and v are in each other's transmission range. In this chapter, we assume edges are undirected (bidirectional), which means two linked nodes are able to transmit and receive data from each other. We further assume the each node has the same amount of traffic.

3.3.2 Terminologies

The *load-balance* factor is our major concern in this work. Thus, finding an appropriate measurement to evaluate load-balance is the key to solve the LBCDS problem. We use *p-norm* to measure load-balance of a CDS. The definition of *p-norm* is given as follows:

Definition 3.1 p-norm ($|\Theta|_p$). *The* p-norm *of an* $n \times 1$ *vector* $\Theta = (\theta_1, \theta_2, \cdots, \theta_n)$ *is:*

$$|\Theta|_p = \left(\sum_{i=1}^{n} |\theta_i|^p\right)^{\frac{1}{p}} \tag{3.1}$$

The authors in [50] stated that *p-norm* shows interesting properties for different values of p. If p is close to 1, the information routes resemble the geometric shortest paths from the sources to the sinks. For $p = 2$, the information flow shows an analogy to the electrostatics field, which can be used to measure the load balance among θ_i. The smaller the *p-norm* value is, the more load balanced the interested feature vector Θ is.

In this chapter, we use node degree, denoted by d_i, as the feature vector Θ in Equation (3.1) to measure the load balance of a CDS, since the degree of each node is a potential indicator of traffic load. Thus, the definition of *CDS p-norm* is given as follows:

Definition 3.2　CDS p-norm ($|\mathbb{B}|_p$). *For a WSN represented by graph* $\mathbb{G}(\mathbb{V}, \mathbb{E})$ *and a CDS* $\mathbb{B} = \{v_1, v_2, \cdots, v_m\}$, *the* CDS p-norm *of an* $m \times 1$ *vector* $\mathbb{D} = (d_1, d_2, \cdots, d_m)$ *is:*

$$|\mathbb{B}|_p = (\sum_{i=1}^{m} |d_i - \bar{d}|^p)^{\frac{1}{p}}$$

where m is the number of dominators in the set $\mathbb{B}$, d_i *represents the node degree of each dominator in the set* $\mathbb{B}$, *and* $\bar{d}$ *is the mean degree of* $\mathbb{G}$.

We use the WSN shown in Figure 3.1 to illustrate how to calculate the *CDS p-norm*. For simplicity, we use $p = 2$ in this chapter. Without specific explanation, p and 2 are interchangeable in this chapter. Two different CDSs for the same network are identified in Figure 3.1. The degree of node v_i is denoted by d_i in Figure 3.1. From the topology shown in Figure 3.1, we can get $\bar{d} = 3$. Therefore, the *CDS p-norm* of the CDS shown in Figure 3.1(a) is $\sqrt{9}$. Similarly, in Figure 3.1(b), the *CDS p-norm* value is $\sqrt{2}$. Clearly, $\sqrt{2} < \sqrt{9}$, which implies that the CDS in Figure 3.1(b) is more load balanced than the CDS in Figure 3.1(a). The result also matches the observation we mentioned in Section 3.1.

When constructing an LBCDS, it is important to allocate dominatees to each dominator. In a traditional/naive way, such as the work in [17], each dominatee is allocated to the neighboring dominator with the smallest ID. Obviously, the load-balance factor is not taken into account. In some environment, the dominator with the smallest ID, which is chosen by majority dominatees, tends to have heavier workload than the other dominators. Therefore, neither node ID nor node degree can reflect workload precisely. In a WSN with a CDS as the VB, only the dominator and dominatee links contribute to workload. Based on this observation, we define the following concepts:

Definition 3.3　Dominatee allocation scheme ($\mathscr{A}$). *For a WSN represented by graph* $\mathbb{G}(\mathbb{V}, \mathbb{E})$ *and a CDS* $\mathbb{B} = \{v_1, v_2, \cdots, v_m\}$, *we need to find m disjoint sets on* $\mathbb{V}$, *i.e,* $\mathbb{A}(v_1), \mathbb{A}(v_2), \cdots, \mathbb{A}(v_m)$, *such that:*

1. *Each set* $\mathbb{A}(v_i)$ $(1 \leq i \leq m)$ *contains exactly one dominator* v_i.

2. $\bigcup_{i=1}^{m} \mathbb{A}(v_i) = \mathbb{V}$, *and* $\mathbb{A}(v_i) \bigcap \mathbb{A}(v_j) = \emptyset$ $(1 \leq i \neq j \leq m)$.

3. $\forall v_u \in \mathbb{A}(v_i)$ $(1 \leq i \leq m)$ *and* $v_u \neq v_i$, *such that* $(v_u, v_i) \in \mathbb{E}$.

A dominatee allocation scheme *is:*

$$\mathscr{A} = \{\mathbb{A}(v_i) \mid \forall v_i \in \mathbb{B}, 1 \leq i \leq m\}$$

Definition 3.4 Valid degree (*d′*). *The valid degree of dominator v_i is the number of its allocated dominatees, i.e., $\forall v_i \in \mathbb{B}, d'_i = |\mathbb{A}(v_i)|$, where $|\mathbb{A}(v_i)| - 1$ represents the number of dominatees in the set $\mathbb{A}(v_i)$.*

In this chapter, we use the *allocation scheme p-norm* to measure the load balance of different dominatee allocation schemes, in which the *valid degree* of each dominator is used as the information vector Θ. The definition of the *allocation scheme p-norm* is given as follows:

Definition 3.5 Allocation Scheme p-norm ($|\mathscr{A}|_p$). *For a WSN represented by graph $\mathbb{G}(\mathbb{V}, \mathbb{E})$, a CDS $\mathbb{B} = \{v_1, v_2, \cdots, v_m\}$, and a dominatee allocation scheme $\mathscr{A}$, the* Allocation Scheme p-norm *is:*

$$|\mathscr{A}|_p = (\sum_{i=1}^{m} |d'_i - \mathcal{E}|^P)^{\frac{1}{p}}$$

where d'_i represents the valid degree of each dominator in the set $\mathbb{B}$, and $\mathcal{E} = \frac{n-m}{m}$ is the expected allocated dominatees on each dominator.

Figure 3.1(b) and Figure 3.1(c) illustrate an imbalanced and a balanced dominatee allocation scheme respectively. The *valid degree* of dominator v_i is denoted by d'_i in Figure 3.1. From the topology shown in Figure 3.1 (b) and (c), we can get $\mathcal{E} = \frac{5}{3}$. Therefore, the *allocation scheme p-norm* of the dominatee allocation scheme shown in Figure 3.1(b) is $\sqrt{2.67}$. Similarly, in Figure 3.1(c), the *allocation scheme p-norm* is $\sqrt{0.67}$. Clearly, $\sqrt{0.67} < \sqrt{2.67}$, which implies the dominatee allocation scheme shown in Figure 3.1(c) is more load-balanced than the scheme shown in Figure 3.1(b). The result further confirms the observation we mentioned in Section 3.1.

3.3.3 Problem Definition

Definition 3.6 Load-balanced CDS (LBCDS) problem. *For a WSN represented by graph $\mathbb{G}(\mathbb{V}, \mathbb{E})$, the* LBCDS *problem is to find a minimum-sized node set $\mathbb{B} \subseteq \mathbb{V}$ and a dominatee allocation scheme $\mathscr{A}$, such that:*

1. $\mathbb{G}[\mathbb{B}] = (\mathbb{B}, \mathbb{E}')$, *where $\mathbb{E}' = \{e | e = (u, v), u \in \mathbb{B}, v \in \mathbb{B}, (u, v) \in \mathbb{E})\}$, is connected.*

2. $\forall u \in \mathbb{V}$ *and $u \notin \mathbb{B}, \exists v \in \mathbb{B}$, such that $(u, v) \in \mathbb{E}$.*

3. $min\{|\mathbb{B}|_p, |\mathscr{A}|_p\}$.

We claim that the LBCDS problem is NP-hard, since it still belongs to the MCDS problem. Based on Definition 3.6, the key issue of the LBCDS problem is to seek a tradeoff between the minimum-sized CDS, the load-balance of a constructed CDS, and a dominatee allocation scheme. GAs are population-based search algorithms, which simulate biological evolution processes and have successfully solved a wide range of NP-hard optimization problems [56][57]. Additionally, GAs have shown themselves to be very good at discovering good solutions with a reasonable amount of time and computation effort. In the following, a novel GA algorithm, named LBCDS-GA, is proposed to solve the LBCDS problem.

3.4 LBCDS-GA Algorithm

In the following subsections, we first provide some basics of the GA optimization approach, and then present the detailed design of the RMCDS-GA algorithm for the RMCDS problem.

3.4.1 GA Overview

GAs are adaptive heuristic search algorithms based on the evolutionary ideas of natural selection and genetics [65]. In nature, over many generations, natural populations evolve according to the principles of natural selection and *survival of the fittest*. By mimicking this process, GAs work with a population of *chromosomes*, each representing a possible solution to a given problem. Each chromosome is assigned a *fitness score* according to how good a solution to the problem it is. The fittest chromosomes are given opportunities to *reproduce* by *crossover* with other chromosomes in the population. This produces new chromosomes as *offsprings*, which share some features taken from each *parent*. The least fittest chromosomes of the population are less likely to be selected for reproduction, and so they *die out*. A whole new population of possible solutions is thus produced by selecting the best chromosomes from the current *generation*, and mating them to produce a new set of chromosomes. This new generation contains a higher proportion of the characteristics possessed by the good chromosomes of the previous generation. In this way, over many generations, good characteristics are spread throughout the population. If the GA has been designed well, the population will converge to an optimal solution to the problem. In the following part of this section, we will design and explain LBCDS-GA step by step.

3.4.2 Representation of Chromosomes

A chromosome is a possible solution of the LBCDS problem. Hence, when designing the encoding scheme of chromosomes, we need to identify dominators and dominatees in a chromosome and a dominatee allocation scheme in a chromosome as well. For convenience, the set of neighboring dominators of each dominatee $v_s \in \mathbb{W}$ is denoted by $\mathbb{H}(v_s) = \{v_r | v_r \in \mathbb{B}, (v_r, v_s) \in \mathbb{E}\}$. In the proposed LBCDS-GA, each node is mapped to a gene in the chromosome. A gene value g_i indicates whether the sensor represented by this gene is a dominator or not. If the sensor is a dominatee, the corresponding gene value represents the allocated dominator. Hence, a generation of chromosomes with gene values is denoted as: $\mathbb{C}^g = \{C_j^g \mid 1 \leq j \leq k, C_j^g = (g_1, g_2, \cdots, g_i, \cdots, g_n)\}$, where k is the number of the chromosomes in each generation of population, and for $1 \leq i \leq n$,

$$g_i = \begin{cases} 1, v_i \in \mathbb{B}. \\ \forall v_t \in \mathbb{H}(v_i), v_i \in \mathbb{W}. \end{cases}$$

Additionally, beyond the aforementioned gene value, there is a meta-gene value G_i to store $\mathbb{H}(v_s)$, $\forall v_s \in \mathbb{W}$. Thus, a generation of chromosomes with meta-gene values is denoted as: $\mathbb{C}^G = \{C_j^G \mid 1 \leq j \leq k, C_j^G = (G_1, G_2, \cdots, G_i, \cdots, G_n)\}$, and for $1 \leq i \leq n$,

$$G_i = \begin{cases} 1, v_i \in \mathbb{B}. \\ \mathbb{H}(v_i), v_i \in \mathbb{W}. \end{cases}$$

Through the above description we know, as long as we choose a specific node from each node set $\mathbb{H}(v_i), \forall v_i \in \mathbb{W}$, we can easily generate C_j^g from C_j^G. Additionally, all the nodes with $g_i/G_i = 1$ form a CDS $\mathbb{B} = \{v_i \mid g_i/G_i = 1, 1 \leq i \leq n\}$. An example WSN is shown in Figure 3.1(c) to illustrate the encoding scheme. There are eight nodes and the CDS is $\mathbb{B} = \{v_3, v_6, v_7\}$. Moreover, according to the topology shown in Figure 3.1, $\forall v_i \in \mathbb{W}$, we can get $\mathbb{H}(v_i)$ easily. Thus, the eight nodes can be encoded using eight meta-genes in a chromosome, e.g., $C^G = (\{v_3\}, \{v_3\}, 1, \{v_3, v_6, v_7\}, \{v_6\}, 1, 1, \{v_7\})$ shown in Figure 3.2. Based on the dominatee allocation scheme shown in Figure 3.1(c), i.e., dominatee v_4 is allocated to dominator v_6, the chromosome with eight genes is $C^g = (\{v_3\}, \{v_3\}, 1, \{v_6\}, \{v_6\}, 1, 1, \{v_7\})$. In conclusion, C_j^G stores all neighboring dominators of each dominatee, while the corresponding C_j^g records one CDS and one specific dominatee allocation scheme.

	G_1	G_2	G_3	G_4	G_5	G_6	G_7	G_8
C^G	$\{v_3\}$	$\{v_3\}$	1	$\{v_3,v_6,v_7\}$	$\{v_6\}$	1	1	$\{v_7\}$
	g_1	g_2	g_3	g_4	g_5	g_6	g_7	g_8
C^g	v_3	v_3	1	v_6	v_6	1	1	v_7

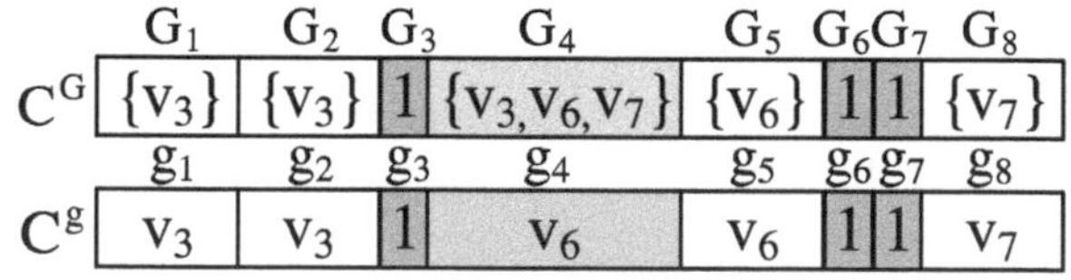

Figure 3.2: A chromosome with meta-genes and genes.

3.4.3 Population Initialization

GAs differ from most optimization techniques because of their global searching effectuated by one population of solutions rather than from one single solution. Hence, a GA search starts with the creation of the first generation, i.e., a population with k chromosomes denoted by P_1. This step is called population initialization. A general method to initialize the population is to explore the genetic diversity. That is, for each chromosome, all dominators are randomly generated. However, the dominators must form a CDS. Therefore we start to create the first chromosome C_1 by running an existing MCDS method, e.g., the latest MCDS construction algorithm [17], and then generate the population with k chromosomes by modifying C_1. We call the procedure, generating the whole population by modifying one specific chromosome, inheritance population initialization (IPI). The IPI algorithm is summarized as follows:

If the number of the generated chromosomes in P_1 is less than k, run the following steps till k non-duplicated chromosomes are generated.

1) Let $t = 1$.

2) In the CDS $\mathbb{B}_t$ represented by the chromosome C_t, start from node v_u with the smallest ID (ID used here is only to create a sequence for generating new chromosomes. Any other features who can rank the nodes also can be applied here.) in $\mathbb{B}_t$, and add one neighboring dominatee by the order of its ID into the CDS each time. $\mathbb{B}_s = \mathbb{B}_t \bigcup \{v_i \mid \forall v_i \in \mathbb{U}(v_u), v_i \text{ has the smallest ID among the nodes in the set } \mathbb{U}(v_u)\}$. Encoding $\mathbb{B}_s$ as a chromosome, add it into P_1. Repeat the procedure, till all nodes in the set $\mathbb{U}(v_u)$ are added into $\mathbb{B}$. For example, in Figure 3.1(a), the CDS $\mathbb{B}_1 = \{v_4, v_7\}$ of the shown WSN is given. Thus, the encoded chromosome with meta-genes is $C_1^G = (\{v_4\}, \{v_4\}, \{v_4\}, 1, \{v_4\}, \{v_4, v_7\}, 1, \{v_7\})$. The node with the smallest ID is v_4 in $\mathbb{B}_1$. Therefore, the chromosomes from C_2^G to C_6^G are generated by adding one node from the set $\mathbb{U}(v_4) = \{v_1, v_2, v_3, v_5, v_6\}$ each time.

3) Move to the node with the second smallest ID in CDS $\mathbb{B}_t$, doing the same procedure as described in step 2), till every node in $\mathbb{B}_t$ is checked. As shown in Figure 3.1(a), $\mathbb{U}(v_7) = \{v_6, v_8\}$. By eliminating the duplicates, the chromosome C_7^G is created by adding v_8.

4) If all the dominators in the current $\mathbb{B}_t$ are checked, move to the next CDS by setting $t = t + 1$, and repeat steps from 2) to 4).

Since each dominatee has two choices: to change to a dominator or to remain as a dominatee, consequently, there are $2^{n-|\mathbb{B}|}$ possible ways to create new chromosomes. Usually, k is much smaller than $2^{n-|\mathbb{B}|}$. Hence, the first population P_1 can be easily generated.

There are several merits that need to be pointed out here when using the IPI algorithm to generate P_1. First, we can guarantee that each chromosome in P_1 is a feasible solution (i.e., a CDS) of the LBCDS problem. Second, the critical nodes (cut nodes) are chosen to be dominators. When reproducing new offsprings, the critical nodes are still dominators in the new chromosomes in the successive generations, which can help guarantee the connectivity of a CDS.

3.4.4 Fitness Function

Given a solution, its quality should be accurately evaluated by the fitness score, which is determined by the fitness function. In our algorithm, we aim to find a minimum-sized CDS $\mathbb{B}$ with minimum $|\mathbb{B}|_p$ and $|\mathscr{A}|_p$ values. Therefore, the fitness function of a chromosome C_i^g is defined as:

$$
\begin{cases}
f(C_i^g) = \dfrac{n - |\mathbb{B}|}{w_1 |\mathbb{B}|_p + w_2 |\mathscr{A}|_p} \\
w_1 + w_2 = 1,\ 0 < w_1, w_2 < 1
\end{cases}
\tag{3.2}
$$

The purpose of doing a linear combination of $|\mathbb{B}|_p$ and $|\mathscr{A}|_p$ values in Equation (3.2) is that a user can change the weight of $|\mathbb{B}|_p$ and $|\mathscr{A}|_p$ values dynamically and easily. The denominator in Equation (3.2) needs to be minimized (the smaller the *p-norm* value, the more load-balanced the interested feature vector), while the numerator needs to be maximized (since we seek an MCDS). As a result, the fitness function value needs to be maximized.

3.4.5 Selection Scheme

During the evolutionary process, election plays an important role in improving the average quality of the population by passing the high quality chromosomes to the next generation. Therefore, the selection operator needs to be carefully formulated to ensure that better chromosomes (higher fitness scores) of the population have a greater probability of being selected for mating, but that worse chromosomes of the population still have a small probability of being selected. Having some probability of choosing worse members is important to ensure that the search process is global and does not simply converge to the nearest local optimum. We adopt *rank selection* (RS) to select parent chromosomes. In order to prevent very fit chromosomes from gaining dominance early at the expense of less fit ones, which would reduce the population's genetic diversity, we set the rank value of each chromosome to be $R_i = log(1 + f(C_i^g))$. Thus, RS stochastically selects chromosomes based on R_i. A real-valued interval, S, is determined as the sum of the chromosomes' expected selection probabilities $P_i = \frac{R_i}{\sum_{j=1}^{k} R_j}$, thus, $S = \sum_{i=1}^{k} P_i$. Chromosomes are then mapped one-to-one into contiguous intervals in the range $[0, S]$. To select a chromosome, a random number is generated in the interval $[0, S]$ and the chromosome whose segment spans the random number is selected. This process is repeated until a desired number of chromosomes have been selected.

3.4.6 Genetic Operations

The performance of a GA relies heavily on two basic genetic operators, *crossover* and *mutation*. Crossover exchanges parts of the current solutions (the parent chromosomes selected by the RS scheme) in order to find better ones. Mutation flips the values of genes, which helps a GA keep away from local optimum. The type and implementation of these two operators depend on the encoding scheme and also on the application. In the LBCDS problem, we can adopt classical operations, however, the new obtained solutions may not be valid (the dominator set represented by the chromosome is not a CDS) after implementing the crossover and mutation operations. Therefore, a correction mechanism needs to be preformed to guarantee the validity of all the new generated offspring solutions.

3.4.6.1 Crossover

The purpose of crossover operations is to produce more valid CDSs represented by the new generated chromosomes. At this stage, we do not need to care dominatee allocations. Therefore, when performing crossover operations, we can logically assume all gene values of dominatees are 0, i.e., $g_i = 0, \forall v_i \in \mathbb{W}$. After the new CDS is created, we can easily fill in all meta-gene values based on its original topology.

In the LBCDS-GA algorithm, we adopt three crossover operators called single-point crossover, two-point crossover, and uniform crossover respectively. With a crossover probability p_c, each time we use the RS scheme to select two chromosomes C_i^g and C_j^g as parents to perform one of the three crossover operators randomly. We use Figure 3.3 to illustrate the three crossover operations.

Suppose that two parent chromosomes (00010011) and (00100110) are selected by the RS scheme from the population. By the slngle-point crossover (shown in Figure 3.3(a)), the genes from the randomly generated crossover point $P = 6$ to the end of the two chromosomes exchange with each other to get (00010110) and (00010111). After crossing, the first offspring (00010110) is a valid solution. However, the other one (00100011) is not valid, thus we need to perform the *correction mechanism*. The mechanism starts with scanning each gene on the offspring chromosome, denoted by C_o^g, till the end of the chromosome. If the value of the current scanned gene is 0, i.e., $g_i = 0$ and the gene value is different from the original chromosome, denoted by C_s^g, without doing crossover and mutation operations, then change the gene value to 1. Whenever the DS represented by the corrected chromosome is a CDS, stop the mechanism. Otherwise, keep repeating the process till the end of C_o^g is reached. The idea behind the correction mechanism is that the DS represented by C_s^g is a CDS. If C_o^g is not valid, then add the dominators represented by C_s^g into the DS represented by C_o^g one by one. Finally, the corrected chromosome must be valid. For example, for the specific invalid offspring chromosome (00100011), when scanning the gene at position P, i.e. $g_6 = 0$, we find the value of g_6 is different after crossing. Therefore, we correct it by setting $g_6 = 1$. Then the corrected chromosome (00010111) is now a valid solution. Consequently, the correction mechanism stops and we get two valid offspring chromosomes (00010110) and (00010111). The correction mechanism is the same for crossover and mutation operations.

By the two-point crossover (shown in Figure 3.3(b)), the two crossover points are randomly generated which are $P_L = 3$ and $P_R = 6$; and then the genes between P_L and P_R of the two parent chromosomes are exchanged with each other. The two

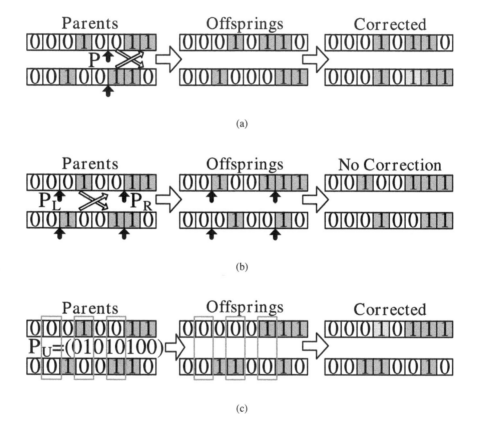

Figure 3.3: Illustration of crossover operations: (a) single-point crossover; (b) two-point crossover; (c) uniform crossover.

offsprings are (00100111) and (00010010) respectively. Since both of the offspring chromosomes are valid, we do not need to do any correction.

For the uniform crossover (shown in Figure 3.3(c)), the vector of uniform crossover P_U is randomly generated, which is $P_U = (01010100)$, indicating that g_2, g_4, and g_6 of the two parent chromosomes exchange with each other. Hence the two offsprings are (00000111) and (00110010). Since the first offspring is not a valid solution, we need to perform the correction mechanism mentioned before, and the corrected chromosome becomes (00110010), which is a valid solution.

3.4.6.2 Gene Mutation

The population undergoes the gene mutation operation after the crossover operation is performed. With a mutation probability p_m, we scan each gene g_i on the offspring chromosomes. If the mutation operation needs to be implemented, the value of the gene flips, i.e., 0 becomes 1, and 1 becomes 0. The correction mechanism mentioned before needs to be preformed if the mutated chromosomes are not valid.

3.4.7 Meta-gene Mutation

Unlike traditional GAs, in LBCDS-GA, we perform an additional operation named meta-gene mutation on k chromosomes in each generation. As mentioned before, the purpose of crossover operations is to produce more valid CDSs represented by the new offspring chromosomes. Moreover, the gene mutation operation after the crossover operation helps a GA keep away from local optimum. In summary, the aforementioned crossover and gene mutation operations only provide the chance to increase diversity of possible CDSs, however, nothing is aimed to create the diversity of dominatee allocation schemes. In fact, to address the LBCDS problem, we need to find a load-balanced CDS and load-balancedly allocate dominatees to dominators. Therefore, meta-gene mutation is proposed in LBCDS-GA to generate more possible dominatee allocation schemes.

As known, as long as choosing a specific node from each node set $\mathbb{H}(v_i), \forall v_i \in \mathbb{W}$, we can easily generate C_j^g from C_j^G. Thus, the procedure to determine gene values from meta-gene values is the procedure to specify a dominatee allocation scheme. According to the observation, we design the following described meta-gene mutation. The original population without doing crossover and gene mutation operations will undergo the meta-gene mutation operation. If the number of neighboring dominators of a dominatee v_i is greater than 1, i.e., $|\mathbb{H}(v_i)| \geq 2$, then randomly pick a node from the set $\mathbb{H}(v_i)$ with a probability p_i. For example, the CDS shown in Figure 3.1(b), and (c) is encoded as the chromosome with meta-genes $(\{v_3\}, \{v_3\}, 1, \{v_3, v_6, v_7\}, \{v_6\}, 1, 1, \{v_7\})$, which is shown in Figure 3.2. Since $G_4 = \mathbb{H}(v_4) = \{v_3, v_6, v_7\}, |\mathbb{H}(v_4)| \geq 2$, we then randomly pick one dominator from the set $\mathbb{H}(v_4)$ with a probability p_i. If v_3 is selected from $\mathbb{H}(v_4)$, it means dominatee v_4 is allocated to dominator v_3. The dominatee allocation scheme is shown in Figure 3.1(b), encoding as the chromosome with genes $(\{v_3\}, \{v_3\}, 1, \{v_3\}, \{v_6\}, 1, 1, \{v_7\})$. Similarly, if dominatee v_4 is allocated to dominator v_6, the dominatee allocation scheme is shown in Figure 3.1(c), encoding as the chromosome with genes $(\{v_3\}, \{v_3\}, 1, \{v_6\}, \{v_6\}, 1, 1, \{v_7\})$.

To easily understand the traditional gene mutation and our proposed meta-gene mutation on chromosomes, we conclude the differences as follows:

1) The gene mutation operation is bit-wised, while the meta-gene mutation is performed at some positions i satisfying the condition $|\mathbb{H}(v_i)| \geq 2$, and $v_i \in \mathbb{W}$.

2) The gene mutation flips the logic gene values, i.e., 0 becomes 1, and 1 becomes 0. In contrast, the meta-gene mutation only flips meta-gene values at some specific positions i, i.e., randomly pick one node from the set $G_i = \mathbb{H}(v_i)$.

3) The purpose of gene mutation is to create diversity of all possible CDSs, while the purpose of meta-gene mutation is to provide more different dominatee allocation schemes. Since constructing a load-balanced CDS and load-balancedly allocating dominatees to dominators are two critical challenges to solve the LBCDS problem, neither of the gene mutations and meta-gene mutations can be ignored in LBCDS-GA.

3.4.8 Replacement Policy

The last step of LBCDS-GA is to create a new population using an appropriate replacement policy. From crossover and gene mutation operations, we can get k offspring chromosomes. In addition, we can get other k chromosomes from the meta-gene mutation operation. In LBCDS-GA, we utilize the best k chromosomes (i.e., the chromosome with the highest fitness score) among those $2k$ chromosomes to generate a new population. However, when creating new population by crossover, gene mutation, and meta-gene mutation, there is a chance to lose the fittest chromosome. Therefore, an elitism strategy, in which the best chromosome (or a few best chromosomes) is retained in the next generation's population, is used to avoid losing the best candidates.

The LBCDS-GA stops and returns the current fittest solution until the number of total generations K is reached or the best fitness score does not change for continuous l generations. Figure 3.4 shows the flow chart of the LBCDS-GA algorithm.

Table 3.1: GA Parameters and Rules

Population size (k)	50
Number of total generations (K)	100
Selection scheme	Rank Selection
Replacement policy	Elitism
Crossover probability (p_c)	1
Gene mutation probability (p_m)	0.2
Meta-gene mutation probability (p_i)	1

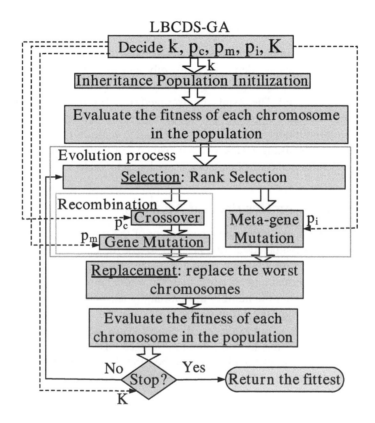

Figure 3.4: Procedure of LBCDS-GA.

3.5   Performance Evaluation

In the simulations, the results of LBCDS-GA are compared with the MCDS construction algorithm in [17] denoted by MIS, which is the latest and best MIS-based CDS construction algorithm. We compare the two algorithms in terms of the size of the constructed CDS, *CDS p-norm*, *allocation scheme p-norm*, the fitness score, network lifetime (which is defined as the time duration until the first dominator runs out of energy), and the average remaining energy over the whole network.

3.5.1   Simulation Environment

We build our own simulator where all nodes have the same transmission range of 50 *m* and all nodes are deployed uniformly and randomly in a square area of 300 *m* × 300 *m*. *n* is incremented from 100 to 1000 by 100. For a certain *n*, 100 instances are generated. The results are averaged over 100 instances. Moreover, we use the CDS-based data aggregation as the communication mode. The simulated energy consumption model is that every node has the same initial 1000 unit energy. Receiving and transmitting a packet both consume 1 of unit energy. Additionally, the particular GA rules and control parameters are listed in Table 3.1.

3.5.2   Simulation Results and Analysis

In Figure 3.5, the X-axis represents the number of the sensor nodes n, while the Y-axis represents the evaluated factors, i.e., the size of the constructed CDS $|\mathbb{B}|$, *CDS p-norm* $|\mathbb{B}|_p$, *allocation scheme p-norm* $|\mathscr{A}|_p$, the fitness score f, network lifetime T, and the average remaining energy E over the whole network respectively.

Figure 3.5(a) shows the number of dominators $|\mathbb{B}|$ of the constructed CDSs by using LBCDS-GA and MIS. From Figure 3.5(a), we can see that, with the increase of the number of the sensor nodes n, $|\mathbb{B}|$ almost keeps stable for the MIS scheme. This is because MIS aims to find a minimum-sized CDS and the area covered by the deployed sensors does not change. On the other hand, for LBCDS-GA, $|\mathbb{B}|$ increases when n increases. This is because the objective of LBCDS-GA is to balance energy consumption on each dominator. If more nodes are chosen as dominators, the energy consumption can be distributed to more components. More importantly, the more dominators, the more possibilities to create diversity of dominatee allocation schemes. Therefore, the load-balanced objective can be achieved.

Figure 3.5(b) shows the *CDS p-norm* $|\mathbb{B}|_p$ values of the constructed CDSs by using LBCDS-GA and MIS. With the increase of n, $|\mathbb{B}|_p$ increases correspondingly for both schemes. This is because when n increases, the area covered by the deployed sensors does not change. Therefore, the density of sensors increases, which means the degree of each dominator increases correspondingly. According to Definition 3.2, larger degrees of dominators imply larger subitems, thus $|\mathbb{B}|_p$ of both schemes increase. Moreover, the $|\mathbb{B}|_p$ of LBCDS-GA is larger than that of MIS. This is because we need more nodes to build an LBCDS (shown in Figure 3.5(a)). More dominators

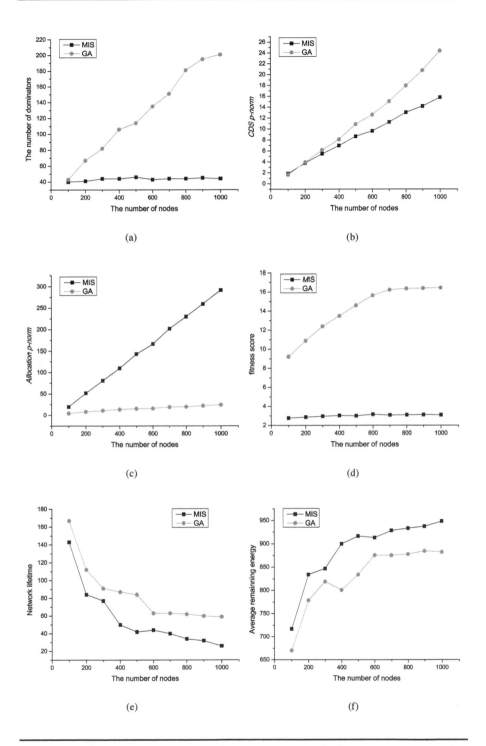

Figure 3.5: Simulation results: (a) fitness score; (b) *CDS p-norm*; (c) *allocation scheme p-norm*; (d) the number of dominators; (e) network lifetime; (f) average remaining energy.

imply more subitems based on Definition 3.2 and thus, the $|\mathbb{B}|_p$ of LBCDS-GA is larger.

Figure 3.5(c) shows the *allocation scheme p-norm* $|\mathscr{A}|_p$ values of the constructed CDSs by using LBCDS-GA and MIS. As mentioned before, the smaller the $|\mathscr{A}|_p$ value, the more load-balanced the dominatee allocation scheme $\mathscr{A}$. With the increase of n, $|\mathscr{A}|_p$ increases quickly for the MIS scheme. This is because, in MIS, dominatees are always allocated to the dominator with the smallest ID. The results also imply that $\mathscr{A}$ of MIS becomes more and more imbalanced when n is getting larger. Nevertheless, for LBCDS-GA, $|\mathscr{A}|_p$ keeps almost the same, which means no matter how large the size of the set $\mathbb{B}$ is, LBCDS-GA always can find a load-balanced $\mathscr{A}$. Additionally, with the increase of n, the difference of $|\mathscr{A}|_p$ values between the two schemes become more and more obvious. This indicates LBCDS-GA becomes more and more effective to find an LBCDS in large scale WSNs.

Figure 3.5(d) shows fitness scores f of the constructed CDSs by using LBCDS-GA and MIS. As mentioned before, the higher the fitness score is, the better quality the solution has. From Figure 3.5(d), we can see that, with the increase of n, f does not change too much for MIS. However, for LBCDS-GA, f increases quickly. The results imply LBCDS-GA can find a more load-balanced CDS than MIS. This is because the MIS scheme does not consider the load-balance factor when building a CDS and allocating dominatees to dominators. Additionally, it is apparent that LBCDS-GA has more benefits when n becomes large.

Figure 3.5(e) shows network lifetime of the two schemes. From Figure 3.5(e), we know that the network lifetime decreases for both schemes with n increasing, since the WSN becomes denser and denser. Additionally, we can see LBCDS-GA prolongs network lifetime by 65% on average compared with MIS. In some extreme cases, such as $n = 1000$, network lifetime is extended by 100% compared with MIS. The result demonstrates that constructing an LBCDS and load-balancedly allocating dominatees to dominators can improve network lifetime significantly.

Figure 3.5(f) shows the average remaining energy E over the whole network of the two schemes. With the increase of n, E increases for both schemes. As the WSN becomes denser and denser, a lot of redundant sensors exist in the WSN. From Figure 3.5(f), we know that LBCDS-GA has less remaining energy than MIS. This is because LBCDS-GA considers the load-balance factor when building a CDS and allocating dominatees to dominators. Thus, the lifetime of the whole network is extended, which means the remaining energy of each node is less than MIS. This also indicates that constructing an LBCDS can balance the energy comsumption on each sensor node, making the lifetime of the whole network prolonged considerably.

3.6 Conclusion

In this chapter, we propose a novel concept load-balanced CDS (LBCDS), which is an MCDS with the minimum $|\mathbb{B}|_p$ and $|\mathscr{A}|_p$ values in order to assure that the workload among each dominator is balanced and load-balancedly allocates dominatees to each dominator. We claim that constructing an LBCDS is an NP-hard problem and propose an effective algorithm named LBCDS-GA to address the problem. The extensive simulation results demonstrate that using an LBCDS as a virtual backbone can balance the energy consumption among dominators. Consequently network lifetime is extended significantly. Particularly, when the node number changes from 100 to 1000, our proposed method prolongs network lifetime by 65% on average compared with the latest MCDS construction algorithm [17].

Chapter 4

Approximation Algorithms for Load-balanced Virtual Backbone Construction in Wireless Sensor Networks

CONTENTS

4.1   Introduction

Wireless sensor networks (WSNs) are composed of spatially distributed wireless sensor nodes, each having very limited communication and sensing range, computing ability, storage capability, and energy resources. They are deployed for monitoring and controlling systems where human intervention is not desirable or feasible. Therefore, WSNs are widely used in many military and civilian applications, such as battlefield surveillance, health care applications, environment and habitat monitoring, and traffic control [52]. Compared with traditional computer networks, WSNs have no fixed or pre-defined infrastructure, resulting in the difficulty to achieve routing scalability and efficiency [53]. To better improve the performance and increase the efficiency of routing protocols, a connected dominating set (CDS) has become a well known approach to form a virtual backbone (VB) in WSNs. A dominating set (DS) is defined as a subset of nodes in a WSN such that each node in the network is either in the set or adjacent to some node in the set. If the induced graph by the nodes in a DS is connected, then this DS is called a CDS. The nodes in a CDS are called *backbone nodes*, otherwise, *non-backbone nodes*. In a WSN with a CDS as its VB, non-backbone nodes may forward their data only to their neighboring backbone nodes. With the help of a CDS, the average message burden of a WSN could be reduced so that routing becomes much easier and can adapt quickly to network topology changes [54]. Furthermore, using backbone nodes as forwarding nodes can efficiently reduce energy consumption, which is also a critical concern in WSNs. In addition to routing protocols, a CDS-based VB has many other applications in WSNs, such as topology control [14], coverage [16], data collection [10][66], and data aggregation [17][67]. Clearly, the benefits of a CDS-based VB can be magnified by making its size (the number of the nodes in the CDS) smaller. In general, the smaller the CDS is, the less communication and storage overhead are incurred. Hence, it is desirable to build a minimum-sized CDS (MCDS)-based VB.

Since VBs bring substantial benefits to WSNs, ever since the idea of employing a CDS-based VB for WSNs was introduced [19], a huge amount of effort has been made to construct different CDS-based VBs for different applications, especially MCDS-based VBs. In the seminal work [20], Guha and Kuller first modeled the problem of constructing the minimum-sized CDS as the MCDS problem in a general graph, which is a well-known NP-hard problem [21]. After that, to make a CDS more resilient in mobile WSNs, the fault tolerance of a CDS-based VB is considered. In [23], k-connected and m-dominated sets are introduced as a generalized abstraction of a fault-tolerance VB. In [25], the authors proposed a minimum routing cost connected dominating set (MOC-CDS) problem, which aims to find a minimum-sized CDS while assuring that any routing path through this CDS is the shortest in WSNs. Additionally, the authors investigated the problem of constructing a quality CDS in terms of size, diameter, and average backbone path length (ABPL) in [26].

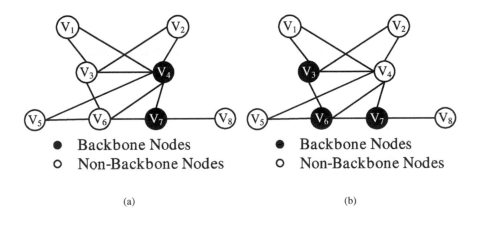

- ● Backbone Nodes
- ○ Non-Backbone Nodes

- ● Backbone Nodes
- ○ Non-Backbone Nodes

(a)   (b)

Figure 4.1: Illustration of a regular VB and a load-balanced VB: (a) regular VB; (b) load-balanced VB.

Unfortunately, all the aforementioned works did not consider the *load-balance* factor when they construct a VB. For instance, when the MCDS-based VB is used in the network shown in Figure 4.1(a), backbone node v_4 is adjacent to five different non-backbone nodes, whereas, backbone node v_7 only connects to two non-backbone nodes. If every non-backbone node has the same amount of data to be transferred through the neighboring backbone node at a fixed data rate, then the number of neighboring non-backbone nodes of each backbone node is a potential indicator of its traffic load. Hence, backbone node v_4 may deplete its energy much faster than backbone node v_7. A counter-example is shown in Figure 4.1(b), the set $\{v_3, v_6, v_7\}$ is served as a VB. Compared with the VB constructed in Figure 4.1(a), the numbers of neighboring non-backbone nodes of all the backbone nodes in Figure 4.1(b) are very similar. On the other hand, the criterion to allocate a non-backbone node to a neighboring backbone node is also critical to balance traffic load on each backbone node. An illustration of the allocation schemes for non-backbone nodes is

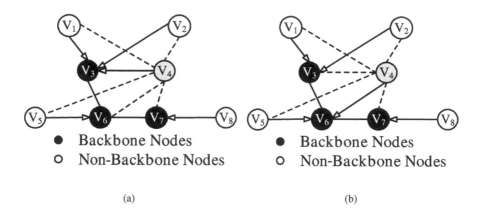

(a)

(b)

Figure 4.2: Illustration of a regular allocation versus a load-balanced allocation: (a) regular allocation; (b) load-balanced allocation.

depicted in Figure 4.2, in which arrow lines represent that the non-backbone nodes are allocated to the arrow pointed backbone nodes, while the dashed lines represent the communication links in the original network topological graph. Although the potential traffic load on each backbone node is evenly distributed in the VB constructed in Figure 4.2, different allocation schemes for non-backbone nodes might break the balance. In Figure 4.2(a), v_4 is allocated to backbone node v_3, while in Figure 4.2(b), v_4 is allocated to backbone node v_6. Apparently, backbone node v_3 has more traffic load than backbone nodes v_6 and v_7 in Figure 4.2(a). However, traffic loads are balanced among backbone nodes in Figure 4.2(b). Moreover, if the workloads on backbone nodes are not balanced, some heavy-duty backbone nodes will deplete their energy quickly. Then, the whole network might be disconnected. Intuitively, compared with the WSN shown in Figure 4.2(a), the VB and the allocation scheme shown in Figure 4.2(b) can extend network lifetime notably. In summary, constructing a load-balanced VB (LBVB) and then load-balancedly allocate non-backbone nodes to backbone nodes are equally important when considering the load-balance factor. Neither of these two aspects can be ignored.

To benefit from the CDS-based VB in WSNs and also take the load-balance factor into consideration, few attempts have been carried out to construct a VB in this manner [68]. In our previous work [68], we proposed a genetic algorithm-based method to build a load-balanced CDS (LBCDS) in WSNs. However, there is no performance ratio analysis in that chapter. In this research, we first investigate how to construct an LBVB. It is well known that in graph theory, a maximal independent set (MIS)[1] is also a DS. Therefore, we construct an LBVB with two steps. The first step is

---

[1] MIS can be defined formally as follows: given a graph $G = (V, E)$, an independent set (IS) is a subset $I \subset V$ such that for any two vertex $v_1, v_2 \in I$, they are not adjacent, i.e., $(v_1, v_2) \notin E$. An IS is called an MIS if we add one more arbitrary node to this subset, the new subset will not be an IS anymore.

to find a minmax degree MIS (MDMIS), and the second step is to make this MIS connected. Subsequently, we explore how to load-balancedly allocate non-backbone nodes to backbone nodes, followed by comprehensive performance ratio analysis. Particularly, our contributions mainly include three aspects as follows:

1. The LBVB problem is an NP-complete problem and therefore cannot be solved in polynomial time unless P = NP. Hence, we solve the LBVB problem with two steps. First, we propose an approximation algorithm by using linear relaxation and random rounding techniques to solve the minmax degree maximal independent set (MDMIS) problem. By analysis, this algorithm yields a solution upper bounded by $O(\ln(n))OPT_{MDMIS}$, where OPT_{MDMIS} is the optimal result of MDMIS, and n is number of sensors in a WSN. Subsequently, the minimal set of nodes are found to make the MDMIS connected. The upper bound on the size of the constructed LBVB is analyzed as well.

2. The problem of load-balancedly allocating non-backbone nodes to backbone nodes is NP-hard and thus it is formulated as an equivalent binary programming. Subsequently, we present a randomized approximation algorithm, which produces a solution with the traffic load on each backbone node upper bounded by $O(\log(n))OPT_{MVBA}$, where OPT_{MVBA} is the optimal result.

3. We also conduct extensive simulations to validate our proposed algorithms. The simulation results show that the constructed LBVB and the allocation scheme for non-backbone nodes can extend network lifetime significantly compared with the state-of-the-art algorithms. Particularly, our proposed algorithms prolong network lifetime by 69% on average compared with the latest and best MCDS-based VB [17], and by 47% on average compared with LBCDS [68].

The rest of this chapter is organized as follows: in Section 4.2, we review some related works on CDS-based VBs. In Section 4.3, we introduce the network model and formally define LBVB and the load-balancedly non-backbone node allocation problem. The design of algorithms and analysis for LBVB and the non backbone nodes allocation problem are presented in Sections 4.4 and 4.5, respectively. The simulation results are presented in Section 4.6 to validate our proposed algorithms. Finally, the chapter is concluded in Section 4.7.

4.2   Related Work

The idea of using a CDS as a VB was first proposed by Ephremides et al. in 1987 [19]. Since then, many algorithms that construct CDSs have been reported [20].

In the following section, we use n to denote the number of sensors in a WSN, Δ to denote the maximum degree of nodes in the graph representing a WSN, and *opt* to denote the size of any optimal MCDS.

4.2.1 Subtraction-based Algorithms for CDS-based VBs

Wu and Li first proposed a distributed algorithm in [33] to obtain a CDS. The CDS construction procedure consists of two stages. In the first stage, each node collects its neighboring information by exchanging message with the one-hop neighbors. If a node finds that there is a direct link between any pair of its one-hop neighbors, it removes itself from the CDS. In the second stage, additional heuristic rules are applied to further reduce the size of the CDS. Wu's algorithm [33] uses Rule 1 and Rule 2, where a node is removed from the CDS, if all its neighbors are covered by its one or two direct neighbors. Later, Dai's [34] work generalizes this as Rule k, in which coverage is defined by an arbitrary number of connected neighbors.

4.2.2 Addition-based Algorithms Using Single Leader for CDS-based VBs

Single-leader construction algorithms for CDS-based VBs use one initiator to initialize the algorithm. Usually, a base station can be the initiator for constructing CDSs in WSNs. In these algorithms, a spanning tree rooted at the initiator is first constructed, and then MISs are identified layer by layer. Finally, a set of connectors to connect the MISs is ascertained to form a CDS. Wan [37] presented an ID-based distributed algorithm to construct a CDS using a single initiator. For UDGs, Wan et al.'s [37] approach guarantees that the approximation factor on the size of a CDS is at most $4\,opt + 1$. Later, many attempts tried to improve this approximation factor on the size of a CDS based on this idea. Wu et al. reported an algorithm with approximation factor of $3.8\,opt + 1.2$ [38]. Yao et al. improved it to $3.67\,opt + 1.33$ [39]. The factor is further improved to $3.478\,opt + 4.874$ [40] and $3.4306\,opt + 4.8185$ [41], subsequently. The current best is that the size of an MIS is at most $3.399\,opt + 4.874$ [42].

4.2.3 Addition-based Algorithms Using Multiple Leader for CDS-based VBs

Algorithms with multiple leaders do not require an initiator to construct a CDS-based VB. Alzoubi et al.'s technique [61] first constructs an MIS using a distributed approach without a leader or tree construction and then interconnects these MIS nodes to get a CDS. Li et al. proposed a distributed algorithm called r-CDS in [62], whose performance ratio is 172. r-CDS is a completely distributed one-phase algorithm where each node only needs to know the connectivity information within its two-hop-away neighborhood. An MIS is constructed based on each node's r value which is defined as the number of this node's two-hop-away neighbors minus the degree of this node. The nodes with smaller r values are preferred to serve as MIS nodes. Adjih et al. [63] presented an approach for constructing an MCDS based on multi-point relays (MPR). However, there is no approximation analysis of the algorithm yet. Recently, in [64], another distributed algorithm was proposed whose performance ratio is 147. This algorithm contains three steps. Step one constructs a forest in which each tree is rooted at a node with the minimum ID among its 1-hop away neighbors. Step two collects the neighborhood information, which is used in Step three to connect neighboring trees.

4.2.4 Other Algorithms

Because VBs can benefit WSNs, a variety of other factors are considered when constructing VBs. More than one CDS can be found for each WSN. To conserve energy, all CDSs are constructed and each one serves as the VB duty cycled in [43]. For the sake of fault tolerance, k-connect m-dominating sets [23] are constructed, where k-connectivity means that between any pair of backbone nodes there exist at least k independent paths, and m-dominating represents that every non-backbone node has at least m adjacent backbone node neighbors. To minimize delivery delay, a special VB called the minimum routing cost CDS (MOC-CDS) [25] is proposed, where each pair of nodes in MOC-CDS has the shortest path. Moreover, the work [26] considers size, diameter, and average backbone path length (ABPL) of a CDS in order to construct a VB with better quality. A genetic algorithm-based LBCDS construction algorithm is proposed in [68] without performance ratio analysis.

4.2.5 Remarks

Most of the above mentioned existing works are considered to construct an MCDS-based VB, involve a k-connect m-dominating CDS-based VB, a minimum routing cost CDS-based VB, or a bounded-diameter CDS-based VB. Unfortunately, they do not consider the load-balance factor when constructing a VB except for the LBCDS investigated in [68]. In contrast, in this chapter, we first illustrate that a traditional MCDS cannot prolong network lifetime by reducing the communication cost. Instead, it actually increases the workload imbalance among backbone nodes, which leads to the reduction of network lifetime. Based on this observation, we then study to build an LBVB and load-balancedly allocate non-backbone nodes to backbone nodes. Approximation algorithms are proposed to address these problems in Section 4.4 and 4.5 followed by comprehensive theoretical analysis.

4.3 Problem Formulation

4.3.1 Network Model

We assume a static connected WSN and all the nodes in the WSN have the same transmission range. Hence, we model a WSN as an undirected graph $\mathbb{G} = (\mathbb{V}, \mathbb{E})$, where $\mathbb{V}$ is the set of n sensor nodes, denoted by v_i ($1 \leq i \leq n$). i is called the node ID of v_i in the chapter; $\mathbb{E}$ represents the link set. $\forall u, v \in \mathbb{V}, u \neq v$, there exists a link (u, v) in $\mathbb{E}$ if and only if u and v are in each other's transmission range. In this chapter, we assume links are undirected (bidirectional), which means two linked nodes are able to transmit and receive data from each other. Moreover, the degree of a node v_i is denoted by d_i, and Δ denotes the maximum degree in the network graph $\mathbb{G}$.

4.3.2 Problem Definition

As we mentioned in Section 4.1, we will solve the LBVB problem in two steps. The first step constructs a minmax degree maximal independent set (MDMIS), and the second step selects additional nodes which together with the nodes in the MDMIS induce a connected topology *LBVB*. In this subsection, we first formally define the MDMIS problem, followed by the problem definition of LBVB.

Definition 4.1   Minmax degree maximal independent set (MDMIS) problem. *For a WSN represented by graph* $\mathbb{G}(\mathbb{V}, \mathbb{E})$, *the* MDMIS *problem is to find a node set* $\mathbb{D} \subseteq \mathbb{V}$ *such that:*

1) $\forall u \in \mathbb{V}$ and $u \notin \mathbb{D}$, $\exists\, v \in \mathbb{D}$, such that $(u, v) \in \mathbb{E}$.

2) $\forall u \in \mathbb{D}$, $\forall v \in \mathbb{D}$, and $u \neq v$, such that $(u, v) \notin \mathbb{E}$.

3) There exists no proper subset or superset of $\mathbb{D}$ satisfying the above two conditions.

4) Minimize $\max\{d_i \mid \forall v_i \in \mathbb{D}\}$.

Taking the load-balance factor into consideration, we are seeking an MIS in which the maximum degree of the nodes in the constructed MIS is minimized. In other words, the potential traffic load on each node in the MIS is as balanced as possible. Now, we are ready to define the LBVB problem.

Definition 4.2   Load-balanced virtual backbone (LBVB) problem. *For a WSN represented by graph* $\mathbb{G}(\mathbb{V}, \mathbb{E})$ *and an* MDMIS $\mathbb{D}$, *the* LBVB *problem is to find a node set* $\mathbb{C} \subseteq \mathbb{V} \backslash \mathbb{D}$ *such that:*

1) The induced subgraph $G[\mathbb{D} \bigcup \mathbb{C}]$ of $\mathbb{G}$ is connected.

2) Minimize $|\mathbb{C}|$, where $|\mathbb{C}|$ is the size of set $\mathbb{C}$.

For convenience, the nodes in set $\mathbb{D}$ are called *independent nodes*, whereas, the nodes in set $\mathbb{C}$ are called *MIS connectors*. Moreover, $\mathbb{B} = \mathbb{D} \bigcup \mathbb{C}$ is an *LBVB* of $\mathbb{G}$. Specifically speaking, $\forall v_i \in \mathbb{B}$, v_i is a *backbone node*.

Constructing an LBVB is a part of the work to balance traffic load on each backbone node. One more important task needs to be resolved is how to allocate non-backbone nodes to its neighboring backbone nodes. The formal definition of the non-backbone node allocation scheme is given as follows:

Definition 4.3   Non-backbone node allocation scheme ($\mathscr{A}$). *For a WSN represented* *by graph* $\mathbb{G}(\mathbb{V},\mathbb{E})$ *and a VB* $\mathbb{B} = \{v_1, v_2, \cdots, v_m\}$, *we need to find m disjoint sets on* $\mathbb{V}$, *denoted by* $\mathbb{A}(v_1), \mathbb{A}(v_2), \cdots, \mathbb{A}(v_m)$, *such that:*

1) Each set $\mathbb{A}(v_i)$ $(1 \leq i \leq m)$ contains exactly one backbone node v_i.

2) $\bigcup_{i=1}^{m} \mathbb{A}(v_i) = \mathbb{V}$, and $\mathbb{A}(v_i) \cap \mathbb{A}(v_j) = \emptyset$ $(1 \leq i \neq j \leq m)$.

3) $\forall v_u \in \mathbb{A}(v_i)$ $(1 \leq i \leq m)$ and $v_u \neq v_i$, such that $(v_u, v_i) \in \mathbb{E}$.

A *non-backbone node allocation scheme* is:

$$\mathscr{A} = \{\mathbb{A}(v_i) \mid \forall v_i \in \mathbb{B}, 1 \leq i \leq m\}.$$

The potential traffic load indicator on each backbone node is the degree of the node, i.e., d_i, for $\forall v_i \in \mathbb{B}$. However, d_i is not the actual traffic load. The actual traffic load can only be determined when a non-backbone node allocation scheme $\mathscr{A}$ is decided. In other words, the number of allocated non-backbone nodes is an indicator of the actual traffic load on each backbone node. According to this observation, we give the following definition:

Definition 4.4 Valid degree (d'). *The* valid degree *of a backbone node* v_i *is the number of its allocated non-backbone nodes, i.e.,* $\forall v_i \in \mathbb{B}, d_i' = |\mathbb{A}(v_i)| - 1$, *where* $|\mathbb{A}(v_i)|$ *represents the cardinality of set* $\mathbb{A}(v_i)$.

Finally, we are dedicated to find a load-balanced non-backbone node allocation scheme $\mathscr{A}$, namely, the maximum *valid degree* of all the backbone nodes is minimized under $\mathscr{A}$.

Definition 4.5 Minmax valid-degree non-backbone node allocation (MVBA) problem. *For a WSN represented by graph* $\mathbb{G}(\mathbb{V},\mathbb{E})$ *and an LBVB* $\mathbb{B} = \{v_1, v_2, \cdots, v_m\}$, *the MVBA problem is to find a backbone allocation scheme* $\mathscr{A}^*$, *such that* $\max\{d_i' \mid \forall v_i \in \mathbb{B}\}$ *is minimized under* $\mathscr{A}^*$.

4.4 Load-Balanced Virtual Backbone Problem

In this section, we first introduce how to solve the MDMIS problem. Since finding an MIS is a well-known NP-complete problem [69] in graph theory, the LBVB problem is NP-complete as well. Next, we formulate the MDMIS problem as an integer nonlinear programming (INP). Subsequently, we show how to obtain an $O(\ln(n))$ approximation solution by using linear programming (LP) relaxation techniques. Finally, we present how to find a minimal set of MIS connectors to form an LBVB $\mathbb{B}$.

4.4.1   INP Formulation of MDMIS

Consider a WSN described by graph $\mathbb{G} = (\mathbb{V}, \mathbb{E})$. First, we define the one-hop neighborhood of a node v_i and then extend it to the r-hop neighborhood.

Definition 4.6   One-hop neighborhood ($\mathbb{N}_1(v_i)$). $\forall v_i \in \mathbb{V}$, *the* 1-*hop neighborhood of node* v_i *is defined as:*

$$\mathbb{N}_1(v_i) = \{v_j \mid v_j \in \mathbb{V}, e_{ij} = (v_i, v_j) \in \mathbb{E}\}.$$

The physical meaning of one-hop neighborhood is the set of the nodes that can be directly reached from node v_i.

Definition 4.7   *r-Hop neighborhood.* $\forall v_i \in \mathbb{V}$, the r-hop neighborhood of node v_i is defined as:

$$\mathbb{N}_r(v_i) = \mathbb{N}_{r-1}(v_i) \cup \{v_k \mid \exists v_j \in \mathbb{N}_{r-1}(v_i), v_k \in \mathbb{N}_1(v_j), v_k \notin \bigcup_{i=1}^{r-1} \mathbb{N}_i(v_i)\}.$$

The physical meaning of the r-hop neighborhood is that the set of the nodes that can be reached from node v_i by passing maximum r number of links.

Next, we formally model the MDMIS problem as an integer nonlinear program (INP).

DS property constraint. As we mentioned earlier, an MIS is also a DS. Hence, we should formulate the DS constraint for the MDMIS problem. For convenience, we assign a decision variable x_i for each sensor $v_i \in \mathbb{V}$, which is allowed to have 0/1 value. This variable sets to 1 *iff* the node is an independent node, i.e., $\forall v_i \in \mathbb{D}, x_i = 1$. Otherwise, it sets to 0. The DS property states that each non-independent node must reside within the one-hop neighborhood of at least one independent node. We therefore have:

$$x_i + \sum_{v_j \in \mathbb{N}_1(v_i)} x_j \geq 1, \forall v_i \in \mathbb{V}.$$

IS property constraint. Since the solution of the MDMIS problem is at least an IS, the IS property is also a constraint of MDMIS. The IS property indicates that no two independent nodes are adjacent, i.e., $\forall v_i, v_j \in \mathbb{D}, (v_i, v_j) \notin \mathbb{E}$. In other words, we have:

$$\sum_{v_j \in \mathbb{N}_1(v_i)} x_i \cdot x_j = 0, \forall v_i \in \mathbb{V}.$$

Consequently, the objective of the MDMIS problem is to minimize the maximum degree of all the independent nodes. We denote z as the objective of the MDMIS problem, i.e., $z = \max_{v_i \in \mathbb{D}}(d_i)$. Mathematically, the MDMIS problem can be formulated as an integer nonlinear programming INP_{MDMIS} as follows:

$$\min \quad z = \max\{d_i \mid \forall v_i \in \mathbb{D}\}$$
$$s.t. \quad x_i + \sum_{v_j \in \mathbb{N}_1(v_i)} x_j \geq 1;$$
$$\sum_{v_j \in \mathbb{N}_1(v_i)} x_i \cdot x_j = 0; \qquad \qquad (INP_{MDMIS})$$
$$x_i, x_j \in \{0,1\}, \forall v_i, v_j \in \mathbb{V}.$$

Since the *IS property constraint* is quadratic, the formulated integer programming INP_{MDMIS} is not linear. To linearize INP_{MDMIS}, the quadratic constraint is eliminated by applying the techniques proposed in [70]. More specifically, the product $x_i \cdot x_j$ is replaced by a new binary variable χ_{ij}, on which several additional constraints are imposed. As a consequence, we can reformulate INP_{MDMIS} exactly to an integer linear programming ILP_{MDMIS} by introducing the following linear constraints:

$$\sum_{v_j \in \mathbb{N}_1(v_i)} \chi_{ij} = 0;$$
$$x_i \geq \chi_{ij};$$
$$x_j \geq \chi_{ij};$$
$$x_i + x_j - 1 \leq \chi_{ij};$$
$$\chi_{ij} \in \{0,1\}, \forall v_i, v_j \in \mathbb{V}.$$

For convenience, we assign a random variable l_{ij} for each edge in the graph $\mathbb{G}$ modeled from a WSN, i.e., $l_{ij} = \begin{cases} 1, & if (v_i, v_j) \in \mathbb{E}. \\ 0, & otherwise. \end{cases}$

Thus, we obtain that $d_i = \sum_{v_j \in \mathbb{N}_1(v_i)} x_i l_{ij}, \forall v_i \in \mathbb{D}$. Moreover, by relaxing the conditions $x_j \in \{0,1\}$, and $\chi_{ij} \in \{0,1\}$ to $x_j \in [0,1]$, and $\chi_{ij} \in [0,1]$, correspondingly, we obtain the following relaxed linear programming LP^*_{MDMIS}:

$$\min z = \max\{1, \max\{d_i = \sum_{v_j \in \mathbb{N}_1(v_i)} x_i l_{ij} \mid \forall v_i \in \mathbb{V}\}\}$$
$$s.t. \quad x_i + \sum_{v_j \in \mathbb{N}_1(v_i)} x_j \geq 1;$$
$$\sum_{v_j \in \mathbb{N}_1(v_i)} \chi_{ij} = 0;$$
$$x_i \geq \chi_{ij}; \qquad \qquad (LP^*_{MDMIS})$$
$$x_j \geq \chi_{ij};$$
$$x_i + x_j - 1 \leq \chi_{ij};$$
$$x_i, x_j, \chi_{ij} \in [0,1], \forall v_i, v_j \in \mathbb{V}.$$

4.4.2 Approximation Algorithm

Due to the relaxation enlarged the optimization space, the solution of LP^*_{MDMIS} corresponds to a lower bound to the objective of INP_{MDMIS}. Given an instance of MDMIS modeled by the integer nonlinear programming INP_{MDMIS}, the sketch of the proposed approximation algorithm (see Algorithm 4) is summarized as follows: first, solve the relaxed linear programming LP^*_{MDMIS} to get an optimal fractional solution, denoted by $(\mathbf{x}^*, z^*)$, where $\mathbf{x}^* = <x^*_1, x^*_2, \cdots, x^*_n>$, and then round x^*_i to integers $\hat{x}_i$ according to five steps:

1. Sort sensor nodes by the x^*_i value (where $1 \le i \le n$) in the decreasing order (line 2).

2. Set all $\hat{x}_i$ to be 0 (lines 3-5).

3. Start from the first node in the sorted node array A (line 8). If there is no node selected as an independent node in v_i's one-hop neighborhood (line 11), then let $\hat{x}_i = 1$ with probability $p_i = x^*_i$ (line 12).

4. Repeat step 3 till reaching the end of array A (lines 9-15).

5. Repeat steps 3 and 4 for $\beta = \frac{3\ln(n)}{\min\{x^*_i \mid v_i \in \mathbb{V}, x^*_i > 0\}}$ times (lines 7-17).

Next the correctness of our proposed approximation algorithm (Algorithm 4) is proven, followed by the performance ratio analysis. Before showing the correctness of Algorithm 4, two important lemmas are derived as follows:

Lemma 4.1
For a WSN represented by $\mathbb{G} = (\mathbb{V}, \mathbb{E})$, if a subset $\mathbb{S} \subseteq \mathbb{V}$ is a DS and meanwhile $\mathbb{S}$ is also an IS, then this subset $\mathbb{S}$ is an MIS of $\mathbb{G}$.

Proof 4.1   If $\mathbb{S} \subseteq \mathbb{V}$ is a DS of $\mathbb{G}$, it implies that $\forall v_i \in \mathbb{V} \backslash \mathbb{S}$, there exists at least one node $v_j \in \mathbb{S}$ in v_i's one-hop neighborhood. Moreover, if $\mathbb{S}$ is also an IS, it implies that no two nodes in $\mathbb{S}$ are adjacent, i.e., $\forall v_s, v_t \in \mathbb{S}, (v_s, v_t) \notin \mathbb{E}$.
Suppose $\mathbb{S}$ is **not** an MIS. In other words, we can find at least one more node that does not violate the DS property and the IS property of $\mathbb{S}$ to be added into $\mathbb{S}$. Suppose v_k is such a node. Based on the DS property, we know that $\exists v_j \in \mathbb{S}$ and $v_j \in \mathbb{N}_1(v_k)$. According to the hypothesis, $v_k \in \mathbb{S}$, and considering the fact that $v_j \in \mathbb{N}_1(v_k)$, we conclude there are two nodes (v_j and v_k) adjacent in $\mathbb{S}$ (i.e., $(v_j, v_k) \in \mathbb{E}$), which contradicts the IS property. Hence, the hypothesis is false and Lemma 7.1 is true. □

Lemma 4.2
The set $\mathbb{D} = \{v_i \mid \hat{x}_i = 1, 1 \le i \le n\}$, where $\hat{x}_i$ is derived from Algorithm 4, is a DS almost surely.

Proof 4.2   Suppose $\forall v_i \in \mathbb{V}, |\mathbb{N}_1(v_i)| = k_i$, where $|\mathbb{N}_1(v_i)|$ is the size of set $\mathbb{N}_1(v_i)$. Let the random variable X_i denote the event that no node in the set $\mathbb{N}_1(v_i) \bigcup \{v_i\}$

Algorithm 4: : Approximation Algorithm for MDMIS

Require: A WSN represented by graph $\mathbb{G} = (\mathbb{V}, \mathbb{E})$; Node degree d_i.

1: Solve LP^*_{MDMIS}. Let $(\mathbf{x}^*, z^*)$ be the optimum solution, where
   $\mathbf{x}^* = <x_1^*, x_2^*, \cdots, x_n^*>, z^* = \max(1, \sum\limits_{v_j \in \mathbb{N}_1(v_i)} x_i^* l_{ij})$.

2: Sort all the sensor nodes by the x_i^* value in the decreasing order. The sorted node ID i is stored in the array denoted by $A[n]$.

3: **for** $i = 1$ **to** n **do**

4:   $\widehat{x_i} = 0$.

5: **end for**

6: $counter = 0$.

7: **while** $counter \leq \beta$, where $\beta = \frac{3\ln(n)}{\min\{x_i^* | v_i \in \mathbb{V}, x_i^* > 0\}}$ **do**

8:   $k = 0$.

9:   **while** $k < n$ **do**

10:    $i = A[k]$.

11:    **if** $\forall v_j \in \mathbb{N}_1(v_i), \widehat{x_j} = 0$, **then**

12:      $\widehat{x_i} = 1$ with probability $p_i = x_i^*$.

13:    **end if**

14:    $k = k + 1$.

15:   **end while**

16:   $counter = counter + 1$.

17: **end while**

18: **return** $(\widehat{\mathbf{x}}, \widehat{z} = \max(1, d_i = \sum\limits_{v_j \in \mathbb{N}_1(v_i)} \widehat{x_i} l_{ij}))$.

---

is selected as an independent node. Additionally, we denote $\mathscr{X} = \max\{x_j^*\}$, for $v_j \in \mathbb{N}_1(v_i) \bigcup \{v_i\}$. For the probability of X_i happening, we have

$$P(X_i) = (1 - p_1)^\beta (1 - p_2)^\beta \cdots (1 - p_{k_i})^\beta \tag{4.1}$$

$$= [(1 - p_1)(1 - p_2) \cdots (1 - p_{k_i})]^\beta \tag{4.2}$$

$$\leq (1 - \mathscr{X})^\beta \tag{4.3}$$

$$\leq (1 - \min\{x_i^* | v_i \in \mathbb{V}, x_i^* > 0\})^\beta \tag{4.4}$$

$$\leq (e^{-\min\{x_i^* | v_i \in \mathbb{V}, x_i^* > 0\}})^\beta \tag{4.5}$$

$$= e^{-\frac{3\ln(n)}{\min\{x_i^* | v_i \in \mathbb{V}, x_i^* > 0\}}(\min\{x_i^* | v_i \in \mathbb{V}, x_i^* > 0\})} = e^{-3\ln(n)} \tag{4.6}$$

$$= \frac{1}{n^3} \tag{4.7}$$

The inequality 4.4 follows the fact that $\mathscr{X} \geq \min\{x_i^*\}$. The inequality 4.5 results from the inequality $1 - x \leq e^{-x}$, $\forall x \in [0, 1]$. Since $\sum\limits_{n>0} \frac{1}{n^3}$ is a particular case of the Riemann zeta function, and it is bounded, i.e., $\sum\limits_{n>0} \frac{1}{n^3} < \infty$ by the result of the Basel problem. Thus, according to the Borel-Cantelli lemma, $P(X_i) \sim 0$, which implies there exists one independent node in the set $\mathbb{N}_1(v_i) \bigcup \{v_i\}$ almost surely. Hence, it is almost certain that the set $\mathbb{D} = \{v_i \mid \widehat{x}_i = 1, 1 \leq i \leq n\}$ derived from Algorithm 4 is a DS. Then, it is reasonable that we consider $\mathbb{D}$ as a DS of $\mathbb{G}$ in the following.[2] Hence $\mathbb{D}$ holds the DS property almost surely. □

Theorem 4.1
The set $\mathbb{D} = \{v_i \mid \widehat{x}_i = 1, 1 \leq i \leq n\}$, where $\widehat{x}_i$ is derived from Algorithm 4, is an MIS.

Proof 4.3   According to lines 11-13 of Algorithm 4, no two nodes can both be set as independent nodes in the one-hop neighborhood. This guarantees the IS property of $\mathbb{D}$, i.e., $\forall v_i, v_j \in \mathbb{D}, (v_i, v_j) \notin \mathbb{E}$. Moreover, $\mathbb{D}$ is a DS as proven in Lemma 7.2. Hence, based on Lemma 7.1, we conclude that $\mathbb{D}$ is an MIS. □

From Theorem 4.1, the solution of our proposed approximation Algorithm 4 is an MIS. Subsequently, we analyze the approximation factor of Algorithm 4 in Theorem 4.2.

Theorem 4.2
Let OPT_{MDMIS} denote the optimal solution of the MDMIS problem. The proposed algorithm yields a solution of performance $O(\ln(n))OPT_{MDMIS}$.

Proof 4.4
The expected d_i of the independent node v_i found by Algorithm 4 is as follows:

$$E[\sum_{v_j \in \mathbb{N}_1(v_i)} \widehat{x}_i l_{ij}] \leq \sum_{v_j \in \mathbb{N}_1(v_i)} E[\widehat{x}_i]E[l_{ij}] \tag{4.8}$$

$$\leq \sum_{v_j \in \mathbb{N}_1(v_i)} (\beta x_i^*)E[l_{ij}] \tag{4.9}$$

$$= \beta \sum_{v_j \in \mathbb{N}_1(v_i)} x_i^* E[l_{ij}] \tag{4.10}$$

$$\leq \beta z^* \tag{4.11}$$

The inequality (4.8) holds because $\widehat{x}_i$ and l_{ij} are independent. The inequality (4.9) holds because the procedure, setting $\widehat{x}_i = 1$ with probability p_i, is repeated β times. By the union bound, we get $Pr[\widehat{x}_i = 1] = Pr[\bigcup_{t \leq \beta} \widehat{x}_i = 1 \text{ at round } t] \leq \beta x_i^*$. This implies $E(\widehat{x}_i) \leq \beta x_i^*$. The inequality (4.11) follows from the fact that $\sum_{v_j \in \mathbb{N}_1(v_i)} x_i^* \cdot E[l_{ij}] \leq \max\{d_i \mid v_i \in \mathbb{D}\} = z^*$.

Applying the Chernoff bound, we obtain the bound:

---

[2]It is almost impossible that $\mathbb{D}$ is not a DS of $\mathbb{G}$. If not, we repeat the entire rounding process.

$$Pr[\sum_{v_j \in \mathbb{N}_1(v_i)} \widehat{x}_i l_{ij} \geq (1+\mu)\beta z^*] \leq (\frac{e^{\mu}}{(1+\mu)^{1+\mu}})^{\beta z^*} \tag{4.12}$$

for arbitrary $\mu > 0$. To simplify this bound, let $\mu = e - 1$, we get

$$Pr[\sum_{v_j \in \mathbb{N}_1(v_i)} \widehat{x}_i l_{ij} \geq (1+\mu)\beta z^*] \leq (\frac{e^{e-1}}{e^e})^{\beta z^*} \tag{4.13}$$

$$\leq e^{-\beta} \tag{4.14}$$

$$= e^{-\frac{3\ln(n)}{\min\{x_i^*\}}} \tag{4.15}$$

$$\leq e^{-3\ln(n)} = \frac{1}{n^3}. \tag{4.16}$$

The inequality (4.14) holds since $z^* = \max\{1, \max\{d_i = \sum_{v_j \in \mathbb{N}_1(v_i)} x_i l_{ij} \mid \forall v_i \in \mathbb{V}\}\} \geq 1$. Applying the union bound, we get the probability that some independent node has a degree larger than $(1+\mu)\beta z^*$,

$$Pr[\widehat{z} \geq (1+\mu)\beta z^*] \leq n\frac{1}{n^3} = \frac{1}{n^2}. \tag{4.17}$$

Again, since $\sum_{n>0} \frac{1}{n^2}$ is a particular case of the Riemann zeta function, which is bounded. Thus, according to the Borel-Cantelli lemma, $P[\widehat{z} \geq (1+\mu)\beta z^*] \sim 0$.

According to the probability of Inequalities (4.7) and (4.17), we get

Pr [some node is selected to be an independent node in 1-hop neighborhood $\bigcap \widehat{z} \leq (1+\mu)\beta z^*] = 1 \cdot (1 - \frac{1}{n^2}) \sim 1$, when $n \sim \infty$. *where* $\mu = e-1$. □

4.4.3 Connected Virtual Backbone

To solve the LBVB problem, one more step is needed after constructing an MDMIS, which is to make the MDMIS connected. Next, we introduce how to find a minimal set of MIS connectors to connect the MDMIS.

We first divide the MDMIS $\mathbb{D}$ into disjoint node sets according to the following criterion:

$$\mathbb{D}_0 = \{v_i \mid v_i \text{ has the smallest node ID among all the nodes in } \mathbb{D}\}$$

$$\mathbb{D}_t = \{v_i \mid v_i \in \mathbb{D}, \exists v_j \in \mathbb{D}_{t-1}, v_i \in \mathbb{N}_2(v_j), v_i \notin \bigcup_{k=0}^{t-1} \mathbb{D}_k\}$$

The independent node with smallest node ID is put into $\mathbb{D}_0$. Clearly, $|\mathbb{D}_0| = 1$. All the independent nodes in the two-hop neighborhood of the nodes in $\mathbb{D}_{l-1}$ (not in $\bigcup_{k=0}^{l-1} \mathbb{D}_k$) are put into $\mathbb{D}_l$. l is called the *level* of the independent node in $\mathbb{D}_l$, i.e., $\mathbb{D}_l$ represents the set of independent nodes of level l in $\mathbb{G}$ with respect to the node in $\mathbb{D}_0$. Additionally, suppose the maximum level of an independent node is L. For each $0 \leq i \leq L-1$, let $\mathbb{S}_i$ be the set of the nodes adjacent to at least one node in $\mathbb{D}_i$ and at least one node in $\mathbb{D}_{i+1}$. Subsequently, compute a minimal set of nodes $\mathbb{C}_i \subseteq \mathbb{S}_i$ to cover node set $\mathbb{D}_{i+1}$. Let $\mathbb{C} = \bigcup_{i=0}^{L-1} \mathbb{C}_i$ and therefore $\mathbb{B} = \mathbb{D} \bigcup \mathbb{C}$ is a *Load Balanced Virtual Backbone* of the original graph $\mathbb{G}$.

We use the WSN shown in Figure 4.3 (a) as an example to explain the construction process of an LBVB. In Figure 4.3 (a) and each circle represents a sensor node. In the first step, it solves the MDMIS problem by Algorithm 4 to obtain $\mathbb{D}$ which is shown in Figure 4.3 (b) by black circles. In $\mathbb{D}$, suppose v_i is the node with the smallest node ID. Then, the number besides each independent node is the level of that node with respect to v_i. In the second phase, we choose the appropriate MIS connectors ($\mathbb{C}$), shown by gray nodes in Figure 4.3 (c), to connect all the nodes in $\mathbb{D}$ to form an LBVB ($\mathbb{B}$). Next, we analyze the number of backbone nodes $|\mathbb{B}|$ produced by our algorithm.

Theorem 4.3
The number of backbone nodes $|\mathbb{B}| \leq 2|\mathbb{D}|$.

Proof 4.5   According to the above proposed algorithm, each MIS connector connects the independent nodes in $\mathbb{D}_i$ and $\mathbb{D}_{i+1}$. Hence,

$$|\mathbb{C}| = |\bigcup_{i=0}^{L-1} \mathbb{C}_i| \leq \sum_{i=0}^{L-1} |\mathbb{D}_{i+1}| \leq |\mathbb{D}|.$$

Finally, we get

$$|\mathbb{B}| = |\mathbb{D} \bigcup \mathbb{C}| \leq 2|\mathbb{D}|.$$

$\square$

4.5   MinMax Valid-degree Non-backbone Node Allocation

The MVBA problem is NP-hard, which can be proven by reduction from the optimization version of the set cover problem. Subsequently, we formulate the MVBA as an integer linear programming (ILP) problem. Then, we present an approximation algorithm by applying the linear relaxation and random rounding technique.

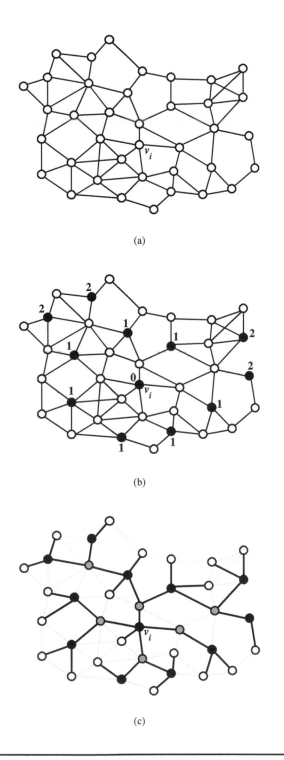

Figure 4.3: Illustration of LBVB construction process.

4.5.1 ILP Formulation of MVBA

According to Definition 4.5, the MVBA problem can be modeled by a binary problem with a linear objective function. We define a binary variable b_i to indicate whether the sensor v_i is a backbone node or not. b_i sets to be 1 *iff* the sensor v_i is a backbone node. Otherwise, b_i sets to be 0 *iff* the sensor v_i is a non-backbone node. Additionally, we assign a random variable a_{ij} for each edge connecting a backbone node v_i and a non-backbone node v_j on the graph modeled from a WSN. Furthermore,

$$a_{ij} = \begin{cases} 1, & \text{if non-backbone node } v_j \text{ is allocated to backbone node } v_i. \\ 0, & \text{otherwise.} \end{cases}$$

Consequently, the MVBA problem can be formulated as an integer linear programming ILP_{MVBA} as follows:

$$
\begin{aligned}
\min \quad & y = \max\{d_i' \mid \forall v_i \in \mathbb{B}\} \\
s.t. \quad & \sum_{v_i \in \mathbb{N}_1(v_j)} b_i a_{ij} = 1, \ \forall v_j \notin \mathbb{B}; \\
& a_{ij} \in \{0,1\}.
\end{aligned}
\qquad (ILP_{MVBA})
$$

The objective function y is the maximum valid degree (d') of all the backbone nodes. The first constraint states that each non-backbone node can be allocated to only one backbone node, whereas the second constraint indicates that a_{ij} is a binary variable. By relaxing variable $a_{ij} \in \{0,1\}$ to $a_{ij} \in [0,1]$, we get the relaxed formulation which falls into a standard linear programming (LP) problem, denoted by LP_{MVBA}^* as follows:

$$
\begin{aligned}
\min \quad & y = \max\{1, \max\{ \sum_{v_j \in \mathbb{N}_1(v_i)} b_i a_{ij} \mid \forall v_i \in \mathbb{B}\}\} \\
s.t. \quad & \sum_{v_i \in \mathbb{N}_1(v_j)} b_i a_{ij} = 1, \ \forall v_j \notin \mathbb{B}; \\
& a_{ij} \in [0,1].
\end{aligned}
\qquad (LP_{MVBA}^*)
$$

Due to the relaxation enlarged the optimization space, the solution of LP_{MVBA}^* corresponds to a lower bound to the objective of ILP_{MVBA}.

4.5.2 Randomized Approximation Algorithm

Given an instance of MVBA modeled by the ILP ILP_{MVBA}, the sketch of the randomized approximation algorithm (see Algorithm 5) is summarized as follows: first, solve the relaxed linear programming LP_{MVBA}^* to get an optimal fractional solution, denoted by $(\mathbf{a}^*, y^*)$, where $\mathbf{a}^* = <a_{11}^*, \cdots, a_{1n}^*, a_{21}^*, \cdots, a_{2n}^*, \cdots, a_{m1}^*, \cdots, a_{mn}^*>$, and then round a_{ij}^* to integers $\widehat{a_{ij}}$ by a random rounding procedure, which consists of four steps:

1. Set all $\widehat{a_{ij}}$ to be 0 (line 2).

2. Let $\widehat{a_{ij}} = 1$ with probability a_{ij}^* and execute this step for α times (lines 3-6), where $\alpha = \frac{3\log(n)}{\min\{a_{ij}^* | a_{ij}^* > 0\}}$.

3. Let $\widehat{a_{ij}} = 1$ with probability $\frac{1}{\Delta}$ (line 7).

4. To ensure $(\widehat{a_{ij}}, \widehat{y})$ is a feasible solution to ILP_{MVBA}, repeat steps 2 and 3 until every non-backbone node is assigned a backbone node.

Algorithm 5: : Approximation Algorithm for MVBA

Require: A WSN represented by graph $\mathbb{G} = (\mathbb{V}, \mathbb{E})$.
1: Solve LP_{MVBA}^*. Let $(\mathbf{a}^*, y^*)$ be the optimum solution.
2: $\widehat{a_{ij}} = 0$.
3: **while** $k \leq \alpha$, where $\alpha = \frac{3\log(n)}{\min\{a_{ij}^* | a_{ij}^* > 0\}}$ **do**
4:   **if** $\sum\limits_{v_i \in \mathbb{N}_1(v_j)} b_i \widehat{a_{ij}} \neq 1$ **then**
5:     $\widehat{a_{ij}} = 1$ with probability a_{ij}^*
6:   **end if**
7:   $k = k + 1$
8: **end while**
9: **if** $((v_i, v_j) \in \mathbb{E})$ and $(v_i \in \mathbb{B})$ and $(\sum\limits_{v_i \in \mathbb{N}_1(v_j)} b_i \widehat{a_{ij}} \neq 1)$ **then**
10:   $\widehat{a_{ij}} = 1$ with probability $\frac{1}{\Delta}$.
11: **end if**
12: **repeat**
13:   line 3 - 8
14: **until** $\sum\limits_{v_i \in \mathbb{N}_1(v_j)} b_i \widehat{a_{ij}} = 1$
15: **return** $(\widehat{a}, \widehat{y} = \max(1, \sum\limits_{v_j \in \mathbb{N}_1(v_i)} b_i \widehat{a_{ij}}))$.

Subsequently, we analyze the approximation factor of Algorithm 5 in Theorem 4.4.

Theorem 4.4

Let OPT_{MVBA} denote the optimal solution of the MVBA problem. The proposed algorithm yields an optimal fractional solution of $O(\log(n))OPT_{MVBA}$.

Proof 4.6   Considering any backbone node v_i and non-backbone node v_j, the expected valid degree of v_i is as follows:

$$E[\sum_{v_j \in \mathbb{N}_1(v_i)} b_i \widehat{a_{ij}}] \tag{4.18}$$

$$= \sum_{v_j \in \mathbb{N}_1(v_i)} b_i E[\widehat{a_{ij}}] \tag{4.19}$$

$$\leq \sum_{v_j \in \mathbb{N}_1(v_i)} b_i(\alpha a_{ij}^* + \frac{1}{\Delta}) \tag{4.20}$$

$$= \alpha \sum_{v_j \in \mathbb{N}_1(v_i)} b_i a_{ij}^* + \frac{1}{\Delta} \sum_{v_j \in \mathbb{N}_1(v_i)} b_i \tag{4.21}$$

$$\leq \alpha y^* + 1 \tag{4.22}$$

$$= \alpha(y^* + \varphi), \text{where } \varphi = \frac{1}{\alpha} \tag{4.23}$$

The Equation (4.19) holds because b_i and $\widehat{a_{ij}}$ are independent. The inequality (4.20) holds since the random rounding technique used in Algorithm 5. Applying the *union bound*, we have the probability, that a non-backbone node v_j is allocated to a backbone node v_i (i.e., $\widehat{a_{ij}} = 1$) when the random rounding shown in Algorithm 5 is done, is: $Pr[\widehat{a_{ij}} = 1] = Pr[\bigcup_{k \leq \alpha+1} \widehat{a_{ij}} = 1 \text{ at iteration } k] \leq \alpha a_{ij}^* + \frac{1}{\Delta}$. This implies $E[\widehat{a_{ij}}] \leq \alpha a_{ij}^* + \frac{1}{\Delta}$. The inequality (4.22) holds because there is at most Δ non-backbone nodes in a backbone node's one-hop neighborhood.

Applying the Chernoff bound, we obtain the following bound:

$$Pr[\sum_{v_j \in \mathbb{N}_1(v_i)} b_i \widehat{a_{ij}} \geq (1+\mu)\alpha(y^*+\varphi)] \leq (\frac{e^\mu}{(1+\mu)^{1+\mu}})^{\alpha(y^*+\varphi)}$$

for arbitrary $\mu > 0$. To simplify this bound, suppose $\mu = e - 1$, then

$$Pr[\sum_{v_j \in \mathbb{N}_1(v_i)} b_i \widehat{a_{ij}} \geq (1+\mu)\alpha(y^*+\varphi)]$$
$$\leq (\frac{e^\mu}{(1+\mu)^{1+\mu}})^{\alpha(y^*+\varphi)} = (\frac{e^{e-1}}{e^e})^{\alpha(y^*+\varphi)}$$
$$= e^{-\alpha(y^*+\varphi)}.$$

Since $y^* = \max\{1, \max\{\sum_{v_j \in \mathbb{N}_1(v_i)} b_i a_{ij} \mid \forall v_i \in \mathbb{B}\}\} \geq 1$, we have

$$Pr[\sum_{v_j \in \mathbb{N}_1(v_i)} b_i \widehat{a_{ij}} \geq (1+\mu)\alpha(y^*+\varphi)]$$
$$\leq e^{-\alpha(y^*+\varphi)} \leq e^{-\alpha} \tag{4.24}$$
$$\leq e^{-\frac{3\log(n)}{\min\{a_{ij}^* \mid a_{ij}^* > 0\}}} \leq \frac{1}{n^3}.$$

According to inequality (4.24), summing over all backbone nodes $v_i \in \mathbb{B}$, we obtain the probability that some backbone node has a *valid degree* larger than $(1 + \mu)\alpha(y^* + \varphi)$ as follows:

$$Pr[\hat{y} \geq (1+\mu)\alpha(y^* + \varphi)] = n\frac{1}{n^3} = \frac{1}{n^2}. \qquad (4.25)$$

Again, $\sum_{n>0} \frac{1}{n^2}$ is a particular case of the Riemann zeta function, which is bounded. Thus, according to the Borel-Cantelli lemma, $Pr[\hat{y} \geq (1+\mu)\alpha(y^* + \varphi)] \sim 0$, where $\alpha = \frac{3\log(n)}{\min\{a_{ij}^*|a_{ij}^*>0\}}$. This implies that Algorithm 5 yields a solution of performance $O(\log(n))y^*$, where y^* is the optimal solution.

Subsequently, we consider the probability that a non-backbone node $v_j \in \mathbb{V}\backslash\mathbb{B}$ is not allocated to a backbone node in its one-hop neighborhood at iteration j. That is,

$$\prod_{k=1}^{\alpha} \prod_{v_i \in \mathbb{N}_1(v_j),\ v_i \in \mathbb{B}} Pr[b_i \widehat{a_{ij}} = 0 \text{ at iteration } k] \qquad (4.26)$$

$$\leq \prod_{k=1, a_{ij}^*>0}^{\alpha} (1 - b_i a_{ij}^*) \qquad (4.27)$$

$$\leq e^{b_i a_{ij}^* \cdot \alpha} \leq \frac{1}{n^3}. \qquad (4.28)$$

The inequality (4.28) results from the inequality $(1-x) \leq e^{-x}$, $\forall x \in [0,1]$. Now, the probability that any non-backbone node is not allocated to a backbone node in its one-hop neighborhood after executing the algorithm is

$$Pr[\text{Some non-backbone node has no neighboring backbone node}]$$
$$\leq n \cdot n^{-3} = \frac{1}{n^2} \qquad (4.29)$$

Based on the probability in inequality (4.25) and Inequality (4.29), we have

$$Pr\ [\text{each non-backbone node is allocated to a backbone node} \bigcap \hat{y} \leq (1+\mu) \\ \alpha(y^* + \varphi)]$$
$$\geq (1 - \frac{1}{n^2})(1 - \frac{1}{n^2}) \sim 1, \text{ when } n \sim \infty$$
$$(4.30)$$

for arbitrary $\mu > 0$. □

From Theorem 4.4, the soultion of our proposed random approximation Algorithm 5 yields a solution upper bounded by $O(\log n)OPT_{MVBA}$. Moreover, this bound can be verified in polynomial time.

4.6 Performance Evaluation

In the simulations, the results of LBVB are compared with the latest and best MCDS construction algorithm [17] denoted by MCDS, and the LBCDS-GA algorithm proposed in [68] denoted by GA. We compare the three algorithms in terms of the number of backbone nodes, network lifetime, which is defined as the time duration until the first backbone node runs out of energy, and the remaining energy over the whole network.

4.6.1 Simulation Environment

We build our own simulator where all the nodes have the same transmission range and all the nodes are deployed uniformly and randomly in a square area. For each specific setting, 100 instances are generated. The results are averaged over these 100 instances (all results are rounded to integers). Moreover, we use the VB-based data aggregation as the communication mode. The simulated energy consumption model is that every node has the same initial 1000 units of energy. Receiving and transmitting a packet both consume 1 unit of energy. In the simulation, we consider the following tunable parameters: the node transmission range, the total number of nodes deployed in the square area, and the side length of the square area. Subsequently, we show the simulation results in three different scenarios.

4.6.2 Scenario 1: Change Total Number of Nodes

In this scenario, all nodes have the same transmission range of 50 *m* and all nodes are deployed uniformly and randomly in a square area of 300 *m* × 300 *m*. The number of nodes is incremented from 50 to 100 by 10. The simulation results are shown in Figure 4.4, where the *X*-axis represents the number of nodes, while the *Y*-axis represents the number of backbone nodes, network lifetime, and the remaining energy over the whole network, respectively.

From Figure 4.4(a), we can see that, with the increase of the number of the sensor nodes, the number of backbone nodes almost keeps stable (from 34 to 39) for all the three algorithms (MCDS, LBVB, and GA). This is because the network deployed area keeps fixed, i.e., if the network deployed area keeps unchanged, the density of the WSN has little effect on the size of the constructed VB.

Despite few changes on the number of backbone nodes, different non-backbone node allocation schemes do affect network lifetime as shown in Figure 4.4(b). From Figure 4.4(b), we know that the network lifetime decreases for all algorithms with the number of nodes increasing. This is because we exploit data aggregation communication mode in a more crowded network. Intuitively, the denser the network is, the more neighbors of each backbone node. Moreover, the number of backbone nodes keeps stable. Hence, with the number of neighbors increasing, the aggregated data on each backbone node becomes heavier. Therefore, the network lifetime decreases for all the three algorithms. Additionally, for a specific network size, LBVB outperforms GA and MCDS. To be specific, LBVB prolongs network lifetime by 69% on average compared to MCDS, and by 47% on average compared to GA. The results demonstrate that load-balancedly allocating non backbone nodes to backbone nodes can improve network lifetime significantly. On the other hand, LBVB outperforms GA, since GA searches the solution in a limited searching space (locally). The local

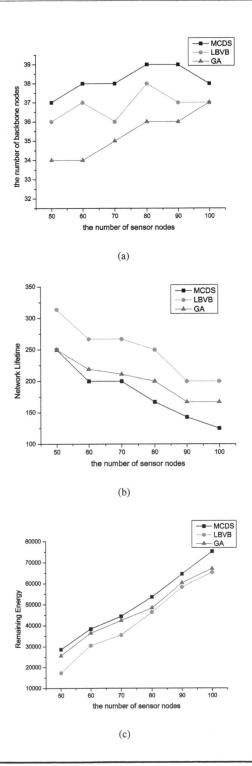

(a)

(b)

(c)

Figure 4.4: Simulation results for Scenario 1: (a) The number of backbone nodes; (b) network lifetime; (c) remaining energy.

optimal solution found by GA might not be the same as the global optimal solution. The results shown in Figure 4.4(b) indicate our proposed algorithms can find a solution which is closer to the optimal solution than GA.

Figure 4.4(c) shows the remaining energy over the whole network of the three algorithms. With the increase of the number of nodes, the remaining energy increases for all algorithms. As the WSN becomes denser and denser, a lot of redundant sensor nodes exist in the WSN. For a specific network size, we observe that LBVB has less remaining energy than GA and MCDS. This is because LBVB considers the load-balance factor when building a VB and allocating non-backbone nodes to backbone nodes. Thus, the lifetime of the whole network is extended, which means the remaining energy of the network is less than MCDS. On the other hand, LBVB prolongs network lifetime compared to GA as shown in Figure 4.4(b), hence LBVB has less remaining energy than GA. In summary, Figure 4.4 indicates that constructing an LBVB can balance the energy consumption on each backbone node, and make the lifetime of the whole network prolonged considerably.

4.6.3   Scenario 2: Change Side Length of Square Area

In this scenario, all nodes have the same transmission range of 20 *m* and 100 nodes are deployed uniformly and randomly in a square area. The side length of the square area is incremented from 100 to 150 by 10. The simulation results are presented in Figure 4.5, where the *X*-axis represents the side length of the square area, while the *Y*-axis represents the number of backbone nodes, network lifetime, and the remaining energy over the whole network respectively.

From Figure 4.5(a), we can see that, with the increase of the area of the network deployed region, the number of backbone nodes increases for all the three algorithms (MCDS, LBVB, and GA). This is because the WSN becomes thinner, so that more backbone nodes are needed to maintain the connectivity of the constructed VB. There is no obvious trend by which the algorithm might produce more backbone nodes when constructing a VB, which implies that the sizes of the constructed VBs are all considered for all the three algorithms.

Different non-backbone node allocation schemes still affect network lifetime as shown in Figure 4.4(b). We know that the network lifetime increases for all algorithms with the side length of the deployed area increasing. It is obvious that the network becomes thinner with the side length of the deployed area increasing. As to a data aggregation communication mode, the aggregated data becomes less on each backbone node when the network becomes thinner. Hence, network lifetime is increasing for all the three algorithms. Additionally, LBVB outperforms GA and MCDS. More specifically, LBVB prolongs network lifetime by 42% on average compared with MCDS, and by 20% on average compared with GA. The reasons are the same as analyzed in Scenario 1.

Figure 4.5(c) shows the remaining energy over the whole network of the three algorithms. With the increase of the side length of the deployed area, the remaining energy decreases for all algorithms. As the WSN becomes thinner and thinner, more nodes are selected as backbone nodes to maintain the connectivity of the constructed

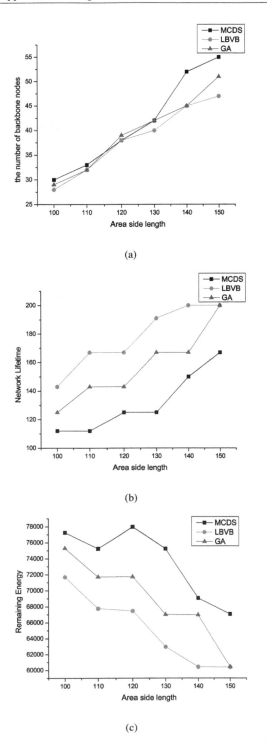

Figure 4.5: Simulation results for Scenario 2: (a) The number of backbone nodes; (b) network lifetime; (c) remaining energy.

VB. Additionally, the traffic load on each backbone node is less as mentioned earlier, hence, the remaining energy decreases with the area of the deployed area increasing. For a specific side length, we observe that LBVB has less remaining energy than GA and MCDS. The reasons are the same as analyzed in Scenario 1.

4.6.4   Scenario 3: Change Node Transmission Range

In this scenario, 100 nodes are deployed uniformly and randomly in a square area of 300 m × 300 m. The node transmission range is incremented from 40 to 65 by 5. The simulation results are recorded in Figure 4.6, where the X-axis represents the node transmission range, while the Y-axis represents the number of backbone nodes, network lifetime, and the remaining energy over the whole network respectively.

From Figure 4.6(a), we can see that, with the increase of the node transmission range, the number of backbone nodes decreases for all the three algorithms (MCDS, LBVB, and GA). This is because there are more nodes in the circle with the node transmission range as the radius, when the node transmission range increases. This is equivalent to the network becoming more denser. Hence, the connectivity of the constructed VB can still be maintained even using fewer backbone nodes. There is still no obvious trend by which the algorithm might produce more backbone nodes when constructing a VB.

From Figure 4.6(b), we know that the network lifetime decreases for all algorithms with the node transmission range increasing. This is because more data need to be aggregated on each backbone node, as the node transmission range increases. Similar results show that LBVB outperforms GA and MCDS for a specific node transmission range. Moreover, LBVB prolongs network lifetime by 25% on average compared with MCDS, and by 6% on average compared with GA. The reasons are the same as analyzed in Scenario 1.

From Figure 4.6(c), we observe that with the increase of the node transmission range, the remaining energy increases for all algorithms. A bunch of redundant sensors exist in the more crowded network, thus the remaining energy increases for all the three algorithms. For a specific node transmission range, LBVB have less remaining energy than GA and MCDS. The reasons are the same as analyzed in Scenario 1.

4.7   Conclusion

In this chapter, we address three fundamental problems of constructing a load-balanced VB in a WSN. More specifically, we solve the LBVB problem which is NP-complete with two steps. First, the MDMIS problem aims to find the optimal MIS such that the maximum degree of all the independent nodes is minimized. To solve this problem, a near optimal approximation algorithm is proposed, which yields an $O(\ln(n))$ approximation factor. Subsequently, the minimal set of MIS connectors make the MDMIS connected. The upper bound on the number of backbone nodes is

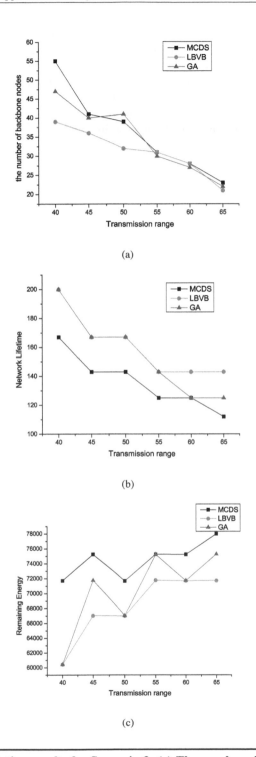

Figure 4.6: Simulation results for Scenario 3: (a) The number of backbone nodes; (b) network lifetime; (c) remaining energy.

analyzed in this chapter as well. In the end, the MVBA problem is dedicated to allocating non-backbone nodes to proper backbone nodes with an objective to minimize the maximum valid degree of all the backbone nodes, which is a NP-hard problem. To solve this problem, we propose an approximation algorithm by using linear relaxing and random rounding techniques, which yields a solution of $O(\log(n))$ approximation factor of traffic load on each backbone node. Simulations show that the proposed algorithms can extend network lifetime significantly.

Chapter 5

A Genetic Algorithm with Immigrants Schemes for Constructing σ-Reliable MCDSs in Probabilistic Wireless Sensor Networks

CONTENTS

5.1   Introduction

Wireless sensor networks are deployed for monitoring and controlling of systems where human intervention is not desirable or possible. Different from wired networks, the topology of wireless sensor networks may change from time to time, and the energy of nodes is very limited and irreplaceable. Therefore, designing an energy-efficient communication scheme for wireless sensor networks is one of the most important issues that has a significant impact on the network performance. The effectiveness of many communication primitives for wireless sensor networks, such as routing [8][9][10], multicast/broadcast [12][13][71], and service discovery [72], relies heavily on the availability of a *virtual backbone* (VB). A *connected dominating set* (CDS) typically serves as a VB of a wireless sensor network. A CDS is defined as a subset of nodes in a wireless sensor network such that each node in the network is either in the set or adjacent to some node in the set, and the induced graph by the nodes in the set is connected. The nodes in a CDS are called *dominators*, or *dominatees*. In a wireless sensor network with a CDS as its VB, dominatees only forward their data to their connected dominators. In addition to communication schemes, a CDS has many other applications in wireless sensor networks, such as topology control [14], coverage [15][16], data aggregation [17], data collection [10][73][74], and query scheduling [18]. Clearly, the benefits of a CDS can be magnified by making its size (the number of the nodes in the CDS) smaller. In general, the smaller the CDS is, the less communication and storage overhead are incurred. Hence, it is desirable to build a minimum-sized CDS (MCDS).

Ever since the idea of employing a CDS for wireless sensor networks was introduced [19], a huge amount of effort has been made to find different CDSs for different applications, especially MCDSs. In the seminal work [20], Guha and Kuller first modeled the problem of constructing the smallest CDS as the MCDS problem in a general graph, which is a well-known NP-hard problem [21]. Subsequently, many polynomial-time approximation algorithms for MCDS construction have been proposed in recent literature. The subtraction-based CDS algorithms begin with a set of all the nodes in a network, then some nodes are removed by pre-defined rules to obtain a CDS. The work in [33] and [34] reports such algorithms. The addition-based CDS algorithms start from a subset of nodes (usually disconnected), then include additional nodes to form a CDS. Two typical such algorithms are [75] and [60].

In all the above mentioned methods, wireless sensor networks are modeled using the *deterministic network model* (DNM). Under this model, there is a transmission radius of each node. According to this radius, specific pairs of nodes are always connected to be neighbors if their physical distance is less than this radius, while the rest of the pairs are always disconnected. The unit disk graph (UDG) model is a special case of the DNM model if all nodes have the same transmission radius. When all nodes are connected to each other, via a single-hop or multi-hop path, a wireless sensor network is said to have full connectivity. In most real applications, however, the DNM cannot fully characterize the behavior of wireless links. This is mainly due to the *transitional region phenomenon* which has been revealed by many empirical studies [76][2][3][4][5]. Beyond the "always connected" region, there is a *transitional region* where a pair of nodes are probabilistically connected. Such pairs of nodes are not fully connected but reachable via the so called *lossy links* [3][5]. As reported in [3][5], there are often many more lossy links than fully connected links in a wireless sensor network. Additionally, in a specific setup [6], more than 90% of the network links are *lossy links*. Therefore, their impact can hardly be neglected.

The employment of lossy links in wireless sensor networks is not straightforward, since when the lossy links are employed, the wireless sensor network may have no guarantee of full network connectivity. When data transmissions are conducted over such topologies, they may degrade the node-to-node delivery ratio. Usually a wireless sensor network has large node density and high data redundancy, thus this certain degraded performance may be acceptable for many wireless sensor network applications. Therefore, as long as an expected percentage of the nodes can be reached, that is the node-to-node delivery ratio satisfies some preset requirement, lossy links are tolerable in a wireless sensor network. In other words, full network connectivity is not always a necessity. Some applications can trade full network connectivity for a higher energy efficiency and larger network capacity [6].

In order to well characterize a wireless sensor network with lossy links, we propose a new network model called the *probabilistic network model* (PNM). Under this model, in addition to transmission radius, there is a *transmission success ratio* (TSR) associated with each link connecting a pair of nodes, which is used to indicate the probability that one node can successfully directly deliver a package to another. Obviously, the core issue under the PNM is how to guarantee the node-to-node delivery ratio of all possible node pairs satisfying the user requirement, in other words, how to guarantee the transmission quality (TQ). For constructing a MCDS under the PNM model, we propose *CDS reliability* to measure its TQ. Given a PNM, *CDS reliability* is defined as the minimum node-to-node delivery ratio between any pair of dominators. Thus, how to find a reliable MCDS under the PNM is the major concern of this chapter. The objective is to seek a MCDS whose reliability satisfies a certain application-dependent threshold denoted by σ (e.g., $\sigma = 80\%$). If $\sigma = 100\%$, finding a reliable MCDS under the PNM is the same as the traditional MCDS problem under the DNM. However, a traditional MCDS algorithm may not find a reliable MCDS under the PNM mode. A counter-example is depicted in Figure 5.2(a). By the latest algorithm proposed in [17], a spanning tree rooted at a specified initiator is first constructed, and then maximal independent sets (MISs) are identified layer by layer. Finally a set of connectors to connect the MISs is ascertained to form a CDS. According to the topology shown in Figure 5.2(a), the constructed CDS by [17] using s_4 as the initiator is $D = \{s_4, s_7, s_8\}$, whose reliability is 0.1. If the threshold $\sigma = 0.7$, the CDS D does not satisfy the constraint at all. The objective of our work is to find a MCDS whose reliability is greater than or equal to σ. One example of the satisfied reliable MCDS is $D' = \{s_3, s_6, s_7\}$ in Figure 5.2(a).

The key challenge finding a reliable MCDS under the PNM is the computation of the *CDS reliability*. It is known that given a network topology, the calculation of the node-to-node delivery ratio is NP-hard when network broadcast is used. Indeed, according to the reliability theory [77], the node-to-node delivery ratio is not practically computable unless the network topology is basically series-parallel, namely, the graph representing a wireless sensor network can be reduced to a single edge by series and parallel replacements. Nevertheless, most network topologies are not series-parallel structures. Thus, instead of computing the accurate *CDS reliability*, we design a greedy based algorithm to approximate the *CDS reliability*. Another challenge is to find a minimum-sized CDS, which is also a NP-hard problem [21]. Intuitively, the smaller the CDS is, the lower the reliability of the CDS is. The key issue then becomes how to find a proper trade-off between the minimum-sized CDS and the *CDS reliability* while satisfying the optimization constraint (i.e., the CDS reliability is no less than the threshold σ). To address this problem, we explore the genetic algorithm (GA) optimization approach. GAs are numerical search tools which operate according to the procedures that resemble the principles of natural selection and genetics [55]. Because of their flexibility and widespread applicability, GAs have been successfully used in a wide variety of problems in several areas of wireless sensor networks [78][79][56][57].

To the best of our knowledge, this work is the first one attempting to construct a MCDS under the PNM for wireless sensor networks. Particularly, the main contributions of this chapter are summarized as follows:

1. We identify and highlight the use of lossy links when constructing a CDS for wireless sensor networks. Based on the lossy links employed in a wireless sensor network, we propose a novel *probabilistic network model* by introducing the TSR of each node pair.

2. In order to measure the quality of a CDS under the PNM, we define a new metric called *CDS reliability*. We then propose a greedy-based algorithm to approximately calculate the *CDS reliability*.

3. We propose a GA with immigrants schemes to build a reliable MCDS under the PNM. Based on the characteristics of GAs and the searching heuristic,

lots of possible solutions are searched and the best one is selected as the final result. We adapt and investigate several genetic algorithms for the construction of a reliable MCDS under the PNM. First, we design the components of the standard GA specifically for our problem. Then, we include several immigrants schemes into the GA to enhance its searching capacity for the optimal solution for our problem.

4. We also conduct extensive simulations to validate our proposed algorithms. The simulation results show that, compared with the traditional MCDS algorithms, our algorithm can obtain a more reliable CDS without increasing the size of a CDS.

The rest of this chapter is organized as follows: in Section 5.2, we review some related works on MCDSs. In Section 5.3, we introduce the PNM and define the reliable MCDS problem. The design of a GA for the reliable MCDS problem under the PNM is presented in Section 5.4. We illustrate the GAs with immigrants schemes in Section 5.5 and conduct extensive simulations to validate our proposed algorithm. Finally, the chapter is concluded in Section 5.6.

5.2 Related Work

In this section, we first briefly review the related works of constructing MCDSs under the DNM. Followed by some related literatures under the PNM. Throughout, we will use n to denote the number of nodes in a wireless sensor network, Δ to denote the maximum nodes in the graph associated with a wireless sensor network, and opt to denote the size of any optimal CDS.

5.2.1 MCDS under DNM

The idea of using a CDS as a virtual backbone was first proposed by Ephremides et al. in 1987 [19]. Since then, many algorithms that construct CDSs have been reported and can be classified into the following four categories based on the network information they use:

■ Centralized algorithms;

■ Subtraction-based localized algorithms;

■ Distributed algorithms using single leader;

■ Distributed algorithms using multiple leaders.

5.2.1.1 Centralized Algorithms for CDSs

Guha et al. [20] first gave two centralized greedy algorithms with performance ratios of $2(H(\Delta) + 1)$ and $H(\Delta) + 2$ respectively, where H is a harmonic function. The greedy function is based on the number of white neighbors of each node. At each step the one(s) with the largest such number become dominator(s).

He et al. [80][81][28][68] considered load-balance factor when constructing a CDS. The authors proposed a greedy algorithm to construct a load-balanced CDS (LBCDS) and then load-balancedly allocate dominatees to dominators (LBAD) in [80][81]. Subsequently, in [28][68], the authors proposed a genetic algorithm (GA) to solve the LBCDS and LBAD problems simultaneously.

5.2.1.2 Subtraction-based Localized Algorithms for CDSs

Wu and Li first proposed a completely localized algorithm in [33] to obtain a CDS. The CDS construction procedure consists of two stages. In the first stage, each node collects its neighboring information by exchanging messages with one-hop neighbors. If a node finds that there is a direct link between any pair of its one-hop neighbors, it removes itself from the CDS. In the second stage, additional heuristic rules are applied to further reduce the size of the CDS. Wu's algorithm [33] uses Rule 1 and Rule 2, where a node is removed from the CDS if all its neighbors are covered by one or two numbers of its direct neighbors. Dai's [34] work generalizes this as Rule k, in which coverage is defined by an arbitrary number of connected neighbors. Dai's algorithm is reduced to Wu's algorithm when k is 1 or 2.

5.2.1.3 Distributed Algorithms for CDSs

A single-leader distributed algorithm for CDSs assumes an initiator to initialize the distributed algorithm. Usually, a base station could be the initiator for construction the CDSs in wireless sensor networks. In these distributed algorithms, a spanning tree rooted at the initiator is first constructed, and then maximal independent sets (MISs) are identified layer by layer, and finally a set of connectors to connect the MISs is ascertained to form a CDS. Wan et al. [75] presented an ID-based distributed algorithm to construct a CDS using a single initiator. For UDGs, Wan et al.'s [75] approach guarantees that the approximation factor on the size of a CDS is at most $8 \, opt + 1$. The algorithm has $O(n)$ time complexity and $O(n \log n)$ message complexity. Subsequently, the approximation factor on the size of a CDS was later improved in another work reported by Cardei et al. [59], in which the authors used the degree-based heuristic and degree-aware optimization to identify Steiner nodes as the connectors in the CDS construction. The approximation factor on the size of a CDS is improved to $8 \, opt$, while this distributed algorithm has $O(n)$ message complexity, and $O(\Delta n)$ time complexity. Later, Li et al. [60] reported a better approximation factor of $5.8 + \ln 4$ by constructing a Steiner tree when connecting all nodes in the MISs.

Distributed algorithms with multiple leaders do not require an initiator to construct a CDS. Alzoubi et al.'s technique [61] first constructs an MIS using a distributed approach without a leader or tree construction and then interconnects these MIS nodes to get a CDS. Li et al. proposed a distributed algorithm r-CDS in [62] whose performance ratio is 172. r-CDS is a completely distributed one-phase algorithm where each node only needs to know the connectivity information within its two-hop-away neighborhood. An MIS is constructed based on each node's r value which is defined as the number of this nodes two-hop-away neighbors minus the degree of this node. The nodes with smaller r values are preferred to serve as MIS nodes. Adjih et al. [63] presented an approach for constructing a MCDS based on multi-point relays (MPRs), but there is no approximation analysis of the algorithm yet. Recently, in [64], another distributed algorithm was proposed whose performance ratio is 147. This algorithm contains three steps. Step 1 constructs a forest in which each tree is rooted at a node with the minimum ID among its one-hop away neighbors. Step 2 collects neighborhood information, which is used in Step 3 to connect neighboring trees.

5.2.2 Related Literature about PNM Model

Traditional routing schemes only consider fully connected links as a path of nodes in a wireless sensor network, and then send data through that sequence of node. Compared with the fully connected links, lossy links only provide probabilistic connectivity. However, there exist more lossy links in a wireless sensor network. Therefore, opportunistic routing schemes (e.g., ExOR [82] and More [83]) proposed lossy links as advantages. The opportunistic routing scheme called ExOR [82] proposed a new unicast routing technique for multi-hop wireless sensor networks. ExOR forwards each packet through a sequence of nodes that can successfully receive the transmission and are close to the destination. ExOR explores package overhearing along lossy links. When a lossy link succeeds, some transmissions can be saved. Later, Chachulski et al. combined random network coding with opportunistic routing to support both unicast and multicast routing in More [83]. The success of opportunistic routing indicates that lossy links provide the potential throughput increase.

Recently, lots of works [6][84][85][86] study the impact of lossy links to the topology control. Ma et al., in [84][85] worked on achieving energy efficiency by turning off redundant nodes and links, while still satisfying the given QoS requirements. Liu et al., investigated how to control the minimal transmission range for each node while the global network reachability satisfies some constraints in [6][86].

5.2.3 Remarks

All the existing works either consider to construct an MCDS under the DNM, design a routing protocol, or investigate the topology control under the PNM. To the best knowledge, of our however, none of them attempt to construct an MCDS under the PNM model, which is more realistic. This is the major motivation of this research work. GAs are a family of computational models inspired by evolution, which have been applied over a broad range of NP-hard optimization problems [78][79][56][57]. Therefore, a GA-based method, namely RMCDS-GA, is proposed in this chapter to construct a reliable MCDS under the PNM. In RMCDS-GA, each possible CDS in a wireless sensor network is represented to be a chromosome (feasible/potential solution), and the fitness function is to evaluate the trade-off between the CDS reliability and the size of the CDS represented by each chromosome.

5.3 Problem Statement

In this section, we give an overview of the reliable MCDS problem under the PNM. We first present the assumptions, and then introduce the PNM. Finally, we give the problem definition and make some remarks.

5.3.1 Assumptions

We assume a static connected wireless sensor network and all nodes in the wireless sensor network have the same transmission range. The transmission success ratio (TSR) associated with each link connecting a pair of nodes is available, which can be obtained by periodic hello messages, or be predicted using link quality index (LQI) [87]. We also assume that the TSR values are fixed. This assumption is reasonable as many empirical studies have shown that LQI is pretty stable in a static environment [88]. Furthermore, no node failure is considered since it is equivalent to a link failure case. No duty cycle is considered either. We do not consider packet collisions or transmission congestion, which are left to the MAC layer. The degradation of the node-to-node delivery ratio is thus only due to the failure of wireless links.

5.3.2 Network Model

Under the *probabilistic network model* (PNM), we model a wireless sensor network as an undirected graph $G(V, E, P(E))$, where V is the set of n nodes, denoted by $s_1, s_2, \ldots, s_n$; E is the set of m lossy links, $\forall u, v \in V$, there exists an edge (u, v) in G if and only if: 1) u and v are in each other's transmission range, $TSR(e = \{u, v\}) > 0$, for each link $e = \{u, v\} \in E$, where $TSR(e)$ indicates the probability that node u can successfully directly deliver a packet to node v; and $P(E) = \{< e, TSR(e) > | e \in E, 0 \leq TSR(e) \leq 1\}$. We assume edges are undirected (bidirectional), which means two linked nodes are able to transmit and receive information from each other with the same TSR value.

Because of the introduction of $TSR(e)$, the traditional definition of the node neighborhood has changed. Hence, we first give the definition of the one-hop neighborhood and then extend it to the r-hop neighborhood.

Definition 5.1 *One-hop neighborhood.* $\forall u \in V$, the one-hop neighborhood of node u is defined as:

$$N_1(u) = \{v | v \in V, TSR(e = \{u, v\})\}$$

The physical meaning of one-hop neighborhood is the set of the nodes that can be directly reached from node u with the probability $TSR(e = \{u, v\})$.

Definition 5.2 *r-Hop neighborhood.* $\forall\, u \in V$, the r-hop neighborhood of node u is defined as:

$$N_r(u) = N_{r-1}(u) \cup \{v | \exists w \in N_{r-1}(u), v \in N_1(w), v \notin \bigcup_{i=1}^{r-1} N_i(u)\}$$

The physical meaning of the r-hop neighborhood is that the set of the nodes that can be reached from node u by passing maximum r number of edges with the probability $TSR(e = \{u, v\}) > 0$.

Definition 5.3 *Node-to-node delivery ratio.* Given a source node u and a destination node v, one path between the node pair can be denoted by the edge permutation $\theta(u, v) = (e_1, e_2, \ldots, e_m)$, and the delivery ratio of the path is denoted by $DR_\theta = \prod_{i=1}^{m} TSR(e_i)$. Furthermore, we use $\Theta(u, v)$ to denote the set of all the possible ways by which node v can be reached from node u. The node-to-node delivery ratio from node u to node v is then defined as:

$$DR^*(u, v) = max\{DR_\theta, \forall\, \theta(u, v) \in \Theta(u, v)\}$$

Clearly, $DR^*(u, v)$ is equivalent to $DR^*(v, u)$.

Definition 5.4 *CDS reliability.* Given a wireless sensor network represented by $G(V, E, P(E))$ under the PNM, and its CDS denoted by D, the reliability of D R_D^* is the minimum node-to-node delivery ratio between any pair of the nodes in the CDS, i.e.,

$$R_D^* = min\{DR^*(u, v), \forall\, u, v \in D, u \neq v\}$$

We use *CDS reliability* to measure the quality of a CDS constructed under the PNM model. By this definition, when a CDS D has a reliability R_D^* satisfying a threshold σ (i.e., $R_D^* \geq \sigma$), we can state that for any pair of the nodes in the CDS the probability that they are connected is no less than the threshold.

According to the reliability theory [77], we know that the computation of the node-to-node delivery ratio is NP-hard. Therefore, the computation of the CDS reliability is also NP-Hard. In summary, we claim that, given a wireless sensor network represented by $G(V, E, P(E))$ under the PNM, a CDS for G denoted by D, and a pre-defined threshold $\sigma \in (0, 1]$, it is NP-hard to verify whether $R_D^* \geq \sigma$.

5.3.3 Problem Definition

After we introduce how to measure the quality of CDSs under the PNM model, we will give the formal definition of the problem we investigate in this chapter.

Definition 5.5 *σ-Reliable MCDS (σ-RMCDS).* Given a wireless sensor network represented by $G(V, E, P(E))$ under the PNM, and a pre-defined threshold $\sigma \in (0, 1]$, the σ-RMCDS problem is to find a minimum-sized node set $D \subseteq V$, such that

1. The induced graph $G[D] = (D, E')$, where $E' = \{e \mid e = (u, v), u \in D, v \in D, (u, v) \in E)\}$, is connected.

2. $\forall u \in V$ and $u \notin D$, $\exists v \in D$, such that $(u, v) \in E$.

3. $R_D^* \geq \sigma$.

We claim that the problem to construct a σ-RMCDS for a wireless sensor network under the PNM model is NP-hard. It is easy to see that the traditional MCDS problem under the DNM is a special case of the σ-RMCDS problem. By setting the TSR values on all edges to 1, we are able to convert the σ-RMCDS problem to the traditional MCDS problem under the DNM. Thus the σ-RMCDS problem belongs to NP. The verification of the σ-RMCDS problem needs to calculate the CDS reliability. It is an NP-hard problem, which is mentioned in Subsection 5.3.2. Therefore, the problem to construct a σ-RMCDS for a wireless sensor network under the PNM is NP-hard.

5.3.4 Remarks

As we already know, computing the node-to-node delivery ratio and the CDS reliability are NP-Hard problems. Therefore, instead of computing the accurate node-to-node delivery ratio, we design a greedy based algorithm to approximate the ratio denoted by $DR(u, v)$. Based on the approximate node-to-node delivery ratio, we then calculate the approximate *CDS reliability* denoted by R_D. When there is no confusion, $DR^*(u, v)$ and $DR(u, v)$, R_D^* and R_D are interchangeable in the chapter.

Based on Definition 5.5, the key issue of the σ-RMCDS problem is to seek a tradeoff between the minimum-sized CDS and the CDS reliability. GAs are population-based search algorithms, which simulate biological evolution processes and have successfully solved a wide range of NP-hard optimization problems [78][79][56][57]. In the following, algorithm RMCDS-GA is proposed to solve the σ-RMCDS problem to search the feasible domain more effectively and reduce the computation time.

5.4 RMCDS-GA Algorithm

In the following sections, we first provide some basics of the GA optimization approach, and then present the detailed design of the RMCDS-GA algorithm for the σ-RMCDS problem.

5.4.1 GA Overview

GAs, first formalized as an optimization method by Holland [65], are search tools modeled after the genetic evolution of natural species. GAs encode a potential solution to a vector of independent variables, called chromosomes. The independent variables consisting of chromosomes are called genes. Each gene encodes one component of the target problem. A binary coding is widely used nowadays. GAs differ from most optimization techniques because of their global searching effectuated by one population of solutions rather than from one single solution. Hence, a GA search starts with the creation of the first generation, a random initial population of chromosomes, i.e., potential solutions to the problem. Then, these individuals in the first generation are evaluated in terms of their "fitness" values, i.e., their corresponding objective function values. Based on their fitness values, a ranking of the individuals in the first generation is dynamically updated. Subsequently, the first generation is allowed to evolve in successive generations through the following steps:

1. Reproduction: selection of a pair of individuals in the current generation as parents. The ranking of the individual in the current generation is used in the selection procedure so that in the long run, the best individuals will have a greater probability of being selected as parents.

2. Recombination: crossover operation and mutation operation:

   (a) Crossover is performed with a crossover probability P_c by selecting a random gene along the length of the parent chromosomes and swapping all the genes of the selected parents chromosomes after that point. The operation generates two new children chromosomes.

   (b) Mutation is performed with a mutation probability P_m by flipping the value of one gene in the chromosomes (e.g., 0 becomes 1, and 1 becomes 0, if binary coding is used).

3. Replacement: utilization of the fittest individual to replace the worst individual of the current generation to create a new generation, so as to maintain the population number k a constant. Every time new children are generated by a GA, the fitness function is evaluated. And then a ranking of the individuals in the current generation is dynamically updated. The ranking is used in the replacement procedures to decide who, among the parents and the children chromosomes, should survive in the next population. This is to resemble the natural principles of the "survival of the fittest".

GAs usually stop when a certain number of total generations denoted by *G* are reached.

Figure 5.1 shows the overview of the RMCDS-GA algorithm.

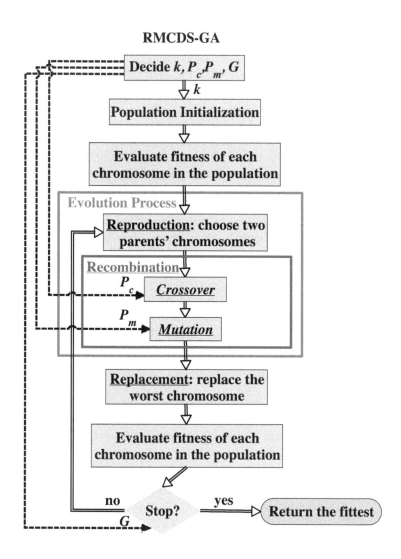

Figure 5.1: Procedure of RMCDS-GA.

One important feature of GAs that must be emphasized here is that the optimization performance of GAs depends mainly on the convergence time of the algorithm. When using GAs, sufficient genetic diversity among solutions in the population should be guaranteed. Lack of such diversity would lead to a reduction of the search space spanned by the GA. Consequently, the GA may prematurely converge to a local minimum because mediocre individuals are selected in the final generation. Alternatively, an excess of genetic diversity, especially at later generations, may lead to a degradation of the optimization performance. In other words, excess genetic diversity may result in very late or even no convergence. In this chapter, genetic diversity is maintained by the crossover, mutation operations and immigrants schemes. In the following part of this section, we will explain RMCDS-GA step by step.

5.4.2 Representation of Chromosomes

In the proposed RMCDS-GA, each node is mapped to a gene in the chromosome. A gene value indicates whether the node represented by this gene is a dominator or not. Hence, a chromosome is denoted as: $C_i = (g_1, g_2, \cdots, g_j, \cdots, g_n)$, where $1 \leq i \leq k$ and k is the number of the chromosomes in the population; $1 \leq j \leq n$ and n is the total number of the nodes in a wireless sensor network.

$$\begin{cases} g_j = 1, \ node \ s_j \ is \ a \ dominator \\ g_j = 0, \ node \ s_j \ is \ a \ dominatee \end{cases}$$

All the nodes with $g_j = 1$ form a CDS denoted by $D = \{s_j | g_j = 1, 1 \leq j \leq n\}$.

An example wireless sensor network under the PNM model is shown in Figure 5.2(a) to illustrate the encoding scheme. There are eight nodes and the CDS is $D = \{s_4, s_7\}$. Thus, the eight nodes can be encoded using eight genes in a chromosome, e.g., $C_1 = (g_1, g_2, \cdots, g_8)$, and then set the values of genes representing the dominators to 1. Finally, the encoded chromosome is $C_1 = (0, 0, 0, 1, 0, 0, 1, 0)$ shown in Figure 5.2(b).

5.4.3 Population Initialization

According to the flowchart of the proposed RMCDS-GA shown in Figure 5.1, after we decide the encoding scheme of the σ-RMCDS problem, the first generation (a population with k chromosomes) should be created. This step is called population initialization in Figure 5.1. A general method to initialize the population is to explore the genetic diversity. That is, for each chromosome, all dominators are randomly generated. However, the dominators must form a CDS. Therefore we start to create the first chromosome by running an existing MCDS method, e.g., Wan's work [17] in which a spanning tree rooted at a specified initiator is first constructed, and then maximal independent sets (MISs) are identified layer by layer; finally a set of connectors to connect the MISs is ascertained to form an MCDS, and then generate the population with k chromosomes by modifying the first chromosome. We call the procedure, generating the whole population by modifying one specific chromosome, inheritance population initialization (IPI).

An example is shown in Figure 5.2(b) to illustrate the IPI process. In Figure 5.2(a), the wireless sensor network and its CDS $D_1 = \{s_4, s_7\}$ are given.

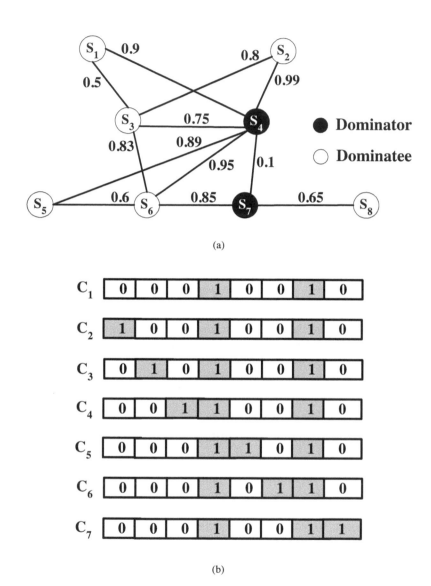

Figure 5.2: (a) A wireless sensor network under the PNM, (b) illustration of population of initialization.

The values on the edges are TSR values and black nodes are dominators. Furthermore, we assume the CDS is constructed by a traditional MCDS method. $C_1 = (0,0,0,1,0,0,1,0)$ represents the CDS generated by Wan's work [17] shown in Figure 5.2(a). Subsequently, we need to generate more chromosomes based on the first chromosome. The IPI algorithm is summarized as follows:

1. Start from the node with the smallest ID, reduce one dominator each time from the original CDS D_1 represented by C_1. If the new obtained node set is still a CDS D_i, then encode it as a chromosome C_i and add it into the initial population. Otherwise, remove the node with the second smallest ID from the original CDS D_1 and make the same checking process as for the node with the smallest ID. Repeat the process till no more new chromosomes can be created. The CDS shown in Figure 5.2(a) is a minimum-sized CDS, i.e., we cannot further reduce its size. Thus we go to step 2.

2. If the size of the original CDS D_1 cannot be reduced, and the number of the generated chromosomes is less than k, then for all the existing chromosomes $C_1, C_2, \cdots, C_i$ we use the following steps till k non-duplicated chromosomes are generated.

   (a) Let $t = 1$.

   (b) In the CDS D_t represented by the chromosome C_t, start from node u with the smallest ID, and add one dominatee node in its one-hop neighborhood $N_1(u)$ by the order of its ID into the CDS each time. If the new obtained node sets form CDSs, then encode them as chromosomes, and add them into the initial population. In Figure 5.2(a), the node with the smallest ID is s_4 in D_1. Therefore, the chromosomes from C_2 to C_6 (shown in Figure 5.2(b)) are generated by adding one node from set $N_1(s_4) = \{s_1, s_2, s_3, s_5, s_6\}$ each time.

   (c) Move to the node with the second smallest ID in CDS D_t till every node in D_t is checked. In Figure 5.2(a), the one-hop neighborhood of the node with the second smallest ID s_7 is $N_1(s_7) = \{s_6, s_8\}$. Since s_6 has already been marked as a dominator, we cannot add it to create a new CDS. By eliminating the duplicates, the chromosome C_7 is created.

   (d) If all the dominators in the current D_t are checked, move to the next CDS by setting $t = t + 1$, repeat the step from 2b) to 2d).

Since each node has two choices: to be a dominator or a dominatee, consequently, there are $2^{n-|D|}$ possible ways to create new chromosomes, where $|D|$ is the size of the CDS denoted by D under the PNM. Usually, k is much smaller than $2^{n-|D|}$. Hence the initial population $C_1, C_2, \cdots, C_k$ can be easily generated.

There are several merits that need to be pointed out here when using the above IPI algorithm to generate the initial population. First, we can guarantee every dominator set represented by a chromosome in the first generation is a CDS, i.e., each chromosome in the initial population is a feasible solution of the σ-RMCDS problem.

Second, the critical nodes (cut nodes), are dominators encoded in each chromosome of the initial population. When performing crossover operations, the critical nodes are still dominators in the new offspring chromosome in the successive generations. The illustrative examples will be shown in Subsection 5.4.6.1. Finally, The IPI stops when k chromosomes are generated. Actually, we can obtain more valid solutions by continuously running the IPI algorithm. As we already know, the population diversity plays an important role on the optimization performance of GAs. Therefore, the extra valid solutions generated by keeping running the IPI algorithm can be used in the replacement process to bring more population diversity in new generations. We will give more detailed description of the replacement scheme in Section 5.5.

5.4.4 Fitness Function

Given a solution, its quality should be accurately evaluated by the fitness value, which is determined by the fitness function. In our algorithm, we aim to find a minimum-sized CDS D whose reliability R_D should be greater than or equal to a preset threshold σ. Therefore, the fitness function of a chromosome C_i in the population is defined as:

$$f(C_i) = \frac{R_D^2}{|D|^2} \tag{5.1}$$

The purpose of raising $|D|$ and R_D to the power of 2 in Equation (5.1) is to enlarge the weight of the size of the CDS D. The denominator in Equation (5.1) needs to be minimized while the numerator needs to be maximized. As a result, the fitness function value will be maximized.

As mentioned in the previous section, precisely calculating the CDS reliability is an NP-hard problem. According to Definition 5.4, we can easily compute the CDS reliability based on the node-to-node delivery ratio of all possible dominator pairs in the CDS. Therefore, we propose a greedy based approximate algorithm to calculate the node-to-node delivery ratio. We adopt a greedy based routing protocol, greedy perimeter stateless routing (GPSR)[89], to find the path between all possible dominator pairs. In this work, we modified the greedy criterion to be the largest TSR values greater than or equal to σ based on GPSR, then we can guarantee that node-to-node delivery ratios between all possible dominator pairs are greater than or equal to σ.

For easier to understand, we first illustrate the idea by an example and then summarize the whole process. For the chromosome C_2 shown in Figure 5.2(b), the CDS represented by C_2 is $D = \{s_1, s_4, s_7\}$, in which there are three possible dominator pairs, i.e. $(s_1, s_4), (s_1, s_7), (s_4, s_7)$. Assume the reliability threshold is $\sigma = 60\%$. Clearly, the TSRs associated with the edges (s_1, s_4) and (s_4, s_7) are both greater than 60% in Figure 5.2, i.e., $TSR(e_1 = \{s_1, s_4\}) = 0.9$, and $TSR(e_2 = \{s_4, s_7\}) = 0.95$. According to the Definition 5.3, we know that $DR(s_1, s_4) = 0.9$ and $DR(s_4, s_7) = 0.95$ respectively. Therefore, the first greedy criterion comes out: the direct edges between sources and destinations with TSR values greater than δ have the highest priority to be chosen as the path between sources and destinations. For dominator pair (s_1, s_7), obviously, there is no direct edge between them. Thus we need to find a multi-hop path between them. The search process starts from the destination s_7. The greedy criterion is based on the TSR values on the edges between s_7 and all its one-hop neighborhood $N_1(s_7) = \{s_4, s_6, s_8\}$. Since $TSR(e_2 = \{s_4, s_7\}) = 0.95 > 0.6$ is the largest TSR value among all the nodes in $N_1(s_7)$, the edge $e_2 = \{s_4, s_7\}$ is chosen. Subsequently, we keep searching from s_4. Apparently, $TSR(e_3 = \{s_2, s_4\}) = 0.99 > 0.6$ is the highest TSR value on the edges from s_4 to all the nodes in $N_1(s_4)$. However, based on the direct edge greedy criterion, there is a direct edge between the source s_1 and the current search node s_4; therefore $e_1 = \{s_1, s_4\}$ is chosen. According to Definition 5.3, $\theta(s_1, s_7) = \{e_1, e_2\}$, $DR(s_1, s_7) = DR_\theta = \prod_{i=1}^{2} TSR(e_i) = 0.9 * 0.95 = 0.855$. Finally, based on Definition 5.4, we know $R(D) = min\{DR(s_1, s_4), DR(s_1, s_7), DR(s_4, s_7)\} = min\{0.9, 0.855, 0.95\} = 0.855$. The fitness of C_2 can then be calculated using Equation (5.1), $f(C_2) = \frac{0.855^2}{3^2} = 0.081225$.

5.4.5 Selection (Reproduction) Scheme

During the evolutionary process, selection plays an important role in improving the average quality of the population by passing the high quality chromosomes to the next generation. The selection operator is carefully formulated to ensure that better chromosomes of the population (with higher fitness values) have a greater probability of being selected for mating, but that worse chromosomes of the population still have a small probability of being selected. Having some probability of choosing worse members is important to ensure that the search process is global and does not simply converge to the nearest local optimum. We adopt *roulette wheel selection* (RWS) since it is simple and effective. RWS stochastically selects individuals based on their fitness values $f(C_i)$. A real-valued interval, S, is determined as the sum of the individuals' expected selection probabilities, i.e. $S = \sum_{i=1}^{k} P_i$, where $P_i = \frac{f(C_i)}{\sum_{j=1}^{k} f(C_j)}$.

Individuals are then mapped one-to-one into contiguous intervals in the range [0, S]. The size of each individual interval corresponds to the fitness value of the associated individual. The circumference of the roulette wheel is the sum of all fitness values of the individuals. The fittest chromosome occupies the largest interval, whereas the least fit has correspondingly smaller interval within the roulette wheel. To select an individual, a random number is generated in the interval $[0, S]$ and the individual whose segment spans the random number is selected. This process is repeated until a desired number of individuals have been selected.

The pseudo-code is shown in Algorithm 6.

We still use the WSN shown in Figure 5.2(a) to illustrate the RWS scheme. The following Table 5.1 lists a sample population of seven individuals (shown in Figure 5.2(b)). These individuals consist of 8-bit chromosomes. The fitness values are calculated by Equation (5.1). We can see from the table: C_1 is the fittest and C_7 is the

---

Algorithm 6: Roulette Wheel Selection

Require: Population number k, each chromosome's fitness value $f(C_i)$.

1: $S = \sum_{i=1}^{k} f(C_i)$;

2: Generate random number r from interval $(0, S)$;

3: Initialize $curS = 0$;

4: **for** $i = 1$ **to** k **do**

5:   $curS += f(C_i)$;

6:   **if** $curS >= r$ **then**

7:    **return** C_i;

8:   **end if**

9: **end for**

---

weakest. Summing these fitness values we can apportion a percentage total of fitness. This gives the strongest individual a value of 35% and the weakest 6%. These percentage fitness values can then be used to configure the roulette wheel (shown in Figure 5.3). The number of times the roulette wheel is spun is equal to size of the population (i.e., k). As can be seen from the way the wheel is now divided, each time the wheel stops this gives the fitter individuals the greatest chance of being selected for the next generation and subsequent mating pool. According to the survival of the fittest in nature selection, individual $C_1 = (00010010)$ will become more prevalent in the general population because it is the fittest and more appropriate for the environment we have put it in.

Table 5.1: Fitness of Seven Chromosomes

No.	Chromosome	$f(C_i)$	% of total
C_1	00010010	$\frac{0.95^2}{2^2} = 0.226$	35
C_2	10010010	$\frac{0.855^2}{3^2} = 0.081$	12
C_3	01010010	$\frac{0.9405^2}{3^2} = 0.098$	15
C_4	00110010	$\frac{0.7125^2}{3^2} = 0.056$	9
C_5	000110101	$\frac{0.8455^2}{3^2} = 0.079$	12
C_6	00010110	$\frac{0.80725^2}{3^2} = 0.072$	11
C_7	00010011	$\frac{0.6175^2}{3^2} = 0.042$	6
	Totals	0.654	100

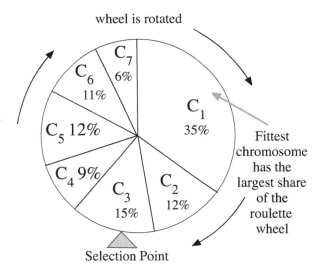

wheel is rotated

Selection Point

Figure 5.3: Roulette wheel selection.

5.4.6 Genetic Operations

The performance of a GA relies heavily on two basic genetic operators, *crossover* and *mutation*. Crossover exchanges parts of the current solutions (the parent chromosomes selected by the RWS scheme) in order to find better ones. Mutation flips the values of genes, which helps a GA keep away from local optimum. The type and implementation of these two operators depend on the encoding scheme and also on the application. In the σ-RMCDS problem, we use the binary coding scheme and all potential solutions must be CDSs. For crossover, we can adopt all classical operations, however, the new obtained solutions may not be valid (the dominator set represented by the chromosome is not a CDS) after implementing the crossover operations. Therefore, a correction mechanism needs to be performed to guarantee validity of all the new generated solutions. Similarly, all traditional mutation operations can be adopted to the σ-RMCDS problem, followed by a correction mechanism.

In this subsection, we introduce three crossover operators and their correction mechanism, followed by a mutation operator and its correction scheme.

5.4.6.1 Crossover

In our algorithm, since a chromosome is expressed by binary codes, we adopt three crossover operators called single-point crossover, two-point crossover, and uniform crossover respectively. With a crossover probability P_c, we use the RWS scheme to select two chromosomes C_i and C_j as parents to perform one of the three crossover operators randomly. We use Figure 5.4 to illustrate the three crossover operations.

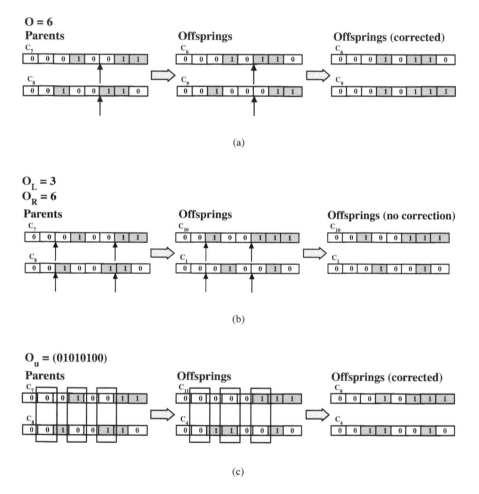

Figure 5.4: Illustration of crossover operations: (a) single-point crossover; (b) two-point crossover; (c) uniform crossover.

Suppose that two parent chromosomes $C_7 = (00010011)$ and $C_8 = (00100110)$ are selected from the population. By the single-point crossover (shown in Figure 5.4(a)), the genes from the crossover point to the end of the two chromosomes exchange with each other to get $C_6 = (00010110)$ and $C_9 = (00010111)$. The crossover point denoted by $O = 6$ is generated randomly. After crossing, the first offspring $C_6 = (00010110)$ is a valid solution. However, the other one $C_9 = (00100011)$ is not valid, thus we need to perform the correction mechanism. The correction starts from the gene in the position of the crossover point O, i.e. g_6. Since g_6 is 1 in the parent chromosome C_8, it changes to 0 after crossing. We correct it by setting $g_6 = 1$. Then $C_9 = (00010111)$ is now a valid solution. In general, we can keep correcting the genes to the end of the chromosome. By the two-point crossover (shown in Figure 5.4(b)), the two crossover points are randomly generated which are $O_L = 3$ and $O_R = 6$; and then the genes between O_L and O_R of the two parent chromosomes are exchanged with each other. The two offsprings are $C_{10} = (00100111)$ and $C_1 = (00010010)$ respectively. Since both of the offspring chromosomes are valid, we do not need to do any correction. As we already know, C_1 is the fittest in the population. This is a good illustration because we can obtain a fitter solution during the evolutionary process through genetic operations. For the uniform crossover (shown in Figure 5.4(c)), the vector of uniform crossover O_U is randomly generated which is $O_U = (01010100)$, indicating that g_2, g_4, and g_6 of the two parent chromosomes exchange with each other. Hence the two offsprings are $C_{11} = (00000111)$ and $C_4 = (00110010)$. Since C_{11} is not a valid solution, we need to perform the correction scheme, and the corrected chromosome becomes to $C_{10} = (00110010)$, which is a valid solution.

5.4.6.2 Mutation

The population will undergo the mutation operation after the crossover operation is performed. With a mutation probability P_m, we scan each gene g_i on the parent chromosomes. If the mutation operation needs to be implemented, the value of the gene flips, i.e., 0 becomes 1 and 1 becomes 0.

An example shown in Figure 5.5 assumes g_3 is mutated in chromosome C_7. The offspring $C_{11} = (00110011)$ is a valid solution, thus no correction needed. While g_6, g_8 are mutated in chromosome C_8, the offspring $C_{12} = (00100011)$ is not a valid solution. Therefore, we perform the similar correction mechanism mentioned in the crossover subsection to make the offspring C_{12} valid by correcting $g_6 = 1$.

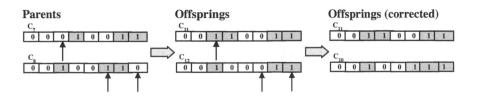

Parents **Offsprings** **Offsprings (corrected)**

Figure 5.5: Illustration of mutation operation.

5.4.6.3 Replacement Policy

The last step of RMCDS-GA is to create a new population using an appropriate replacement policy. Usually, two chromosomes from the evolution process are utilized to replace the two worst chromosomes in the original population for generating a new population. However, when creating new population by crossover and mutation, we have a big chance to lose the fittest chromosome. Therefore, an elitism strategy, in which the best chromosome (or a few best chromosomes) is retained in the next generation's population, is used to avoid losing the best candidates.

The RMCDS-GA stops and returns the current fittest solution until the number of total generations G is achieved or the best fitness value does not change for ten continuous generations. In the RMCDS-GA algorithm, we use G to stop the algorithm.

5.5 Genetic Algorithms with Immigrants Schemes

As mentioned in Section 5.4.1, the optimization performance of GAs depends mainly on the convergence time of the algorithm and appropriate population diversity may result in fast convergence time. In this section we investigate how the immigrants schemes affect the convergence time of the proposed RMCDS-GA algorithm.

In general, to converge at a proper pace is usually what we expect for GAs to find the optimal solutions for many optimization problems. However, for the σ-RMCDS problem, the convergence becomes a challenge. GAs usually require a certain population diversity level to maintain their adaptability. The crossover and mutation correction mechanisms in RMCDS-GA may reduce the population diversity. This slows down the speed of convergence. To address this problem, the random immigrants approach is a quite natural and simple way [90][91][92][93], which is inspired by the flux of immigrants that wander in and out of a population between two generations in nature. It maintains the diversity level of the population through replacing some individuals of the current population with random individuals, called random immigrants, in every generation. As to which individuals in the population should be replaced, usually there are two strategies: replacing random individuals or replacing the worst ones. In this chapter, GA with random immigrants (GARI) uses the second replacement strategy, i.e., utilize random immigrants to replace the worst individuals of the current population. The random immigrants can be obtained by running the IPI algorithm or by randomly running another existing MCDS algorithm. In order to avoid significant disruption of the ongoing search progress by random immigrants, the ratio of the number of random immigrants to the population size denoted by P_{ri} is set to a small value, e.g., $P_{ri} = 0.1$.

However, in some cases, random immigrants may not have any actual effect because individuals in the previous population may still be quite fit in the new population. In this case, random immigrants may thus degrade the performance. Based on the above consideration, GA with the elitism-based immigrants (GAEI), which uses elitism, i.e., the best chromosome (or a few best chromosomes), to create immigrants and replace the worst individuals in the current population, is also used to address the σ-RMCDS problem. The IPI algorithm can be performed to create immigrants from the elitism.

To further investigate the performance of GARI and GAEI, we propose the GA with hybrid immigrants (GAHI). In GAHI, in addition to the $P_{ri} \times k$ immigrants which are randomly created, $P_{ei} \times k$ immigrants are created from the elite of the previous generation, where P_{ei} is the ratio of the number of elitism-based immigrants to the population size. These two sets of immigrants will then replace the worst individuals in the current population.

The pseudo-code for GAEI and GAHI is shown in Algorithm 7.

Algorithm 7: RMCDS-GA with Immigrants Schemes

Require: k, P_{ri}, P_{ei}, G.
1: $g = 0$ {g represents the current generation number};
2: Initialize population $P(0)$ using IPI algorithm;
3: **while** $g < G$ **do**
4:     Calculate the fitness of each chromosome in population $P(g)$;
5:     Select two parent chromosomes in $P(g)$ using RWS selection;
6:     Crossover with P_c;
7:     Mutation with P_m;
8:     Calculate the fitness of each chromosome in interim population $P'(g)$
        {perform elitism-based immigrants};
9:     Generate $P_{ei} \times k$ immigrants by modify $E(g-1)$; {$E(g-1)$ denotes the elite
        in $P(g-1)$};
10:    Calculate the fitness of these immigrants
        {perform hybrid immigrants};
11:    **if** GAHI is used **then**
12:        generate $P_{ri} \times k$ immigrants by modify $E(g-1)$;
13:        calculate the fitness of these immigrants;
14:    **end if**
15:    Replace the worst individuals in $P'(g)$ with the immigrants;
16:    $P(g+1) = P'(g)$;
17:    g++;
18: **end while**
19: **return** the fittest individual in population $P(G)$.

5.6 Performance Evaluation

In the simulations, we implement the traditional GAs without immigrants and the three GAs with immigrants (GARI, GAEI, GAHI) to solve the σ-RMCDS problem.

These algorithms are compared with Wan's work [17] denoted by MIS, which is the latest and best MIS-based CDS construction algorithm.

5.6.1 Simulation Environment

We build our own simulator where all nodes have the same transmission range (10 m) and all nodes are deployed uniformly in a square area. Moreover, a random value between 0.9 and 0.98 is assigned to the TSR value associated to a pair of nodes inside the transmission range, otherwise, a random value between $(0, 0.8]$ is assigned to the TSR value associated to a pair of nodes beyond the transmission range. For a certain n, 100 instances are generated. The results are averaged among 100 instances. Additionally, the particular GA rules and control parameters are listed in Table 5.2.

Table 5.2: GA Parameters and Rules

Population size (k)	20
Number of total generations (G)	100
Selection scheme	Roulette wheel selection
Replacement policy	Elitism
Immigrants schemes	RI, EI, HI
P_{ri}	0.1
P_{ei}	0.1
Crossover probability (P_c)	1
Mutation probability (P_m)	0.001

5.6.2 Simulation Results

In Table 5.3, we show that traditional MCDS construction algorithms cannot solve the σ-RMCDS problem under the PNM, especially for large scale wireless sensor networks. In Table 5.3, we list the number of times that MIS and RMCDS-GA can find a CDS with a reliability greater than or equal to σ by running 100 simulations separately; σ is decreased from 0.6 to 0.4 by 0.1. From Table 5.3, we find that, with increasing n, the number of the times of satisfied CDSs for MIS and RMCDS-GA both decrease. This is because the sizes of CDSs increase which leads to a lower node-to-node delivery ratio. Moreover, RMCDS-GA can guarantee more satisfied CDSs than MIS, especially when $n \geq 200$. In other words, for large scale wireless sensor networks, it is hard to construct a satisfied CDS for MIS since the MIS algorithm does not consider reliability. Additionally, both MIS and RMCDS-GA can find more satisfied CDSs when σ decreases. In conclusion, traditional MCDS construction algorithms do not take reliability into consideration, while RMCDS-GA can find a satisfied reliable MCDS which is more practical in real environments.

Table 5.3: MIS-based CDSs and RMCDS-GA-generated CDSs

n	$\sigma = 0.6$		$\sigma = 0.5$		$\sigma = 0.4$	
	MIS	*GA*	*MIS*	*GA*	*MIS*	*GA*
50	100	100	100	100	100	100
80	94	100	100	100	100	100
120	57	100	98	100	100	100
160	21	100	90	100	100	100
200	5	96	44	100	88	100
250	2	91	12	93	56	100
400	1	90	4	17	10	100

In Table 5.4, R_{MIS} and R_{GA} represent the reliability of a CDS generated by MIS and RMCDS-GA, respectively. $|D_{MIS}|$ and $|D_{GA}|$ represent the size of the CDS constructed by MIS and RMCDS-GA, respectively. In Table 5.4, the reliability of CDSs decreases when the area size increases, since the number of the dominators increases. RMCDS-GA can guarantee to find a more reliable CDS than MIS, i.e., $R_{GA} > R_{MIS}$. More importantly, the sizes of the CDSs obtained by MIS and RMCDS-GA are almost the same. On average, RMCDS-GA can find a CDS with 10% more reliability without increasing the size of a CDS than MIS. In summary, RMCDS-GA does not trade CDS size for CDS reliability.

Table 5.4: *R and |D|* Results of MIS and RMCDS-GA Algorithms

Area (m^2)	n	R_{MIS}	R_{GA}	$|D_{MIS}|$	$|D_{GA}|$
40×40	50	0.65	0.77	17	18
50×50	80	0.59	0.72	24	26
60×60	120	0.51	0.68	33	33
70×70	160	0.46	0.62	40	44
80×80	200	0.44	0.58	51	51
90×90	250	0.39	0.53	63	62
100×100	400	0.32	0.49	78	78

5.7 Conclusion

In this chapter, we have investigated the σ-RMCDS problem using a new network model called PNM. The PNM is based on empirical studies showing that most wireless links are lossy links which only probabilistically connect pairs of nodes. Different from the traditional DNM which assumes that links are either connected or disconnected, the PNM enables the employment of lossy links by introducing the TSR value on each lossy link. In this chapter we focus on constructing a minimum-sized CDS while its reliability satisfies a preset application-dependent threshold. We prove that σ-RMCDS is an NP-hard problem and propose a GA with immigrants schemes to address the problem. The simulation results show that compared to the traditional MCDS algorithm, RMCDS-GA can find a more reliable CDS without increasing the size of a CDS.

Chapter 6

Constructing Load-balanced Virtual Backbones in Probabilistic Wireless Sensor Networks via Multi-Objective Genetic Algorithm

CONTENTS

6.1 Introduction

Wireless sensor networks (WSNs) are emerging as the desired environment for increasing numbers of military and civilian applications, such as disaster control, environment and habitat monitoring, battlefield surveillance, and health care applications [52]. Indeed, WSNs distinguish themselves clearly from wired networks with many features, i.e., absence of a fixed infrastructure, wireless multi-hop communication, limited bandwidth, and irreplaceable battery. Therefore, designing an energy-efficient communication scheme for WSNs is one of the most important issues that has a significant impact on the network performance. Due to the infrastructure-less and dynamic nature in WSNs, most routing protocols in WSNs (i.e., flooding) usually cause a serious broadcasting storm [53]. A connected dominating set (CDS) has been a well known approach for constructing a virtual backbone (VB) to alleviate the broadcasting storm thus improving the performance and increase the efficiency of routing protocols in WSNs. A dominating set (DS) is defined as a subset of nodes in a WSN such that each node in the network is either in the set or adjacent to some nodes in the set. If the induced subgraph by the nodes in a DS is connected, then this DS is called a CDS. The nodes in a CDS are called *dominators* denoted by set $\mathbb{B}$, or *dominatees* denoted by set $\mathbb{W}$. Using CDS as a VB, only nodes in a CDS need to forward the messages. Thus, the average message burden of a WSN could be reduced so that routing becomes much easier and can adapt quickly to network topology changes [54]. Furthermore, using dominators as forwarding nodes can efficiently reduce the energy consumption, which is also a critical concern in WSNs. In addition to routing protocols, a CDS-based VB has many other applications in WSNs, such as topology control [14], coverage [16], data collection [10], and data aggregation [17]. Since every node in a CDS works as a central management agent with heavy load, constructing a CDS with minimized size can greatly help to reduce the transmission interference, the number of control messages, and the storage overhead [54]. Hence, it is desirable to build a minimum-sized CDS (MCDS)-based VB.

Ever since the idea of employing a CDS as a VB for WSNs was introduced [19], a huge amount of approximation algorithms have been proposed to construct an MCDS-based VB, which is a well-known NP-hard problem [21]. In the seminal work [20], Guha and Kuller first modeled the problem of constructing the smallest CDS as the MCDS problem in a general graph, which is a well-known NP-hard problem [21]. Subsequently, many polynomial-time approximation algorithms for constructing an MCDS-based VB have been proposed in recent literature, such as abstraction-based algorithms [33][34], and addition-based algorithms [17]. After that, to make a CDS-based VB more resilient in mobile WSNs, the fault tolerance of a VB is considered. In [23], *k*-connected and *m*-dominated sets are introduced as a generalized abstraction of a fault-tolerance VB. In [25], the authors proposed a minimum routing cost connected dominating set (MOC-CDS), which aims to find a minimum CDS while assuring that any routing path through this CDS is the shortest in WSNs. Additionally, the authors investigate the problem of constructing a qualified CDS in terms of size, diameter, and average backbone path length (ABPL) in [26].

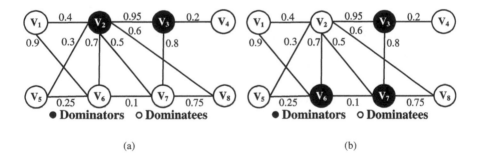

(a)   (b)

Figure 6.1: Illustration of a regular VB and an LBVB.

Unfortunately, all of the above mentioned work is based on the ideal deterministic network model (DNM), where any pair of nodes in a network is either fully connected or completely disconnected. In most real applications, however, the DNM cannot fully characterize the behavior of wireless links due to the *transitional region phenomenon* [5][6]. Beyond the "always connected" region, there is a *transitional region* where a pair of nodes is probabilistically connected via the so called *lossy links* [5]. As reported in [5], there are often many more lossy links than fully connected links in a WSN. Therefore, a more practical network model for WSNs is the probabilistic network model (PNM). Under this model, there is a *delivery ratio* (γ_{ij}) associated with each link connecting a pair of nodes v_i and v_j, which is used to indicate the probability that v_i can successfully deliver a package to v_j (an example is shown in Figure 6.1). For convenience, the WSNs considered under the DNM/PNM are called deterministic/probabilistic WSNs.

On the other hand, all the aforementioned work did not consider the *load-balance* factor in constructing a CDS. Without considering balancing the traffic load among the nodes on each dominator, some heavy loaded nodes may quickly exhaust their energy (such as dominator v_2 shown in Figure 6.1(a)), which might cause network partitions or malfunctions. To benefit from the CDS-based VB in WSNs and also take the load-balance factor into consideration, we constructed VBs in this manner in [80, 28]. We proposed a greedy algorithm to build a load-balanced CDS (LBCDS) and then load-balancedly allocate dominatees (LBAD) based on *expected allocation probability* for deterministic WSNs in [80]. Subsequently, we investigated the LBCDS and LBAD problems simultaneously for deterministic WSNs, and proposed a genetic algorithm to solve it in [28]. However, our previous work studied the load-balanced VB (LBVB) construction problem under the DNM rather than the more practical PNM, i.e., for probabilistic WSNs. Actually, how to measure the traffic load on each node for probabilistic WSNs is different. For example, in Figure 6.1, $\gamma_{27} = 0.5$, which means the probability that v_2 can successfully deliver a packet to v_7 is 50%. Then the expected number of transmissions to guarantee v_2 delivered one packet to v_7 is $\frac{1}{0.5} = 2$. In other words, the less the γ_{ij} value, the more potential traffic load on the link from v_j to v_i (the formal definition of *potential traffic load* is given in Section 6.3). Therefore, in this research, we investigate how to construct an LBVB for probabilistic WSNs, denoted by LBVBP. In order to better control the trade-off between the size of the constructed VB and the balance of traffic loads among all the dominators, we propose a novel multi-objective genetic algorithm (MOGA) to solve this problem. The detailed design, algorithm description and theoretical analysis are presented in Section 6.4. Particularly, the main contributions of this chapter are summarized as follows:

- We identify and highlight the use of lossy links when constructing a VB for probabilistic WSNs. Moreover, in order to measure the load balance of the nodes on a VB under the PNM, we define two new metrics: *potential traffic load*, and *actual traffic load*, which measure the potential traffic load and actual traffic load of each node in the network, respectively.

- In order to measure the load-balance factor of a constructed VB, we define two new metrics *VB p-norm* and *partition p-norm* based on potential traffic load and actual traffic load of each dominator, respectively.

- The LBVB construction problem under PNM (LBVBP) is an NP-hard problem, and we propose an effective multi-objective genetic algorithm (MOGA) to solve it, called LBVBP-MOGA. The convergence of the proposed algorithm is theoretically analyzed in the chapter.

- We also conduct simulations to validate our proposed algorithms. The simulation results show that the constructed LBVB can extend network lifetime by 65% on average compared with the existing state-of-the-art approaches.

The rest of this chapter is organized as follows: in Section 6.2, we review some related literature in Section 6.3, we introduce the network model and formally define the LBVBP construction problem under PNM. The design of the LBVBP-MOGA algorithm and the convergence analysis of the algorithm are presented in Section 6.4. The simulation results are presented in Section 6.5 to validate our proposed algorithm. Finally, the chapter is concluded in Section 6.6.

6.2 Related Work

In this section, we first briefly review the related work of constructing a CDS-based VB under the DNM. And then we summarize some related literatures under the PNM, followed by a literature review of multi-objective genetic algorithms (MO-GAs). Throughout, we will use n to denote the number of sensors in a WSN, and opt to denote the size of any optimal CDS.

6.2.1 CDS-based VBs under DNM

The idea of using a CDS as a VB was first proposed by Ephremides et al. in 1987 [19]. Since then, many algorithms that construct CDSs have been reported. Centralized algorithms were first proposed by Guha et al. [20]. Afterwards, plenty of distributed algorithms were proposed.

Single-leader distributed algorithms for CDS-based VBs use one initiator to initialize the algorithm. Usually, a base station can be the initiator for constructing CDSs in WSNs. In these distributed algorithms, a spanning tree rooted at the initiator is first constructed, and then maximum independent sets (MISs) are identified layer by layer. Finally, a set of connectors to connect the MISs is ascertained to form a CDS. Wan et al. [75] presented an ID-based distributed algorithm to construct a CDS using a single initiator. For UDGs, their [75] approach guarantees that the approximation factor on the size of a CDS is at most $4\,opt + 1$. Later, many attempts tried to improve this approximation factor on the size of a CDS based on this idea. Wu et al. reported an approximation factor of $3.8\,opt + 1.2$ [38]. Yao et al. improved it to $3.67\,opt + 1.33$ [39]. The factor is further improved to $3.478\,opt + 4.874$ [40] and $3.4306\,opt + 4.8185$ [41], subsequently. The current best is that the size of an MIS is at most $3.399\,opt + 4.874$ [42].

Distributed algorithms with multiple leaders do not require an initiator to construct a CDS-based VB. Alzoubi et al.'s technique [61] first constructs an MIS using a distributed approach without a leader or tree construction and then interconnects these MIS nodes to get a CDS. Li et al. proposed a distributed algorithm called r-CDS in [62], whose performance ratio is 172.

Moreover, some work compared the performance of protocols for clustering and virtual backbone formation, such as the the work by Basagni et al. [94].

6.2.2 Related Literature about PNM Model

Traditional routing schemes only considered fully connected links as a path of nodes in a WSN, and then send data through that sequence of nodes. Compared with the fully connected links, lossy links only provide probabilistic connectivity. However, there exist more lossy links in a WSN. Therefore, opportunistic routing schemes (e.g., ExOR [82] and More [83]) proposed to treat lossy links as advantages. The opportunistic routing scheme called ExOR [82] proposed a new unicast routing technique for multi-hop wireless networks. ExOR forwards each packet through a sequence of nodes that can successfully receive the transmission and are close to the destination. ExOR explores package overhearing along lossy links. When a lossy link succeeds, some transmissions can be saved. Later, Chachulski et al. combined random network coding with opportunistic routing to support both unicast and multicast routing in More [83]. The successful of opportunistic routing indicates that lossy links provide the potential throughput increase.

Recently, many works [6, 84, 85, 86] study the impact of lossy links to the topology control. Ma et al. in [84, 85] worked on achieving energy efficiency by turning off redundant nodes and links, while still satisfying the given QoS requirements. Liu et al. investigated how to control the minimal transmission range for each node while the global network reachability satisfies some constraints in [6, 86].

6.2.3 Literature Review of MOGAs

Solving multi-objective problems is a very difficult, since the objectives often conflict across a high-dimensional problem space and the optimization of multi-objective problems may also require extensive computational resources. A good MOGA must satisfy the following two aspects: 1) convergence to the pareto optimal set (the formal definition will be given in Section 6.4) and 2) maintenance of diversity in the solution set [95, 96, 97]. Many methods [98, 99] are designed to reach these goals. Recently, some new evolutionary paradigms have successfully applied to multi-objective problems (MOPs) such as particle swarm optimization [100], artificial immune systems [101], estimation of distribution [102], and scatter search [103].

6.2.4 Remarks

All the above mentioned VB construction works consider VBs for deterministic WSNs but none of them attempt to construct a load-balanced VB under the PNM, which is more realistic. This is the major motivation of this work. In this chapter, we first show an example to illustrate that an unbalanced VB cannot prolong network lifetime by reducing the communication cost. Instead, it actually leads to the reduction of network lifetime. Based on this observation, we then study to build an LBVB for more practical probabilistic WSNs (LBVBP). A multi-objective genetic algorithm (MOGA) is proposed followed by its convergence analysis.

6.3 Network Model and Problem Definition

In this section, we first present the assumptions, and then introduce the PNM. Subsequently, we formally define some terminologies. Finally, we give the problem definition and make some remarks about the proposed problem.

6.3.1 Assumptions

We assume a static connected WSN with the set of n nodes $\mathbb{V} = \{v_0, v_1, \cdots, v_{n-1}\}$. All the nodes in the WSN have the same transmission range. The *delivery ratio* γ_{ij} associated with each link connecting a pair of nodes v_i, v_j is available, which can be obtained by periodic hello messages, or be predicted using link quality index (LQI) [87]. We also assume that the γ_{ij} values are fixed. This assumption is reasonable as many empirical studies have shown that LQI is pretty stable in a static environment [88]. Furthermore, no node failure is considered since it is equivalent to a link failure case. No duty cycle is considered either. We do not consider packet collisions or transmission congestion, which are left to the MAC layer.

6.3.2 Network Model

Under the *probabilistic network model* (PNM), we model a WSN as an undirected graph $\mathbb{G}(\mathbb{V}, \mathbb{E}, \Upsilon(\mathbb{E}))$, where $\mathbb{V}$ is the set of n nodes, denoted by v_i, where $0 \leq i < n$. i is called the node ID of v_i in the chapter. $\mathbb{E}$ is the set of lossy links. $\forall\, v_i, v_j \in \mathbb{V}$, there exists a link (v_i, v_j) in $\mathbb{G}$ if and only if: 1) v_i and v_j are in each other's transmission range, and 2) $\gamma_{ij} > 0$. For each link $(v_i, v_j) \in \mathbb{E}$, γ_{ij} indicates the probability that node v_i can successfully directly deliver a packet to node v_j; and $\Upsilon(\mathbb{E}) = \{\gamma_{ij} \mid (v_i, v_j) \in \mathbb{E}, 0 < \gamma_{ij} \leq 1\}$. We assume the links are undirected (bidirectional), which means two linked nodes are able to transmit and receive information from each other with the same γ_{ij} value.

Because of the introduction of γ_{ij}, the traditional definition of the node neighborhood has changed. Hence, we first give the definition of the one-hop neighborhood and then extend it to the h-hop neighborhood.

Definition 6.1 One-hop neighborhood ($\mathbb{N}_1(v_i)$). $\forall v_i \in \mathbb{V}$ of node v_i is defined as:

$$N_1(v_i) = \{v_j | v_j \in \mathbb{V}, \gamma_{ij} > 0\}.$$

The physical meaning of one-hop neighborhood is the set of the nodes that can be reached from node v_i via 1 hop neighbors with positive probability. In this chapter, we use $|\mathbb{N}_1(v_i)|$ to represent the cardinality of the one-hop neighborhood set of node v_i.

Definition 6.2 *h*-Hop neighborhood($\mathbb{N}_h(v_i)$). $\forall v_i \in \mathbb{V}$ of node v_i is defined as:

$$\mathbb{N}_h(v_i) = \mathbb{N}_{h-1}(v_i) \cup \{v_k \mid \exists v_j \in \mathbb{N}_{h-1}(v_i),$$
$$v_k \in \mathbb{N}_1(v_j), v_k \notin \bigcup_{i=1}^{h-1} \mathbb{N}_i(v_i)\}.$$

The physical meaning of the *h*-hop neighborhood is the set of nodes that can be reached from node v_i by passing maximum *h* number of *lossy links* with positive probability.

6.3.3 Preliminary

Since load balance is the major concern of this work, the measurement of traffic load balance under PNM is critical to solve the LBVBP construction problem. Hence, in this subsection, we first define a novel metric called *potential traffic load* to measure the potential traffic load on each node.

Without knowing the communication protocol, the number of neighboring nodes of a node (i.e., $|\mathbb{N}_1(v_i)|$) is a potential indicator of the traffic load on each node. However, it is not the only factor to indicate the potential traffic load on each node in probabilistic WSNs. As we mentioned in Section 6.1, the less the γ_{ij} value, the more potential traffic load on v_j from v_i. Therefore, a more reasonable and formal definition of the potential traffic load is given as follows:

Definition 6.3 Potential traffic load (ι_i). $\forall v_i \in \mathbb{V}$. *The potential traffic load of v_i is defined as:* $\iota_i = \sum_{v_j \in \mathbb{N}_1(v_i)} \frac{1}{\gamma_{ij}}$.

After knowing the *potential traffic load* of each node, how to measure load-balance of a constructed VB is another challenge. We use *p-norm* to measure load balance in this chapter. The definition of *p-norm* is given as follows:

Definition 6.4 p-norm. *The p-norm of an $n \times 1$ vector $\mathbb{X} = (x_1, x_2, \cdots, x_n)$ is* $|\mathbb{X}|_p = (\sum_{i=1}^n |x_i|^p)^{\frac{1}{p}}$.

The authors in [50] stated that *p*-norm shows interesting properties for different values of *p*. If *p* is close to 1, the information routes resemble the geometric shortest paths from the sources to the sinks. For $p = 2$, the information flow shows an analogy to an electrostatics field, which can be used to measure the load balance among x_i. More importantly, the smaller the *p-norm* value, the more load-balanced the interested feature vector $\mathbb{X}$. For simplicity, we use $p = 2$ in this chapter.

In this chapter, we use *potential traffic load* (*Definition 7.3*) as the feature vector $\mathbb{X}$. According to Definition (6.4), we define the VB p-norm as follows:

Definition 6.5 VB p-norm ($|\mathbb{B}|_p$). *For WSN* $\mathbb{G}(\mathbb{V}, \mathbb{E}, \Upsilon(\mathbb{E}))$, *and a VB* $\mathbb{B} = \{v_1, v_2, \cdots, v_m\}$. *The* VB p-norm *of an* $m \times 1$ *vector* $\mathbb{L} = (l_1, l_2, \cdots, l_m)$ *is:*

$$|\mathbb{B}|_p = (\sum_{i=1, v_i \in \mathbb{B}}^{m} |l_i - \bar{l}|^p)^{\frac{1}{p}}, \tag{6.1}$$

where m is the cardinality of set $\mathbb{B}$, l_i *represents the potential traffic load of each node in set* $\mathbb{B}$, *and* $\bar{l} = (\sum_{j=1, v_j \in \mathbb{B}}^{m} l_j)/m$ *is the average potential traffic load on set* $\mathbb{B}$.

As we mentioned early, the smaller the *VB p-norm* value, the more load-balanced the constructed VB. We still use the WSN shown in Figure 6.1 to illustrate how to use *VB p-norm* to measure the load balance of constructed VBs. Two different VBs (represented by the black nodes) for the same network are identified in Figure 6.1(a), and (b), in which the VB shown in Figure 6.1(a) is a typical minimum-sized CDS-based VB. According to Definition 7.3, in Figure 6.1(a), $l_2 = 20.28$, $l_3 = 8.55$. Hence, $\bar{l} = \frac{l_2 + l_3}{2} = 14.42$. Therefore, based on Equation (6.1), the *VB p-norm* value of the VB shown in Figure 6.1(a) is $\sqrt{68.80}$. Similarly, we can get the *VB p-norm* value of the VB shown in Figure 6.1(b), which is $\sqrt{34.68}$. Clearly, $\sqrt{34.68} < \sqrt{68.80}$, which implies that the VB shown in Figure 6.1(b) is more load-balanced than the VB shown in Figure 6.1(a).

Actually, if one dominatee is adjacent to more than one dominator, one of the adjacent dominators is chosen by the dominatee to perform data transmission. Hence, it is important to load-balancedly allocate dominatees to each dominator to further balance the traffic loads among each dominator. In a traditional/naive way [17], each dominatee is allocated to the neighboring dominator with the smallest ID. Obviously, the load-balance factor is not taken into account. In some environments, the dominator with the smallest ID, which is chosen by majority dominatees, tends to have heavier workload than the other dominators. Therefore, neither node ID nor *potential traffic load* can reflect the actual workload precisely. In a WSN with a CDS as the VB, only the dominator and dominatee links contribute to the actual traffic load. Based on this observation, we define the following concepts:

Definition 6.6 Dominator partition ($\mathscr{P}$). *For a WSN represented by graph* $\mathbb{G}(\mathbb{V}, \mathbb{E}, \Upsilon(\mathbb{E}))$ *and a VB* $\mathbb{B} = \{v_1, v_2, \cdots, v_m\}$, *m disjoint sets are identified on* $\mathbb{V}$, *i.e.,* $\mathbb{P}(v_1), \mathbb{P}(v_2), \cdots, \mathbb{P}(v_m)$, *such that:*

1. *Each set* $\mathbb{P}(v_i)$ $(1 \leq i \leq m)$ *contains exactly one dominator* v_i.

2. $\bigcup_{i=1}^{m} \mathbb{P}(v_i) = \mathbb{V}$, *and* $\mathbb{P}(v_i) \bigcap \mathbb{P}(v_j) = \emptyset$ $(1 \leq i \neq j \leq m)$.

3. $\forall v_u \in \mathbb{P}(v_i)$ $(1 \leq i \leq m)$ *and* $v_u \neq v_i$, $(v_u, v_i) \in \mathbb{E}$.

A dominator partition *is:* $\mathscr{P} = \{\mathbb{P}(v_i) \mid v_i \in \mathbb{B}, 1 \leq i \leq m\}$.

We also use the WSN shown in Figure 6.1 to explain the concept of *dominator partition.* Three different dominator partitions are shown in Figure 6.2, in which only dominator and dominatee links are presented in the figure. According to Definition 6.6, we have $\mathbb{P}(v_3) = \{v_2, v_4\}$, $\mathbb{P}(v_6) = \{v_1, v_5\}$, and $\mathbb{P}(v_7) = \{v_8\}$ for the partition $\mathscr{P} = \{\mathbb{P}(v_3), \mathbb{P}(v_6), \mathbb{P}(v_7)\}$ shown in Figure 6.2(a). Without considering *delivery ratio* on each dominator and dominatee link, it is obvious that the dominator partition shown in Figure 6.2(b) is the most unbalanced of the workloads on each dominator. Moreover, without further information, it is hard to reveal which partition is more balanced than the other shown in Figure 6.2(a), and (c). According to above observations, we define the following concepts and metric to measure the load-balance of a dominator partition.

Definition 6.7   Authorized link set ($\mathbb{L}_i$). $\forall v_i \in \mathbb{B}$, *the authorized link set of dominator v_i is the set of the dominator and dominatee links formed by nodes in $\mathbb{P}(v_i)$*, i.e.,

$$\mathbb{L}_i = \{(v_i, v_j) \mid v_j \in \mathbb{P}(v_i), 1 \leq i \leq m\}.$$

As we have already known, l_i is only the indicator of the potential traffic load on each dominator v_i. The actual traffic load can be determined when a dominator partition is decided. In other words, the *authorized link set* $\mathbb{L}_i$ along with the corresponding *delivery ratio* of each link are the indicators of the actual traffic load on each dominator v_i. According to this observation, we give the following definition:

Definition 6.8   Actual traffic load (l_i). $\forall v_i \in \mathbb{B}$, *the actual traffic load of v_i is defined as:* $l_i = \sum_{(v_i, v_j) \in \mathbb{L}(v_i)} \frac{1}{\gamma_{ij}}$.

In this chapter, we use *partition p-norm* to measure the load-balance of different dominator partitions, in which, the *actual traffic load* l_i of each dominator v_i is used as the feature vector $\mathbb{X}$ shown in Definition 6.4. The definition of the *partition p-norm* is given as follows:

Definition 6.9   Partition p-norm ($|\mathscr{P}|_p$). *For a WSN represented by graph* $\mathbb{G}(\mathbb{V}, \mathbb{E}, \Upsilon(\mathbb{E}))$, *a VB* $\mathbb{B} = \{v_1, v_2, \cdots, v_m\}$, *and a dominator partition $\mathscr{P}$, the partition p-norm is:*

$$|\mathscr{P}|_p = \left(\sum_{i=1, v_i \in \mathbb{B}}^{m} |l_i - \bar{l}|^p \right)^{\frac{1}{p}}. \tag{6.2}$$

where $\bar{l} = (\sum_{j=1, v_j \in \mathbb{B}}^{m} l_j)/m$ *is the average actual traffic load on set* $\mathbb{B}$.

It is worth mentioning that the smaller the *partition p-norm* value, the more load-balanced the dominator partition. Figure 6.2 illustrates unbalanced and balanced dominator partitions. We can use *partition p-norm* to reveal which partition is the most balanced one. In Figure 6.2(a), the *actual traffic loads* of dominators are $l_3 = \frac{1}{\gamma_{32}} + \frac{1}{\gamma_{34}} = 6.05$, $l_6 = \frac{1}{\gamma_{61}} + \frac{1}{\gamma_{65}} = 5.11$, $l_7 = \frac{1}{\gamma_{78}} = 1.33$, and $\bar{l} = \frac{l_3 + l_6 + l_7}{3} = 4.16$.

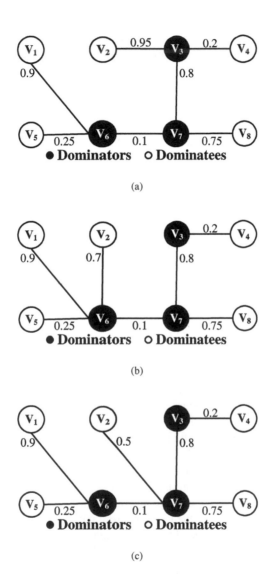

(a)

(b)

(c)

Figure 6.2: Illustration of unbalanced and balanced dominator partitions.

According to Equation (6.2), we obtain the *partition p-norm* of the dominator partition shown in Figure 6.2(a), which is $\sqrt{12.48}$. In the same way, the *partition p-norm* values of the dominator partitions shown in Figure 6.2(b) and (c) are $\sqrt{14.33}$ and $\sqrt{1.99}$, respectively. Clearly, $\sqrt{1.99} < \sqrt{12.48} < \sqrt{14.33}$, which implies the dominator partition shown in Figure 6.2(c) is the most balanced partition among all three dominator partitions shown in Figure 6.2. The partition shown in Figure 6.2(b) is the most unbalanced, which further confirms the observation we mentioned earlier.

6.3.4 Problem Definition

Finally, we are dedicated to construct a load-balanced VB. The physical meaning of an LBVB is that the constructed VB has the minimum *VB p-norm* and minimum *partition p-norm* values under PNM. The formal definition of the LBVBP problem is as follows:

Definition 6.10 Load-balanced VB problem in probabilistic WSNs (LBVBP). *For a WSN represented by graph* $\mathbb{G}(\mathbb{V}, \mathbb{E}, \Upsilon(\mathbb{E}))$, *the LBVBP problem is to find a minimum-sized node set* $\mathbb{B} \subseteq \mathbb{V}$ *and a dominator partition* $\mathscr{P}$, *such that:*

1. $\mathbb{G}[\mathbb{B}] = (\mathbb{B}, \mathbb{E}')$, *where* $\mathbb{E}' = \{e | \ e = (u, v), u \in \mathbb{B}, v \in \mathbb{B}, (u, v) \in \mathbb{E})\}$, *is connected.*

2. $\forall u \in \mathbb{V}$ *and* $u \notin \mathbb{B}$, $\exists v \in \mathbb{B}$, *such that* $(u, v) \in \mathbb{E}$.

3. *Minimize* $\{|\mathbb{B}|_p, |\mathscr{P}|_p\}$.

The LBVBP construction problem is NP-hard, since it still belongs to the MCDS problem. Based on Definition 6.10, the key issue of the LBVBP construction problem is to seek a VB that satisfies multiple constraints, i.e., the minimum size, the minimum *VB p-norm*, and the minimum *partition p-norm*. In reality, the multiple objectives are potentially in conflict. Conflicting objectives result in a set of compromised solutions known as the Pareto-optimal set. Since none of the solutions in this set can be considered as better than the others with respect to all the objectives, the goal of multi-objective optimization problem is to find as many as Pareto-optimal solutions as possible. Multi-objective genetic algorithm (MOGA) is a powerful tool to fulfill the above requirements due to its inherent parallelism and its ability to exploit the similarities among solutions by recombination [95]. Hence, in the following, a novel MOGA algorithm, named LBVBP-MOGA, is proposed to solve the LBVBP construction problem.

6.4 LBVBP-MOGA Algorithm

This section provides the detailed design of the LBVBP-MOGA algorithm. The first part provides an overview of the multiple-objective genetic algorithms (MOGAs) and some notions to be used in the chapter. The second part introduces the LBVBP-MOGA algorithm step-by-step. The last part analyzes the convergence of the proposed LBVBP-MOGA algorithm.

6.4.1 Overview of MOGAs

6.4.1.1 Multi-objective Problem (MOP) Definitions and Overview

Similar definitions for MOPs are given in related literature [95]. We merely introduce three important concepts in this subsection. Without loss of generality, we only consider minimization problems in this chapter, and it is easy to convert a maximization problem into a minimization problem.

Definition 6.11 Multi-objective problem (MOP). *In general, an MOP minimizes* $F(x) = (f_1(x), \cdots, f_k(x))$ *subject to* $g_i(x), i = 1, \cdots, m, x \in \Omega$ *(* Ω *is the decision variable space). An MOP solution minimizes the components of a objective vector* $F(x)$, *where* $x = (x_1, \cdots, x_n)$ *is an n-dimensional decision variable vector from some universe* Ω.

Because of multiple objectives and constraints, a key concept in determining a set of MOP solutions is *Pareto optimality*. To ensure understanding and consistency, we mathematically define *Pareto dominance* and *Pareto optimality* as follows:

Definition 6.12 Pareto dominance $(\preceq)$. *A vector* $u = (u_1, \cdots, u_k)$ *is said to dominate* $v = (v_1, \cdots, v_k)$ *(denoted by* $u \preceq v$*) if and only if* u *is partially less than* v, *i.e.,* $\forall i \in \{1, \cdots, k\}, u_i \leq v_i \bigwedge \exists i \in \{1, \cdots, k\}, u_i < v_i$.

Definition 6.13 Pareto optimality. *A solution* $x \in \Omega$ *is said to be* Pareto optimal *with respect to* Ω *if and only if there is no* $x' \in \Omega$ *for which* $F(x') \preceq F(x)$.

The set of all Pareto optimal decision vectors is called the *Pareto optimal set*. The corresponding set of the objective vectors is called the *Pareto optimal front*.

6.4.1.2 GA Overview

GAs work with a population of *chromosomes*, each representing a possible solution to a given problem. Each chromosome is assigned a *fitness score* according to how good a solution to the problem it is. The fittest chromosomes are given opportunities to *reproduce*, by *crossover* with other chromosomes in the population. This produces new chromosomes as *offsprings*, which share some features taken from each *parent*. The least fit chromosomes of the population are less likely to be selected for reproduction, and so they *die out*. A whole new population of possible solutions is thus produced by selecting the best chromosomes from the current *generation*, and mating them to produce a new set of chromosomes. In this way, over many generations, good characteristics are spread throughout the population. If the GA has been designed well, the population will converge to an optimal solution to the problem. The overview of GAs is pictured in Figure 6.3.

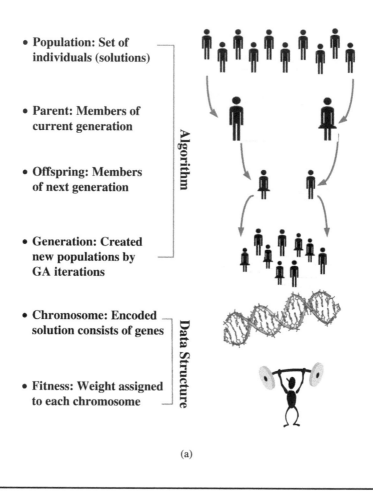

- **Population: Set of individuals (solutions)**

- **Parent: Members of current generation**

- **Offspring: Members of next generation**

- **Generation: Created new populations by GA iterations**

- **Chromosome: Encoded solution consists of genes**

- **Fitness: Weight assigned to each chromosome**

(a)

Figure 6.3: Genetic algorithm overview.

6.4.1.3 MOGA Overview

MOGA is a recently developed algorithmic tool to solve MOPs. MOGAs are very attractive because they have the ability to search partially ordered spaces for several alternative trade-offs. Additionally, an MOGA can track several solutions simultaneously via its population.

According to the definition of Pareto optimality, we can discover distinct differences between MOPs and single objective problems (SOPs). The solutions in the objective space of SOPs are scalar numbers and their relations have only two possibilities: less than and greater than or equal. However, the solutions of MOPs are vectors and their relations have three possibilities: $\mathbf{u} \preceq \mathbf{v}$, $\mathbf{v} \preceq \mathbf{u}$, and nondominated. This difference requires that MOGAs have more complicated fitness assignment rules and replacement rules.

6.4.2 Design of LBVBP-MOGA

In this section, we will design and explain the MOGA algorithm to solve the LBVBP problem called LBVBP-MOGA step by step.

6.4.2.1 Representation of Chromosomes

A chromosome is a possible solution of the LBVBP problem. Hence, when designing the encoding scheme of chromosomes, we need to identify dominators and dominatees in a chromosome and a *dominator partition* in a chromosome as well. For convenience, the set of neighboring dominators of each dominatee $v_s \in W$ is denoted by $\mathbb{D}(v_s) = \{v_r | v_r \in \mathbb{B}, (v_r, v_s) \in \mathbb{E}\}$. In the proposed LBVBP-MOGA, each node is mapped to a gene in the chromosome. A gene value g_i indicates whether the sensor represented by this gene is a dominator or not. If the sensor is a dominator (i.e., $\forall v_i \in \mathbb{B}$), the corresponding gene value is $g_i = 1$. Otherwise, the corresponding gene value is the two-tuples, which represent the allocated dominator of the dominatee v_i, and the neighboring dominators set of the dominatee v_i, respectively. Hence, a generation of chromosomes with gene values is denoted as: $\mathbb{C}^G = \{C_j \mid 1 \leq j \leq G, C_j = (g_1, g_2, \cdots, g_i, \cdots, g_n)\}$, where G is the number of chromosomes in each generation of population, and for $1 \leq i \leq n$,

$$g_i = \begin{cases} 1, \forall v_i \in \mathbb{B}. \\ < \forall v_t \in \mathbb{D}(v_i) \mid \mathbb{D}(v_i) >, \forall v_i \in W. \end{cases}$$

Through the above description we know that, as long as choosing a specific dominator from the neighboring dominator set $\mathbb{D}(v_i), \forall v_i \in W$, we can decide a specific *dominator partition*. Additionally, all the nodes with $g_i = 1$ form a VB $\mathbb{B} = \{v_i \mid g_i = 1, 1 \leq i \leq n\}$. We still use the probabilistic WSN shown in Figure 6.1(b) to illustrate the encoding scheme. There are eight nodes and the VB is $\mathbb{B} = \{v_3, v_6, v_7\}$. Moreover, according to the topology shown in Figure 6.1(b), we can get $\mathbb{D}(v_i), \forall v_i \in W$ easily. Thus, the *dominator partition* shown in Figure 6.2(b) can be encoded using eight genes in a chromosome, i.e., $C_1 = (v_6 \mid \{v_6\}, v_7 \mid \{v_3, v_6, v_7\}, 1, v_3 \mid \{v_3\}, v_6 \mid \{v_6\}, 1, 1, v_7 \mid \{v_7\})$ (shown in Figure 6.4). In conclusion, C_j records one possible VB and one possible dominator partition associated with the VB, while $\mathbb{C}^G$ represents the G different solutions to the LBVBP problem.

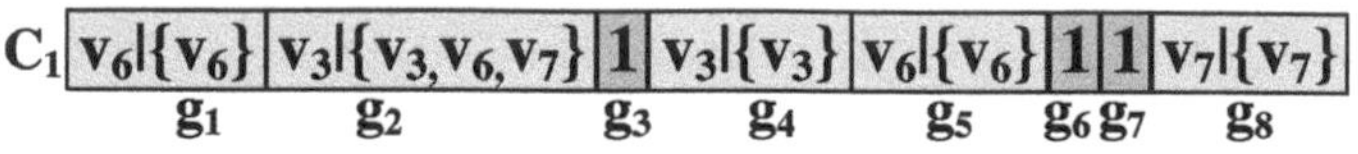

$$C_1 \boxed{v_6|\{v_6\}} \boxed{v_3|\{v_3,v_6,v_7\}} \boxed{1} \boxed{v_3|\{v_3\}} \boxed{v_6|\{v_6\}} \boxed{1}\boxed{1} \boxed{v_7|\{v_7\}}$$
$$\quad\ g_1 \qquad\ g_2 \qquad\qquad g_3 \quad\ g_4 \qquad g_5 \quad g_6 g_7 \qquad g_8$$

Figure 6.4: A chromosome with meta-genes and genes.

6.4.2.2 Population Initialization

GAs differ from most optimization techniques because of their global searching effectuated by one population of solutions rather than from one single solution. Hence, a GA search starts with the creation of the first generation, i.e., a population with G chromosomes denoted by P_1. This step is called population initialization. A general method to initialize the population is to explore the genetic diversity. That is, for each chromosome, all dominators are randomly generated. However, the dominators must form a VB. Therefore we start to create the first chromosome C_1 by running an existing MCDS method, e.g., the latest MCDS construction algorithm [17], and then generate the population with G chromosomes by modifying C_1. We call the procedure, generating the whole population by modifying one specific chromosome, inheritance population initialization (IPI) [28].

6.4.2.3 Fitness Function

Given a solution, its quality should be accurately evaluated by the fitness score, which is determined by the fitness function. In MOGAs, multiple conflict objectives need to be achieved. Hence, in our algorithm, three different fitness functions are defined as follows:

$$\begin{cases} f_1(C_j) = \min\{|\mathbb{B}|\}; \\ f_2(C_j) = \min\{|\mathbb{B}|_p\}; \\ f_3(C_j) = \min\{|\mathscr{P}|_p\}. \end{cases}$$

As we have mentioned, different from the relations of the solutions in SOPs, the relations of the solutions of MOPs have three possibilities. We use the following operator to summarize the relations.

Definition 6.14 Fitter operator.

$$Fitter(C_i,C_j) = \begin{cases} 1 & \mathbf{F}(C_i) \preceq \mathbf{F}(C_j); \\ -1 & \mathbf{F}(C_j) \preceq \mathbf{F}(C_i); \\ 0 & \text{non-dominated} \end{cases} \qquad (6.3)$$

where the objective vector is $\mathbf{F}(C_k) = (f_1(C_k), f_2(C_k), f_3(C_k)), 1 \le k \le G.$

When the objective vector of C_i *Pareto dominates* that of C_j, the *fitter operator* $fitter(C_i, C_j)$ returns 1; when the objective vector of C_i is *Pareto dominated* by that of C_j, the *fitter operator* $fitter(C_i, C_j)$ returns -1; when they are non-dominated, the *fitter operator* $fitter(C_i, C_j)$ returns 0.

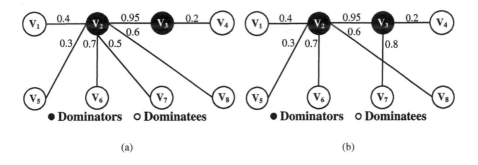

(a)   (b)

Figure 6.5: Illustration of two different dominator partitions.

We will use the feasible solutions (encoded as chromosomes) shown in Figure 6.5 and Figure 6.2 to illustrate the *fitter operator*. The three fitness values of the five different chromosomes are listed in Figure 6.6(a). To be more specific, we draw the five chromosomes in the two-dimensional objective space (f_2 *vs.* f_3) in Figure 6.6(b). From Figure 6.6(b), we know C_2 Pareto dominates C_1, C_5 Pareto dominates C_3, and C_3 Pareto dominates C_4.

6.4.2.4 Selection Scheme and Replacement Policy

During the evolutionary process, election plays an important role in improving the average quality of the population by passing the high quality chromosomes to the next generation. Therefore, in MOPS, the selection operator needs to be more carefully formulated to ensure that better chromosomes (the chromosomes closer to the Pareto optimal set) of the population have a greater probability of being selected for mating. We adopt *dominating tree (DT)* [103] to select parent chromosomes. A DT is a binary tree, in which each node has three fields: *id, left-link, and right-link*. The left-link field links to its left sub-tree whose root node is *dominated* by the node, and the right-link filed links to its right sub-tree whose root node is *non-dominated* by the node. A *sibling chain* of a DT is defined as a chain constituted by its root and the root's right-link nodes. A DT has some useful features [103]:

■ The sibling chain of a DT consists of and only consists of all Pareto optimal nodes in the DT.

■ The root of a DT Pareto dominates all nodes in its left sub-tree.

■ The leftmost node in the DT can be regarded as the "worst" node of the DT.

Chromosomes	f_1	f_2	f_3
C_1 (Fig. 4(a))	2	8.29	4.19
C_2 (Fig. 4(b))	2	8.29	1.89
C_3 (Fig. 2(a))	3	5.89	3.53
C_4 (Fig. 2(b))	3	5.89	3.79
C_5 (Fig. 2(c))	3	5.89	1.41

(a)

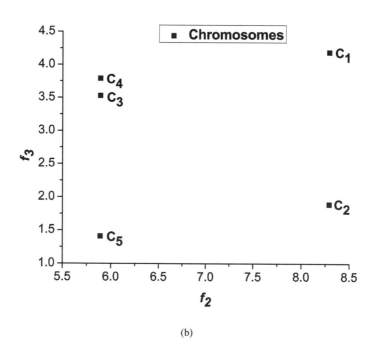

(b)

Figure 6.6: (a) Multiple objective fitness values of five chromosomes; (b) five chromosomes in the two-dimensional objective space.

Figure 6.7 demonstrates a DT consisting of the five encoded chromosomes shown in Figure 6.5 and Figure 6.2. According to the above features, we have:

- C_5 and C_2 (the *sibling chain*) are Pareto optimal nodes in the DT.

- C_5 Pareto dominates C_3 and C_4 in the DT.

- C_4 can be considered as the "worst" node in the DT.

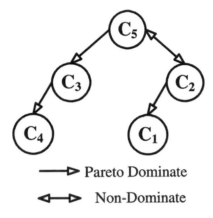

Figure 6.7: Illustration of a dominating tree.

Based on the above description, in each generation, we random select a pair of parent chromosomes in the set of Pareto optimal nodes (i.e., from *sibling chain*). The two new-generated offsprings will be inserted into the DT one by one and the "worst" (left-most) individual will be deleted from the DT each time. This process will be repeated until the stopping criterion (such as, a desired number of total generation is reached) is satisfied.

6.4.2.5 Genetic Operations

The performance of a GA relies heavily on two basic genetic operators, *crossover* and *mutation*. Crossover exchanges parts of the the parent chromosomes in order to find better ones. Mutation flips the values of genes, which helps a GA keep away from local optimum. In the LBVBP problem, we can adopt classical operations; however, the new obtained solutions may not be valid (the dominator set represented by the chromosome is not a CDS) after implementing the crossover and mutation operations. Therefore, a correction mechanism [28] needs to be performed to guarantee the validity of all the new generated offspring solutions.

The purpose of crossover operations is to produce more valid VBs represented by the offspring chromosomes. At this stage, we do not need to care about dominator partitions. Therefore, when performing crossover operations, we can logically assume all gene values of dominatees are 0. In the LBVBP-MOGA algorithm, we adopt three crossover operators called single-point crossover, two-point crossover, and uniform crossover respectively. (shown in Figure 6.8). With a crossover probability p_c, each time we select two chromosomes from the set of poreto optimal nodes as parents to perform one of the three crossover operators randomly. As mentioned earlier, after crossover operation, the new generated offsprings may not be a valid solution (the constructed VB is not a CDS). Thus we need to perform the correction mechanism. The mechanism starts with scanning each gene g_i on the offspring chromosome from the position of the crossover point, till the end of the chromosome. If g_i value is different from the corresponding value of its parent, then the mechanism corrects the value. The processes of the correction mechanism are shown in Figure 6.8(a) and (c).

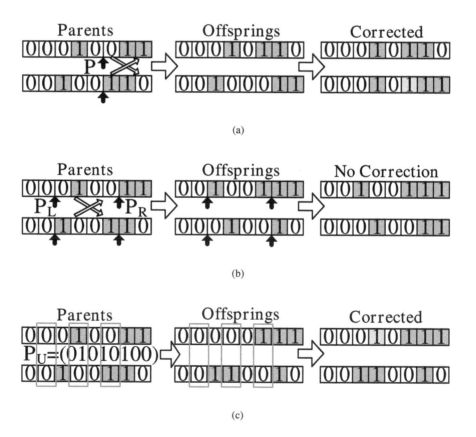

Figure 6.8: Illustration of crossover operations: (a) single-point crossover; (b) two-point crossover; (c) uniform crossover.

Suppose that two parent chromosomes (00010011) and (00100110) are selected by the RS scheme from the population. By the single-point crossover (shown in Figure 6.8(a)), the genes from the randomly generated crossover point $P = 6$ to the end of the two chromosomes exchange with each other to get (00010110) and (00010111). After crossing, the first offspring (00010110) is a valid solution. However, the other one (00100011) is not valid, thus we need to perform the *correction mechanism*. The mechanism starts with scanning each gene on the offspring chromosome, denoted by C_o^g, till the end of the chromosome. If the value of the current scanned gene is 0, i.e., $g_i = 0$ and the gene value is different from the original chromosome, denoted by C_s^g, without doing crossover and mutation operations, we change the gene value to 1. Whenever the DS represented by the corrected chromosome is a CDS, stop the mechanism. Otherwise, keep repeating the process till the end of C_o^g is reached. The idea behind the correction mechanism is that the DS represented by C_s^g is a CDS. If C_o^g is not valid, then add the dominators represented by C_s^g into the DS represented by C_o^g one by one. Finally, the corrected chromosome must be valid. For example, for the specific invalid offspring chromosome (00100011), when scanning the gene at position P, i.e. $g_6 = 0$, we find the value of g_6 is different after crossing. Therefore, we correct it by setting $g_6 = 1$. Then the corrected chromosome (00010111) is now a valid solution. Consequently, the correction mechanism stops and we get two valid offspring chromosomes (00010110) and (00010111). The correction mechanism is the same for crossover and mutation operations.

By the two-point crossover (shown in Figure 6.8(b)), the two crossover points are randomly generated which are $P_L = 3$ and $P_R = 6$; and then the genes between P_L and P_R of the two parent chromosomes are exchanged with each other. The two offsprings are (00100111) and (00010010) respectively. Since both of the offspring chromosomes are valid, we do not need to do any correction.

For the uniform crossover (shown in Figure 6.8(c)), the vector of uniform crossover P_U is randomly generated, which is $P_U = (01010100)$, indicating that g_2, g_4, and g_6 of the two parent chromosomes exchange with each other. Hence the two offsprings are (00000111) and (00110010). Since the first offspring is not a valid solution, we need to perform the correction mechanism mentioned before, and the corrected chromosome becomes (00110010), which is a valid solution.

The population undergoes the gene mutation operation after the crossover operation is performed. With a mutation probability p_m, we scan each gene g_i on the offspring chromosomes. If the mutation operation needs to be implemented, the value of the gene flips, i.e., 0 becomes 1, and 1 becomes 0. The same correction mechanism needs to be performed if the mutated chromosomes are not valid.

In order to increase the diversity of possible dominator partitions, we propose an additional step called *dominatee mutation* in LBVBP-MOGA to generate more feasible dominator partitions.

As known, as long as we choose a specific node from the neighboring dominator set $\mathbb{D}(v_i), \forall v_i \in \mathbb{W}$, we can easily explore a dominator partition. According to the observation, we design the following *dominatee mutation*. The original population without doing crossover and gene mutation operations will undergo the dominatee mutation operation. If the number of neighboring dominators of a dominatee v_i is greater than 1, i.e., $|\mathbb{D}(v_i)| \geq 2$, then we randomly pick a node from the set $\mathbb{D}(v_i)$. We use the VB shown in Figure 6.1(b) to illustrate the dominatee mutation. According to the topology, we get $|\mathbb{D}(v_2)| = |\{v_3, v_6, v_7\}| = 3 > 1$, which satisfies the condition to perform the dominatee mutation. Therefore, we randomly pick one dominator from the set $\mathbb{D}(v_2)$. If v_3 is selected from $\mathbb{D}(v_2)$, it means dominatee v_2 is allocated to dominator v_3. The corresponding dominator partition is shown in Figure 6.2(a). Similarly, if dominatee v_2 is allocated to dominator v_6, or v_2 is allocated to dominator v_7, the dominator partitions are shown in Figure 6.2(b), and (c), respectively. In summary, the process of dominatee mutation for v_2 is shown in Figure 6.9.

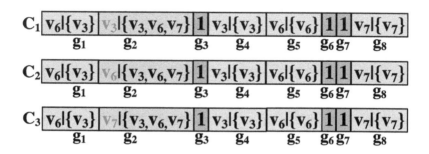

Figure 6.9: Illustration of dominatee mutation.

The pseudo-code of LBVBP-MOGA is shown in Algorithm 8.

6.4.3 Convergence Analysis

If MOGA can find the global optimum (i.e., convergence to the *Pareto optimal set*) of an MOP with probability one, we say that the MOGA converges. In this subsection, we will show our proposed LBVBP-MOGA converges.

Given that $\mathbf{x}$ is a decision variable, $\mathbb{S}$ is the space of all feasible decision variable, $\Phi : \mathbb{S} \rightarrow \mathbb{R}^k (k > 2)$ is a multi-objective fitness function, t is the generation number, and $P(t)$ is the population of the tth generation. Back et al. prove that a GA converges with probability one if it fulfills the following conditions in [104]:

1. $\forall \mathbf{x}, \mathbf{x}'$ is reachable from $\mathbf{x}$ by means of crossover and mutation operations;

2. the population sequence $P(0), P(1), \cdots$ is monotone, i.e.,

$$\forall t : \min\{\Phi(\mathbf{x}(t+1)) \mid \mathbf{x}(t+1) \in P(t+1)\} \leq \min\{\Phi(\mathbf{x}(t)) \mid \mathbf{x}(t) \in P(t)\} \tag{6.4}$$

The above definition is appropriate in the context of single objective GAs. However, in the context of MOGAs, a more general definition of monotonicity is needed. At any given generation t, let $P_{found}(t)$ is the set of Pareto optimal solutions that can be

---

Algorithm 8: : LBVBP-MOGA

Require: p_c, p_m, K {K represents the number of total generations}.

1: $t = 0$ {t represents the current generation number};
2: Initialize population $P(0)$ using the IPI algorithm;
3: **while** $t < K$ **do**
4:     Calculate the fitness vector of each chromosome in population $P(t)$;
5:     Construct *dominating tree (DT) T* for population $P(t)$;
6:     Select two parents chromosomes from $P(t)$ in the set of Pareto optimal nodes {from the *sibling chain of T*};
7:     Perform dominatee mutation;
8:     Crossover with p_c;
9:     Mutation with p_m;
10:     Insert new offsprings to the dominating tree T one by one;
11:     Delete the "worst" (left-most) one from T;
12:     Calculate the fitness vector of each chromosome in interim population $P'(t)$
13:     $P(t+1) = P'(t)$;
14:     t++;
15: **end while**
16: **return** the fittest individual in population $P(K)$.{the root of the constructed DT}

---

found by the MOGA through generation t. Moreover, let $P_{true}(t)$ denote the true set of Pareto optimal solution that can be found through generation t. The definition of monotonicity is given by the condition as follows:

$$P_{found}(t) = \{\mathbf{x} \in P_{true}(t) \mid \forall \mathbf{x}' \in P_{true}(t) \ s.t. \ \mathbf{F}(\mathbf{x}) \preceq \mathbf{F}(\mathbf{x}')\}$$

with $P_{found}(0) = \emptyset$. It can be shown by induction on t that under the above condition, $P_{found}(t)$ consists of the set of solutions evaluated through generation t that are Pareto optimal with respect to the set of all such solutions. Thus, $P_{found}(t+1)$ either retains or improves upon solutions in $P_{found}(t)$. In this manner, the above condition ensures that $P_{found}(t)$ monotonically moves toward to the Pareto optimal set P^* of an MOP.

Theorem 6.1
An MOGA satisfying the following two conditions converges to P^* of an MOP with probability one, i.e., $Prob\{\lim_{t\to\infty} \{P^* = P_{found}(t)\}\} = 1$.

$$\forall x, x' \text{ is reachable from } x \text{ by means of} \atop \text{crossover and mutation operations;}} \tag{6.5}$$

$$P_{found}(t) = \{x \in P_{true}(t) \mid \atop \forall x' \in P_{true}(t) \ s.t. \ F(x) \preceq F(x')\} \tag{6.6}$$

Proof 6.1   An MOGA can be viewed abstractly as a Markov chain consisting of two states. In the first state, $P^* = P_{found}(t)$, and in the second state $P^* \neq P_{found}(t)$. By Equation (6.6), there is zero probability of transiting from the first state to the second state. Thus, the first state is absorbing. By Equation (6.5), there is a non-zero probability of transiting from the second state to the first state. Thus, the second state is transient. Based on the above analysis, we can conclude that an MOGA can be absorbed to the first state (i.e., converges to P^*) within finite states, if the above two conditions (Equation (6.5) and (6.6)) are satisfied. □

Theorem 6.2
The proposed LBVBP-MOGA algorithm is convergent.

Proof 6.2 The offspring chromosomes (i.e., $\mathbf{x}'$) are created only by crossover and mutation operations, which satisfy Equation (6.5). Moreover, according to the proposed selection scheme and replacement policy, in each generation we random select parent chromosomes in the set of Pareto optimal nodes (i.e., $\mathbf{F}(\mathbf{x}) \preceq \mathbf{F}(\mathbf{x}')$). Newly-generated offsprings will be inserted into the DT one by one and the "worst" individual will be deleted from the DT each time. This process satisfies the monotonicity condition, which is shown in Equation (6.6). According to Theorem 6.1, our proposed LBVBP-MOGA is convergent. □

6.5   Performance Evaluation

Since there are no existing works studying the LBVB construction problem for probabilistic WSNs currently, in the simulations, the results of LBVBP-MOGA (denoted by MOGA) are compared with the recently published minimum-sized CDS construction algorithm [17] denoted by MCDS, and the LBCDS-GA algorithm proposed in [28] denoted by GA. We compare the three algorithms in terms of network lifetime, which is defined as the time duration until the first dominator runs out of energy.

6.5.1   Simulation Environment

We build our own simulator where all the nodes have the same transmission range and all the nodes are deployed uniformly and randomly in a square area. For each specific setting, 100 instances are generated. The results are averaged over these 100 instances (all results are rounded to integers). Moreover, a random value between 0.5 and 0.98 is assigned to the delivery ratio (γ_{ij}) value associated to a pair of nodes (v_i and v_j) inside the transmission range, otherwise, a random value between $(0, 0.5)$ is assigned to γ_{ij} associated to a pair of nodes beyond the transmission range. Moreover, we use the VB-based data aggregation as the communication mode. The simulated energy consumption model is that every node has the same initial 1000 units of energy. Receiving and transmitting a packet both consume 1 unit of energy. In the simulation, we consider the following tunable parameters: the node transmission range, the total number of nodes deployed in the square area, and the side length of the square area. Subsequently, we show the simulation results in two different scenarios.

6.5.2 Simulation Results

Figure 6.10 shows the network lifetime of three methods (MOGA, MCDS, GA)) under two different scenarios. From Figure 6.10(a), we know that the network lifetime increases for all the three algorithms with the side length of the deployed area increasing. It is obvious that the density of the network becomes thinner with the side length of the deployed area increasing. As to a data aggregation, the thinner the network is, the fewer neighbors of each dominator. In other words, the aggregated data are less on each dominator when the network becomes thinner. Hence, network lifetime is increasing for all the three algorithms. Additionally, we can see both MOGA and GA outperform MCDS. Furthermore, MOGA prolongs network lifetime by 42% on average compared with MCDS, and by 20% on average compared with GA. The results demonstrate that load-balancedly allocating dominatees to dominators can improve network lifetime significantly. On the other hand, MOGA outperforms GA, since MOGA takes multiple objectives into consideration simultaneously, making the MOGA easier to converge to a global optimum. Additionally, the local optimal solution found by GA might not be the same as the global optimal solution. Hence, the results shown in Figure 6.10(a) indicate our proposed MOGA can find a solution which is closer to the optimal solution than GA.

From Figure 6.10(b), we know that the network lifetime decreases for all algorithms with the node transmission range increasing. The fact is that the network becomes denser with the node transmission range increasing. The denser the network is, the more number of neighbors on each dominator. Since we use data aggregation as the communication mode in the simulations, the aggregated data are increasing on each dominator when the network becomes denser. Hence, network lifetime is decreasing for all the three algorithms. Similar results indicate that both MOGA and GA outperform MCDS. To be specific, MOGA prolongs network lifetime by 25% on average compared with MCDS, and by 6% on average compared with GA. The reasons are the same as analyzed before.

From Figure 6.10(c), we know that the network lifetime decreases for all the three algorithms with the number of nodes increasing. This is because we perform data aggregation in a more crowded network. Intuitively, the denser the network is, the more number of neighbors of each dominator. With the number of neighbors increasing, the aggregated data on each dominator becomes heavier. Hence, the network lifetime decreases for all the three algorithms. Additionally, we can see both MOGA and GA outperform MCDS. Furthermore, MOGA prolongs network lifetime by 69% on average compared with MCDS, and by 47% on average compared with GA. The reasons are the same as analyzed before.

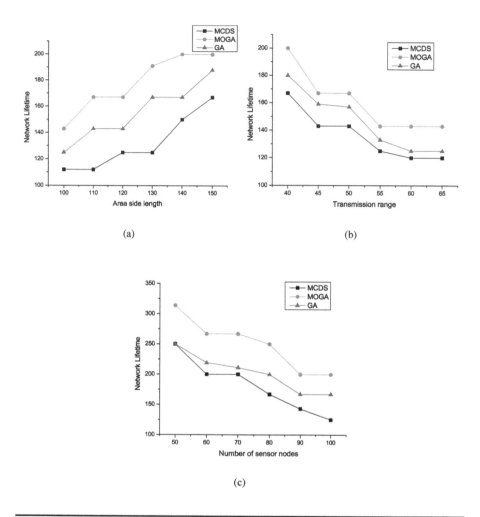

Figure 6.10: Network lifetime: (a) the node transmission range is *20 m*, the number of nodes is *100*, and the side length of the deployed area changes from *100 m* to *150 m*; (b) a square area of *300 m × 300 m*, the number of nodes is *100*, and the node transmission range changes from *40 m* to *65 m*; (c) a square area of *300 m × 300 m*, the node transmission range is *50 m*, and the number of nodes changes from *50* to *100*.

6.6   Conclusion

In this chapter, we address the problem of constructing a load-balanced VB in a probabilistic WSN (LBVBP), which is a minimum-sized CDS with the minimum $|\mathbb{B}|_p$ and $|\mathscr{P}|_p$ values in order to assure that the workload among each dominator is balanced. We propose an effective multi-objective genetic algorithm named LBVBP-MOGA to solve the problem. The simulation results demonstrate that using an LBVB can extend network lifetime significantly. Particularly, when the node number changes from 100 to 1000, our proposed method prolongs network lifetime by 65% on average compared with the state-of-the-art techniques [17][28].

Chapter 7

Constructing Load-balanced Data Aggregation Trees in Probabilistic Wireless Sensor Networks

CONTENTS

7.1   Introduction

In wireless sensor networks (WSNs), sensor nodes periodically sense the monitored environment and send the information to the sink (or base station), at which the gathered/collected information can be further processed for end-user queries. In this *data gathering* process, *data aggregation* can be used to fuse data from different sensors to eliminate redundant transmissions, since the data sensed by different sensors have spatial and temporal correlations [105]. Hence, through this in-network data aggregation technique, the amount of data that needs to be transmitted by a sensor is reduced, which in turn decreases each sensor's energy consumption so that the whole network lifetime is extended. For continuous monitoring applications with a periodical traffic pattern, a tree-based topology is often adopted to gather and aggregate sensing data because of its simplicity. Compared with an arbitrary network topology, a tree-based topology conserves the cost of maintaining a routing table at each node, which is computationally expensive for the sensor nodes with limited resources. For clarification, data gathering trees capable of performing aggregation operations are also referred to as *data aggregation trees* (DATs), which are directed trees rooted at the sink and have a unique directed path from each node to the sink. Additionally, in a DAT, sensing data from different sensors are combined at intermediate sensors according to certain aggregation functions including COUNT, MIN, MAX, SUM, and AVERAGE [106].

Due to the dense sensor deployment, many different DATs can be constructed to relay data from the monitored area to the sink. According to the diverse requirements of different applications, the DAT related works can be roughly classified into three categories: energy-efficient aggregation scheduling [107, 108, 74], minimum-latency aggregation scheduling [109, 17, 11], and maximum-lifetime aggregation scheduling[110, 111, 10]. It is worth mentioning that aggregation scheduling attracts a lot of interest in the current literature. However, unlike most of the existing works which spend lots of effort on aggregation scheduling, we mainly focus on the DAT construction problem.

Furthermore, most of the existing DAT construction works are based on the ideal *deterministic network model* (DNM), where any pair of nodes in a WSN is either connected or disconnected. Under this model, any specific pair of nodes are neighbors if their physical distance is less than the transmission range, while the rest of the pairs are always disconnected. However, in most real applications, the DNM cannot fully characterize the behaviors of wireless links due to the existence of the *transitional region phenomenon* [5]. It is revealed by many empirical studies [5, 3] that, beyond the "always connected" region, there is a *transitional region* where a pair of nodes are probabilistically connected via the so-called *lossy links* [5]. Even without collisions, data transmissions over lossy links cannot be guaranteed. Moreover, as reported in [5], usually there are many more lossy links (sometimes [6] 90% more) than fully connected links in a WSN. Therefore, in order to well characterize WSNs with lossy links, a more practical network model is the *probabilistic network model* (PNM). Under this model, there is a *transmission success ratio* (t_{ij}) associated with each link connecting a pair of nodes v_i and v_j, which is used to indicate the probability that a node can successfully deliver a package to another. An example is shown in Figure 7.1(a), in which the number over each link represents its corresponding transmission success ratio, and v_0 is the sink. For convenience, the WSNs considered under the DNM are called *deterministic WSNs*, whereas the WSNs considered under the PNM are called *probabilistic WSNs*. When $t_{ij} = 1$, DNM can be viewed as a special case of PNM.

On the other hand, all the aforementioned works did not consider the *load-balance* factor when they construct a DAT. Without considering balancing the traffic load among the nodes in a DAT, some heavy-loaded nodes may quickly exhaust their energy, which might cause network partitions or malfunctions. For instance, for aggregating the sensing data from eight different nodes to the sink node v_0, a shortest-path-based DAT for the probabilistic WSN (Figure 7.1(a)) is shown in Figure 7.1(b). The intermediate node v_4 aggregates the sensing data from four different nodes, whereas v_7 only aggregates one sensing data from v_8. For simplicity, if every link shown in Figure 7.1 is always there and every node has the same amount of data to be transferred through the intermediate nodes with a fixed data rate, heavy-loaded v_4 must deplete its energy much faster than v_7. From Figure 7.1(b), we know that the intermediate nodes usually aggregate the sensing data from neighboring nodes in a shortest-path-based DAT. Actually, the number of neighboring nodes of an intermediate node is a potential indicator of the traffic load on each intermediate node. However, it is not the only factor to impact the traffic load on each intermediate node. The criterion to assign a parent node, to which data is aggregated for each node on a DAT, is also critical to balance traffic load on each intermediate node. We refer to the procedure that assigns a unique parent node for each node in the network as the parent node assignment (PNA) in this chapter. Two PNAs different from Figure 7.1(b) are depicted in Figure 7.1(c) and (d). Evidently, with respect to load balance, the PNA shown in Figure 7.1(d) is the best (although it may induce high aggregation delay), which also implies the LBDAT shown in Figure 7.1(d) can extend network lifetime notably compared with the DATs shown in Figure 7.1(b) and Figure 7.1(c), since the traffic load is evenly distributed over all the intermediate nodes.

In summary, the investigated problem in this chapter is distinguished from all the prior works in three aspects. First, most of the current literature investigates the DAT

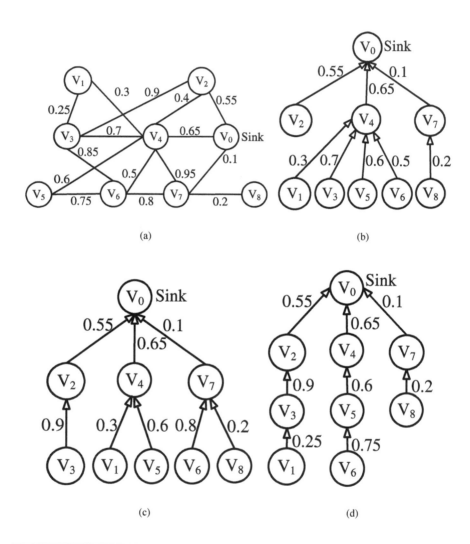

Figure 7.1: A Probabilistic WSN, DATs, and an LBDAT.

construction problem under the DNM, whereas our work is suitable for both DNM and PNM. Second, the *load-balance* factor is not considered when constructing a DAT in most of the aforementioned works. Finally, the DAT construction problem is our major concern, whereas the prior works focus on the aggregation scheduling problem. Therefore, in this chapter, we explore the DAT construction problem under the PNM considering balancing the traffic load among all the nodes in a DAT. To be specific, in this chapter, we construct a load-balanced DAT (LBDAT) under the PNM in three phases. We first investigate how to construct a load-balanced maximal independent set (LBMIS). An MIS can be defined formally as follows: given a graph $\mathbb{G} = (\mathbb{V}, \mathbb{E})$, an independent set (IS) is a subset $\mathbb{I} \subseteq \mathbb{V}$ such that for any two vertex $v_1, v_2 \in \mathbb{I}$, they are not adjacent, i.e., $(v_1, v_2) \notin \mathbb{E}$. An IS is called an MIS if we add one more arbitrary node to this subset, the new subset will not be an IS anymore. After obtaining an LBMIS, we attempt to find a minimum-sized set of nodes called *LBMIS connector set* $\mathbb{C}$ to make this LBMIS $\mathbb{M}$ connected, which is called the connected MIS (CMIS) problem. Finally, we seek a load-balanced parent node assignment (LBPNA). After an LBPNA is decided, by assigning a direction to each link in the constructed tree structure, we obtain an LBDAT. Comprehensive performance ratio analysis is presented as well.

The main contributions of this chapter are summarized as follows:

1. We identify and highlight the use of lossy links when constructing a DAT. Moreover, in order to measure the load balance of the nodes in a DAT under the PNM, we define two new metrics: *potential load* and *actual load*.

2. The LBDAT construction problem is an NP-complete problem and we solve it in three phases. First, we propose an approximation algorithm by using the linear relaxation and random rounding techniques to solve the LBMIS problem, which is an NP-hard problem. Theoretical analysis shows that this algorithm yields a solution upper bounded by $O(\ln(n))opt_{LBMIS}$, where opt_{LBMIS} is the optimal result of LBMIS, and n is the number of sensors in a WSN. Subsequently, a minimum-sized set of nodes are identified to make the LBMIS connected. Finally, to solve LBDAT, we present a randomized approximation algorithm to find an LBPNA. The approximation algorithm produces a solution in which the actual traffic load on each intermediate node is upper bounded by $O(\log(n))opt_{LBPNA}$, where opt_{LBPNA} is the optimal result.

3. We also conduct simulations to validate our proposed algorithms.

The rest of this chapter is organized as follows: in Section 7.3, we introduce the network model and formally define the LBDAT construction problem. The design of algorithms, theoretical analysis of the CMIS problem, and the LBDAT construction problem are presented in Sections 7.4 and 7.5, respectively. The simulation results are presented in Section 7.6 to validate our proposed algorithms. Finally, the chapter is concluded in Section 7.7.

7.2 Related Work

The problem of data gathering and aggregation in WSNs has been extensively investigated in the literature. Moreover, a tree-based topology to periodically aggregate collected data in WSNs is widely adopted because of its simplicity. However, most of existing works concentrated on the aggregation scheduling problem in deterministic WSNs, which is very different from our research problem. To be specific, we focus on constructing an LBDAT to perform data aggregation in probabilistic WSNs in this chapter. Therefore, in this section, we review the most related works. Based on the different user requirements, the existing DAT related works can be roughly divided into three categories: energy-efficient aggregation scheduling [107, 108, 112], minimum-latency aggregation scheduling [109, 17, 113, 114, 115], and maximum-lifetime aggregation scheduling [110, 111, 116, 117, 118].

7.2.1 Energy-efficient Aggregation Scheduling

As to battery powered sensors in WSNs, energy efficiency is always the major concern. Hence, it is important to minimize the total energy consumed by a WSN when designing a DAT. The authors proposed a power efficient data gathering and aggregation protocol (PEDAP) in [107], in which a near optimal minimum energy cost spanning tree is constructed for data aggregation. At first, only the sink node is included in the tree. Then, it keeps selecting nodes not in the tree one by one to join the current tree iteratively. The selected node is the one that can transmit packets to one of the nodes in the current tree with the minimum energy cost. However, PEDAP does not consider each node's energy and thus cannot achieve energy awareness. Therefore power aware PEDAP (PEDAP-PA) is proposed in [107] to improve PEDAP by considering the remaining energy of the sender. Later, the authors tried to construct an energy-balanced minimum degree spanning tree (MDST) in [112]. It starts from an arbitrary tree and tries to balance degrees of nodes in the tree according to their energy. However, a node with fewer children in a DAT does not mean it relays fewer data. Differed from the previous centralized algorithms, in [108], the authors proposed a localized, self organizing, robust, and energy-efficient DAT for a WSN called localized power-efficient data aggregation protocol (L-PEDAP). The proposed approach consists of two phases. In the first phase, it computes a sparse topology over the original graph of the WSN using the one-hop neighborhood information. In the second phase, it constructs a DAT over the edges of the computed sparse topology. Moreover, L-PEDAP is adaptive since it considers the dynamic changes when constructing a routing tree.

7.2.2 Minimum Latency Aggregation Scheduling

The minimum-latency data aggregation problem was proven to be an NP-hard problem in [109]. Moreover, Chen et al. [109] designed a $(\Delta - 1)R$-approximation algorithm based on a shortest path tree for data aggregation, where Δ is the maximum degree of the network graph and R is the network radius. Subsequently, the first-fit algorithm is proposed by Huang [113], in which a connected dominating set (CDS)-based tree is first constructed, and then maximal interference-free set of links is scheduled in each time slot. The latency of Huang's approach is bounded by $23R + \Delta - 18$. However, the already scheduled transmissions could also interference with the candidate links neglected in [113]. Hence, as a successor, Wan [17] developed a $15R + \Delta - 4$ approximation algorithm called sequential aggregation scheduling (SAS) to solve the minimum-latency aggregation schedule (MLAS) problem. Similar to Huang's work, Wan et al. in [17] also divided the aggregation process into the tree construction phase and the scheduling phase. The main difference is that the parents of leaf nodes are dynamically determined during the scheduling process. Subsequently, Xu et al. [114] developed an approximation algorithm with bound $16R' + \Delta - 14$, where R' is the inferior network radius which is smaller than R. Recently, Li et al. proposed a distributed scheduling algorithm named Clu-DDAS based on a novel cluster-based aggregation tree in [115] whose latency bound is $4R' + 2\Delta - 2$.

All the above works devoted efforts to find a *data aggregation schedule* for each link on the constructed DAT which leads to the minimum *data aggregation latency*. Hence, all these studies differed from our work. We mainly focus on the load-balanced tree construction in probabilistic WSNs.

7.2.3 Maximum Lifetime Aggregation Scheduling

Wu et al. [112] proved that constructing an arbitrary aggregation tree with the maximum lifetime is NP-complete. Hence, a huge number of approximation algorithms are proposed to construct a DAT with maximum lifetime. Xue et al. in [110], using linear programming formulation, modeled this problem as a multi-commodity flow problem, where a commodity represents the data generated from a sensor node and delivered to a base station. A fast approximate algorithm is presented, which is able to compute $(1 - \varepsilon)$-approximation to the optimal lifetime for any $\varepsilon > 0$. Lin et al. considered a more general network model in which the transmission power levels of sensors are heterogeneous and adjustable in [111]. The proposed algorithm starts from an arbitrary spanning tree rooted at the base station. Subsequently, one of the heavily loaded nodes is reduced to normalized load by partially rearranging the current tree to create a new tree. The upper bound on the lifetime of the constructed DAT is also presented in [111]. The authors in [116] proposed a combinatorial iterative algorithm for finding an optimal continuous solution to the maximum lifetime data gathering with aggregation (MLDA) problem that consists of up to $n - 1$ aggregation trees and achieves lifetime T_0. They obtained an α-approximate optimal integral solution, where $\alpha = \frac{T_0 - n + 1}{T_0}$, and n is the number of sensors in a WSN. The decentralized lifetime maximizing tree (DLMT) with energy consideration is proposed in [117]. Recently, Luo et al. proposed a distributed shortest-path based DAT in [118]. The authors transformed the problem of maximizing the lifetime of DATs into a general version of semi-matching problem, and showed that the problem can be solved by a min-cost max-flow approach in polynomial time.

7.2.4 Remarks

All the existing works are considered in designing a data aggregation schedule or constructing a DAT under the DNM. However, none of them attempt to construct a load-balanced DAT under the PNM, which is more realistic. This is the major motivation of this research work. Moreover, all the aforementioned works were focused on constructing energy-efficient aggregation scheduling, minimum-latency aggregation scheduling, or maximum-lifetime aggregation scheduling. Unfortunately, they do not consider the load-balance factor when constructing a DAT. In contrast, in this chapter, we first show an example to illustrate that an imbalanced DAT cannot prolong network lifetime by reducing the communication cost. Instead, it actually leads to the reduction of network lifetime. Based on this observation, we then build an LB-DAT for more practical probabilistic WSNs. Approximation algorithms are proposed in the chapter followed by comprehensive theoretical analysis.

7.3 Network Model and Problem Definition

In this section, we give an overview of the LBDAT construction problem under the PNM. We first present the assumptions, and then introduce the PNM. Finally, we give the problem definitions and make some remarks for the proposed problems.

7.3.1 Assumptions

We assume a static connected WSN with the set of n nodes $\mathbb{V}_s = \{v_1, v_2, \cdots, v_n\}$ and one sink node v_0. All the nodes have the same transmission range. The *transmission success ratio* ι_{ij} associated with each link connecting a pair of nodes v_i, v_j is available, which can be obtained by periodic hello messages, or be predicted using link quality index (LQI) [87]. We also assume that the ι_{ij} values are fixed. This assumption is reasonable as many empirical studies have shown that LQI is pretty stable in a static environment [88]. Furthermore, no node failure is considered since it is equivalent to a link failure case. No duty cycle is considered either. We do not consider packet collisions or transmission congestion, which are left to the MAC layer.

We further assume that the n nodes monitor the environment in the deployed area and periodically report the collected data to the sink node v_0 along the LBDAT $\mathbb{T}$ (the formal definition of LBDAT will be given later). Every node produces a data package of B bits during each report interval. Moreover, an intermediate node can aggregate multiple incoming B-bit packets, together with its own package into a single outgoing B-bit package. Furthermore, we assume the data receiving rate of each node v_i is γ_i, and $\mathcal{R}$ denotes the maximum data receiving rate of all the nodes. Finally, the degree of a node v_i is denoted by d_i, whereas δ/Δ denotes the minimum/maximum node degree in the network.

7.3.2   Network Model

Under the *Probabilistic Network Model* (PNM), we model a WSN as an undirected graph $\mathbb{G}(\mathbb{V}, \mathbb{E}, \mathbb{P}(\mathbb{E}))$, where $\mathbb{V} = \mathbb{V}_s \cup \{v_0\}$ is the set of $n+1$ nodes, denoted by v_i, where $0 \leq i \leq n$. i is called the node ID of v_i in the chapter. $\mathbb{E}$ is the set of lossy links. $\forall v_i, v_j \in \mathbb{V}$, there exists a link (v_i, v_j) in $\mathbb{G}$ if and only if: 1) v_i and v_j are in each other's transmission range, and 2) $l_{ij} > 0$. For each link $(v_i, v_j) \in \mathbb{E}$, l_{ij} indicates the probability that node v_i can successfully directly deliver a packet to node v_j; and $\mathbb{P}(\mathbb{E}) = \{l_{ij} \mid (v_i, v_j) \in \mathbb{E}, 0 \leq l_{ij} \leq 1\}$. We assume the links are undirected (bidirectional), which means two linked nodes are able to transmit and receive information from each other with the same l_{ij} value. Because of the introduction of l_{ij}, we define the one-hop neighborhood and the h-hop neighborhood as follows.

Definition 7.1   One-hop neighborhood ($\mathbb{N}_1(v_i)$). $\forall v_i \in \mathbb{V}$ of node v_i is defined as: $N_1(v_i) = \{v_j \mid v_j \in \mathbb{V}, l_{ij} > 0\}$.

The physical meaning of one-hop neighborhood is the set of the nodes that can be directly reached from node v_i.

Definition 7.2 h-Hop neighborhood ($\mathbb{N}_h(v_i)$). $\forall v_i \in \mathbb{V}$ of node v_i is defined as:
$$\mathbb{N}_h(v_i) = \mathbb{N}_{h-1}(v_i) \cup \{v_k \mid \exists v_j \in \mathbb{N}_{h-1}(v_i), v_k \in \mathbb{N}_1(v_j), v_k \notin \bigcup_{i=1}^{h-1} \mathbb{N}_i(v_i)\}.$$

The physical meaning of the h-hop neighborhood is that the set of the nodes that can be reached from node v_i with some probability by passing at most h links.

7.3.3   Problem Definition

Since load balance is the major concern of this work, the measurement of the traffic load balance under the PNM is critical to solve the LBDAT construction problem. Hence, in this subsection, we first define a novel metric called *potential load* to measure the potential traffic load on each node.

As we mentioned in Section 7.1, the number of neighboring nodes of a node (i.e., $|\mathbb{N}_1(v_i)|$) is a potential indicator of the traffic load on each node. However, it is not the only factor to indicate the potential traffic load on each node in probabilistic WSNs. For example, if $l_{ij} = 0.5$, then the expected number of transmissions to guarantee v_i to deliver one packet to v_j is $\frac{1}{0.5} = 2$. The less the l_{ij} value, the more potential traffic load on v_j from v_i. Therefore, a more reasonable and formal definition of the potential load is given as follows:

Definition 7.3   Potential load (ρ_i), $\forall v_i \in \mathbb{V}_s$, of v_i is defined as: $\rho_i = \sum_{v_j \in \mathbb{N}_1(v_i)} \lceil \frac{B}{\gamma_i} \rceil \frac{1}{l_{ij}}$.

We solve the LBDAT construction problem in three phases in this chapter. First, we construct a load-balanced maximal independent set (LBMIS), and then we select additional nodes to connect the nodes in LBMIS, denoted by the connected MIS (CMIS) problem. Finally, we acquire a load-balanced parent node assignment (LBPNA). After LBPNA is determined, by assigning a direction of each link in the constructed tree structure, we obtain an LBDAT. In this subsection, we formally define the LBMIS, CMIS, LBPNA, and LBDAT construction problems sequentially.

Definition 7.4 Load-balanced maximal independent set (LBMIS) problem. For a probabilistic WSN represented by graph $\mathbb{G}(\mathbb{V},\mathbb{E},\mathbb{P}(\mathbb{E}))$, the *LBMIS* problem is to find a node set $\mathbb{M} \subseteq \mathbb{V}$ such that:

1. $v_0 \in \mathbb{M}$.

2. $\forall u \in \mathbb{V}$ and $u \notin \mathbb{M}$, $\exists v \in \mathbb{M}$, such that $(u,v) \in \mathbb{E}$.

3. $\forall u \in \mathbb{M}$, $\forall v \in \mathbb{M}$, and $u \neq v$, such that $(u,v) \notin \mathbb{E}$.

4. There exists no proper subset or superset of $\mathbb{M}$ satisfying the conditions 1, 2, and 3.

5. Maximize $\min\{\rho_i \mid \forall v_i \in \mathbb{M}\}$.[1]

Taking the load-balance factor into consideration, we are seeking an MIS in which the minimum potential load of the nodes in the constructed LBMIS is maximized. In other words, the potential traffic load on each node in the LBMIS is as balanced as possible. Now, we are ready to define the CMIS problem.

Definition 7.5 Connected maximal independent set (CMIS) problem. For a probabilistic WSN represented by graph $\mathbb{G}(\mathbb{V},\mathbb{E},\mathbb{P}(\mathbb{E}))$ and an *LBMIS* $\mathbb{M}$, the *CMIS* problem is to find a node set $\mathbb{C} \subseteq \mathbb{V}\backslash\mathbb{M}$ such that:

1. The induced graph $G[\mathbb{M}\bigcup\mathbb{C}]$ on $\mathbb{G}$ is connected.

2. Minimize $|\mathbb{C}|$, where $|\mathbb{C}|$ is the cardinality of set $\mathbb{C}$.

For convenience, the nodes in set $\mathbb{M}$ are called *independent nodes*, whereas, the nodes in set $\mathbb{C}$ are called *LBMIS connectors*. Moreover, the nodes in the set $\mathbb{G}\backslash(\mathbb{M}\bigcup\mathbb{C})$ are called *leaf nodes*. Furthermore, $\forall v_i \in \mathbb{M}\bigcup\mathbb{C}$, v_i is also called a *non-leaf node*. Hence, the set of non-leaf nodes are denoted by $\mathbb{D} = \mathbb{M}\bigcup\mathbb{C}$.

Constructing a load-balanced connected topology is just one part of the work to build an LBDAT. In order to measure the actual traffic load, one more important task needed to be resolved is how to do parent node assignment for leaf nodes in the network. Since the actual traffic load of each node in a DAT is depended on the number of its children, which are composed of leaf nodes and non-leaf nodes, we

---

[1]Maxmin and minmax can achieve the load-balance objective similarly according to [119] for the specific communication model data aggregation. In this chapter, minmax is also applicable.

give the formal definition of the parent node assignment for leaf nodes to non-leaf nodes as follows:

Definition 7.6　Parent node assignment for leaf nodes ($\mathcal{A}_L$). For a probabilistic WSN represented by graph $\mathbb{G}(\mathbb{V}, \mathbb{E}, \mathbb{P}(\mathbb{E}))$ and a CMIS $\mathbb{D} = \{v_1, v_2, \cdots, v_m\}$, we need to find m disjoint sets on $\mathbb{V}$, denoted by $\mathbb{L}(v_1), \mathbb{L}(v_2), \cdots, \mathbb{L}(v_m)$, such that:

1. Each set $\mathbb{L}(v_i)$ ($1 \leq i \leq m$) contains exactly one non-leaf node v_i.

2. $\bigcup_{i=1}^{m} \mathbb{L}(v_i) = \mathbb{V}$, and $\mathbb{L}(v_i) \bigcap \mathbb{L}(v_j) = \emptyset$ ($1 \leq i \neq j \leq m$).

3. $\forall v_u \in \mathbb{L}(v_i)$ ($1 \leq i \leq m$) and $v_u \neq v_i$, such that $(v_u, v_i) \in \mathbb{E}$.

4. Assign v_i ($1 \leq i \leq m$) as the parent node of the nodes in $\mathbb{L}(v_i) \backslash \{v_i\}$. A *parent node assignment for leaf nodes* is $\mathcal{A}_L = \{\mathbb{L}(v_i) \mid \forall v_i \in \mathbb{D}, 1 \leq i \leq m\}$.

Definition 7.7　Parent node assignment for non-leaf nodes ($\mathcal{A}_I$). For a probabilistic WSN represented by graph $\mathbb{G}(\mathbb{V}, \mathbb{E}, \mathbb{P}(\mathbb{E}))$ and a CMIS $\mathbb{D} = \{v_1, v_2, \cdots, v_m\}$, we need to find m sets on $\mathbb{D}$, denoted by $\mathbb{I}(v_1), \mathbb{I}(v_2), \cdots, \mathbb{I}(v_m)$, such that:

1. $\forall v_i \in \mathbb{M}$, the set $\mathbb{I}(v_i)$ contains exactly one independent node v_i.

2. $\forall v_j \in \mathbb{C}$, the set $\mathbb{I}(v_j)$ contains exactly one LBMIS connector v_j.

3. $\forall v_i \in \mathbb{D}, 1 \leq |\{\mathbb{I}(v_j) \mid v_i \in \mathbb{I}(v_j), j = 1, 2, \cdots, m\}| \leq 2$.

4. $\bigcup_{i=1}^{m} \mathbb{I}(v_i) = \mathbb{D}$.

5. $\forall v_u \in \mathbb{I}(v_i)$ ($1 \leq i \leq m$) and $v_u \neq v_i$, such that $(v_u, v_i) \in \mathbb{E}$.

6. Assign v_i ($1 \leq i \leq m$) as the parent node of the nodes in $\mathbb{I}(v_i) \backslash \{v_i\}$.

A *parent node assignment for non-leaf nodes* is:

$$\mathcal{A}_I = \{\mathbb{I}(v_i) \mid \forall v_i \in \mathbb{D}, 1 \leq i \leq m\}.$$

$\mathcal{A}_L$ and $\mathcal{A}_I$ together are called a parent node assignment (PNA) $\mathcal{A}$. According to the above definitions, as to each set $\mathbb{L}(v_i)$ in $\mathcal{A}_L$, v_i is the *parent node* of the nodes in set $\mathbb{L}(v_i) \backslash \{v_i\}$, whereas, the nodes in set $\mathbb{L}(v_i) \backslash \{v_i\}$ are called the *leaf child nodes* of v_i. Similarly, as to each set $\mathbb{I}(v_i)$ in $\mathcal{A}_I$, v_i is the *parent node* of the nodes in set $\mathbb{I}(v_i) \backslash \{v_i\}$, whereas, the nodes in set $\mathbb{I}(v_i) \backslash \{v_i\}$ are called the *non-leaf child nodes* of v_i. As we have already known, ρ_i is only the indicator of the potential traffic load on each non-leaf node. The actual traffic load only can be determined when a PNA, i.e., $\mathcal{A} = \{\mathcal{A}_L, \mathcal{A}_I\}$, is decided. In other words, the number of leaf children and non-leaf child nodes (i.e., $|\mathbb{L}(v_i)| - 1$ and $|\mathbb{I}(v_i)| - 1$) along with the corresponding t_{ij} are the indicators of the actual traffic load on each non-leaf node v_i. According to this observation, we give the following definition:

Definition 7.8 Actual load (α_i). The actual load of a non-leaf node v_i is: $\forall v_i \in \mathbb{D}, \alpha_i = \sum\limits_{v_j \in \{\mathbb{L}(v_i) \bigcup \mathbb{I}(v_i) \mid i \neq j\}} \lceil \frac{B}{\gamma_i} \rceil \frac{1}{u_{ij}}$.

Load balance is our major concern, hence, when doing parent node assignment, we must take it into consideration. The formal definition of load-balanced parent node assignment is as follows:

Definition 7.9 Load-balanced parent node assignment (LBPNA $\mathcal{A}^*$). For a probabilistic WSN represented by graph $\mathbb{G}(\mathbb{V}, \mathbb{E}, \mathbb{P}(\mathbb{E}))$ and a *CMIS* $\mathbb{D} = \{v_1, v_2, \cdots, v_m\}$, the *LBPNA* problem is to find a parent node assignment $\mathcal{A}^*$ for $\mathbb{V}$, such that: $\min\{\alpha_i \mid \forall v_i \in \mathbb{D}\}$ is maximized under $\mathcal{A}^*$.

After a $\mathcal{A}^*$ is decided, every node in the network has a unique parent node. Hence, a tree structure (LBDAT) is established.

7.3.4 Remarks

Since finding an MIS is a well-known NP-hard problem [21] in graph theory, CMIS is NP-hard as well. Therefore CMIS cannot be solved in polynomial time unless P = NP. Consequently, we propose an approximation algorithm by using linear relaxation and random rounding technique to obtain an approximate solution. Additionally, the key aspect to solve the LBDAT construction problem is to find an LBPNA $\mathcal{A}^*$, and obtaining an LBPNA is NP-complete. Then we formulate it as an equivalent binary programming. We also present a randomized approximation algorithm to find the approximate solution to $\mathcal{A}^*$. After specifying the direction of each link in $\mathcal{A}^*$, we obtain an LBDAT $\mathbb{T}$.

7.4 Connected Maximal Independent Set

In this section, we first introduce how to solve the load-balanced maximal independent set (LBMIS) problem. We formulate the LBMIS problem as an integer nonlinear programming (INP). Subsequently, we show how to obtain an $O(\ln(n))$ approximation solution by using linear programming (LP) relaxation techniques. Finally, we present how to find a minimum-sized set of LBMIS connectors to form a CMIS $\mathbb{D}$.

7.4.1 INP Formulation of LBMIS

For convenience, we assign a decision variable ω_i for each sensor $v_i \in \mathbb{V}$, which is allowed to be 0/1 value. This variable sets to 1 if and only if the node is an independent node, i.e., $\forall v_i \in \mathbb{M}, \omega_i = 1$. Otherwise, it sets to 0.

It is well known that in graph theory, an MIS is also a dominating set (DS). A DS is defined as a subset of nodes in a WSN such that each node in the network is either in the set or adjacent to some node in the set. Hence, we formally model the LBMIS as an integer nonlinear programming (INP) as follows:

Sink node constraint. All aggregated data are reported to the sink node, hence the sink node is deliberately set to be an independent node, i.e., $\omega_0 = 1$.

DS property constraint. Since an MIS is also a DS, we should formulate the DS constraint for the LBMIS problem first. The DS property states that each non-independent node must reside within the one-hop neighborhood of at least one independent node. We therefore have $\omega_i + \sum_{v_j \in \mathbb{N}_1(v_i)} \omega_j \geq 1, \forall v_i \in \mathbb{V}$.

IS property constraint. Since the solution of the LBMIS problem is at least an IS, the IS property is also a constraint of LBMIS. The IS property indicates that no two independent nodes are adjacent, i.e., $\forall v_i, v_j \in \mathbb{M}, (v_i, v_j) \notin \mathbb{E}$. In other words, we have $\sum_{v_j \in \mathbb{N}_1(v_i)} \omega_i \cdot \omega_j = 0, \forall v_i \in \mathbb{V}$.

Consequently, the objective of the LBMIS problem is to maximize the minimum potential load (ρ_i) of all the independent nodes ($\forall v_i \in \mathbb{M}$). We denote v as the objective of the LBMIS problem, i.e., $v = \min_{v_i \in \mathbb{M}} (\rho_i)$. Mathematically, the LBMIS problem can be formulated as an integer nonlinear programming INP_{LBMIS} as follows:

$$
\begin{aligned}
\max \quad & v = \min\{\rho_i \mid \forall v_i \in \mathbb{M}\} \\
s.t. \quad & \omega_0 = 1; \\
& \omega_i + \sum_{v_j \in \mathbb{N}_1(v_i)} \omega_j \geq 1; \\
& \sum_{v_j \in \mathbb{N}_1(v_i)} \omega_i \cdot \omega_j = 0; \\
& \omega_i, \omega_j \in \{0,1\}, \ \forall v_i, v_j \in \mathbb{V}.
\end{aligned}
\qquad (INP_{LBMIS})
$$

Since the IS property constraint is quadratic, the formulated integer programming INP_{LBMIS} is not linear. To linearize INP_{LBMIS}, the quadratic constraint is eliminated by applying the techniques proposed in [70]. More specifically, the product $\omega_i \cdot \omega_j$ is replaced by a new binary variable ϖ_{ij}, on which several additional constraints are imposed. As a consequence, we can reformulate INP_{LBMIS} exactly to an integer linear programming ILP_{LBMIS} by introducing the following linear constraints:

1. $\sum_{v_j \in \mathbb{N}_1(v_i)} \varpi_{ij} = 0$;

2. $\omega_i \geq \varpi_{ij}; \ \omega_j \geq \varpi_{ij}$;

3. $\omega_i + \omega_j - 1 \leq \varpi_{ij}; \ \varpi_{ij} \in \{0,1\}, \ \forall v_i, v_j \in \mathbb{V}$.

According to Definition 7.3, the potential load of an independent node v_i is $\rho_i = \sum_{j:\omega_i t_{ij} > 0} \lceil \frac{B}{\eta} \rceil \frac{1}{t_{ij}}$. Moreover, by relaxing the conditions $\omega_j \in \{0,1\}$ and $\varpi_{ij} \in \{0,1\}$ to $\omega_j \in [0,1]$ and $\varpi_{ij} \in [0,1]$, correspondingly, we obtain the following relaxed linear programming LP^*_{LBMIS}:

$$\max v = \min\left\{\rho_i = \sum_{j:\omega_i l_{ij}>0} \lceil \tfrac{B}{\gamma_i} \rceil \tfrac{1}{l_{ij}} \,\middle|\, \forall v_i \in \mathbb{V}_s\right\}$$

$$
\begin{aligned}
s.t. \quad & \omega_0 = 1; \\
& \omega_i + \sum_{v_j \in \mathbb{N}_1(v_i)} \omega_j \geq 1; \\
& \sum_{v_j \in \mathbb{N}_1(v_i)} \varpi_{ij} = 0; \\
& \omega_i \geq \varpi_{ij}; \quad \omega_j \geq \varpi_{ij}; \\
& \omega_i + \omega_j - 1 \leq \varpi_{ij}; \\
& \omega_i, \omega_j, \varpi_{ij} \in [0,1], \; \forall v_i, v_j \in \mathbb{V}_s.
\end{aligned}
\qquad (LP^*_{LBMIS})
$$

7.4.2 Approximation Algorithm

Due to the relaxation enlarged the optimization space, the solution of LP^*_{LBMIS} corresponds to an upper bound to the objective of INP_{LBMIS}. Given an instance of LBMIS modeled by the integer nonlinear programming INP_{LBMIS}, we propose an approximation algorithm as shown in Algorithm 9 to search for an LBMIS.

Algorithm 9: Approximation Algorithm for LBMIS

---

1   Solve LP^*_{LBMIS}. Let $(\boldsymbol{\omega}^*, v^*)$ be the optimum solution, where
$$\boldsymbol{\omega}^* = <\omega_1^*, \omega_2^*, \cdots, \omega_n^*>, v^* = \min\left\{ \sum_{j:\omega_i^* l_{ij}>0} \lceil \tfrac{B}{\gamma_i} \rceil \tfrac{1}{l_{ij}} \,\middle|\, \forall v_i \in \mathbb{V}\right\};$$

2   Sort all the sensor nodes by the ω_i^* value in the decreasing order. The sorted node ID i is stored in the array denoted by $A[n]$;

3   $\widehat{\omega_0} = 1$;

4   **for** $i = 1$ *to* n **do**

5     $\lfloor \; \widehat{\omega_i} = 0;$

6   $counter = 0$;

7   **while** $counter \leq \tau$, *where* $\tau = \dfrac{7\ln(n)}{\min\{\omega_i^* | v_i \in \mathbb{V}, \omega_i^* > 0\}}$ **do**

8     $k = 0$;

9   **while** $k < n$ **do**

10       $i = A[k]$;

11   **if** $\forall v_j \in \mathbb{N}_1(v_i), \; \widehat{\omega_j} = 0$ **then**

12         $\lfloor \; \widehat{\omega_i} = 1$ with probability $p_i = \omega_i^*$;

13       $k = k + 1$;

14     $counter = counter + 1$;

15   **return** $(\widehat{\boldsymbol{\omega}}, \widehat{v} = \min\{ \sum_{j:\widehat{\omega_i} l_{ij}>0} \lceil \tfrac{B}{\gamma_i} \rceil \tfrac{1}{l_{ij}} \,\middle|\, \forall v_i \in \mathbb{V}\})$

The basic idea of Algorithm 9 is as follows: first, solve the relaxed linear programming LP^*_{LBMIS} to get an optimal fractional solution, denoted by $(\boldsymbol{\omega}^*, v^*)$, where $\boldsymbol{\omega}^* = <\omega_1^*, \omega_2^*, \cdots, \omega_n^*>$, and then round ω_i^* to integer $\widehat{\omega}_i$ according to the six steps shown in lines 2-14 of Algorithm 9.

1. Sort sensor nodes by the ω_i^* value (where $1 \leq i \leq n$) in the decreasing order.

2. Set the sink node to be the independent node, i.e., $\widehat{\omega}_0 = 1$.

3. Set all $\widehat{\omega}_i$ to be 0.

4. Start from the first node in the sorted node array A. If there is no node been selected as an independent node in v_i's 1-hop neighborhood, then let $\widehat{\omega}_i = 1$ with probability $p_i = \omega_i^*$.

5. Repeat step 4) till reach the end of array A.

6. Repeat step 4) and 5) for $\frac{7\ln(n)}{\min\{\omega_i^* | v_i \in \mathbb{V}, \omega_i^* > 0\}}$ times.

Next, the correctness of our proposed approximation algorithm (Algorithm 9) is proven, followed by the performance ratio analysis. Before showing the correctness of Algorithm 9, two important lemmas are given as follows.

Lemma 7.1
For a probabilistic WSN represented by $\mathbb{G} = (\mathbb{V}, \mathbb{E}, \mathbb{P}(\mathbb{E}))$, if a subset $\mathbb{S} \subseteq \mathbb{V}$ is a DS and meanwhile $\mathbb{S}$ is also an IS, then this subset $\mathbb{S}$ is an MIS of $\mathbb{G}$.

Proof: If $\mathbb{S} \subseteq \mathbb{V}$ is a DS of $\mathbb{G}$, it implies that $\forall v_i \in \mathbb{V} \backslash \mathbb{S}$, there exists at least one node $v_j \in \mathbb{S}$ in v_i's 1-hop neighborhood. Moreover, if $\mathbb{S}$ is also an IS, it implies that no two nodes in $\mathbb{S}$ are adjacent, i.e., $\forall v_s, v_t \in \mathbb{S}, (v_s, v_t) \notin \mathbb{E}$.

Suppose $\mathbb{S}$ is **not** an MIS. In other words, we can find at least one more node, that does not violate the DS property and the IS property of $\mathbb{S}$, to be added into $\mathbb{S}$. Suppose v_i is such a node. Based on the DS property, we know that $\exists v_j \in \mathbb{S}$ and $v_j \in \mathbb{N}_1(v_i)$. According to the hypothesis, $v_i \in \mathbb{S}$ and considering the fact that $v_j \in \mathbb{N}_1(v_i)$, we conclude that there are two nodes (v_i and v_j) are adjacent in $\mathbb{S}$ (i.e., $(v_i, v_j) \in \mathbb{E}$), which is contradicted to the IS property. Hence, the hypothesis is false and Lemma 7.1 is true. □

Lemma 7.2
The set $\mathbb{M} = \{v_i \mid \widehat{\omega}_i = 1, 0 \leq i \leq n\}$, where $\widehat{\omega}_i$ is derived from Algorithm 9, is a DS almost surely.

Proof: Suppose $\forall v_i \in \mathbb{V}$, $|\mathbb{N}_1(v_i)| = k_i$, where $|\mathbb{N}_1(v_i)|$ is the cardinality of the set $\mathbb{N}_1(v_i)$. Let the random variable W_i denote the event that no node in the set $\mathbb{N}_1(v_i) \bigcup \{v_i\}$ is selected as an independent node. Additionally, we denote $\mathscr{W} = \max\{\omega_j^* \mid v_j \in \mathbb{N}_1(v_i) \bigcup \{v_i\}\}$. For the probability of W_i happening, we have

$P(W_i) = [(1 - \omega_1)(1 - \omega_2) \cdots (1 - \omega_{k_i})(1 - \omega_i)]^\tau \leq (1 - \mathcal{W})^\tau \leq (1 - \min\{\omega_i^*|v_i \in$

$\mathbb{V}, \omega_i^* > 0\})^\tau \leq (e^{-\min\{\omega_i^*|v_i \in \mathbb{V}, \omega_i^* > 0\}})^\tau = e^{-\frac{7\min\{\omega_i^*|v_i \in \mathbb{V}, \omega_i^* > 0\}\ln(n)}{\min\{\omega_i^*|v_i \in \mathbb{V}, \omega_i^* > 0\}}} \leq e^{-7\ln(n)} = \frac{1}{n^7}.$

Thus, according to the Borel-Cantelli lemma, $P(W_i) \sim 0$, which implies that there exists one independent node in the set $\mathbb{N}_1(v_i) \bigcup \{v_i\}$ almost surely, i.e., it is almost sure that the set $\mathbb{M} = \{v_i \mid \widehat{\omega}_i = 1, 0 \leq i \leq n\}$ derived from Algorithm 9 is a DS. Then, it is reasonable that we consider $\mathbb{M}$ is a DS of $\mathbb{G}$ in the following.[2] □

Based on Lemma 7.1 and Lemma 7.2, the following theorem can be obtained.

Theorem 7.1
The set $\mathbb{M} = \{v_i|\widehat{\omega}_i = 1, 0 \leq i \leq n\}$, where $\widehat{\omega}_i$ is derived from Algorithm 9, is an MIS.

Proof: According to Algorithm 9, no two nodes can both be set as independent nodes in the one-hop neighborhood. This guarantees the IS property of $\mathbb{M}$, i.e., $\forall v_i, v_j \in \mathbb{M}, (v_i, v_j) \notin \mathbb{E}$. Moreover, $\mathbb{M}$ is a DS as proven in Lemma 7.2. Hence, based on Lemma 7.1, we conclude that $\mathbb{M}$ is an MIS. □

From Theorem 7.1, we know that the solution of Algorithm 9 is an MIS. Subsequently, we analyze the approximation factor of Algorithm 9 in Theorem 7.2.

Theorem 7.2
Let opt_{LBMIS} denote the optimal solution of the LBMIS problem. The proposed algorithm yields a solution of $O(\ln(n))opt_{LBMIS}$.

Proof: The expected ρ_i of the independent node v_i found by Algorithm 9 is

$$E[\sum_{j:\widehat{\omega}_i l_{ij} > 0} \lceil \frac{B}{\gamma_i} \rceil \frac{1}{l_{ij}}] = \lceil \frac{B}{\gamma_i} \rceil \sum_{v_j \in \mathbb{N}_1(v_i)} \frac{E[\widehat{\omega}_i]}{l_{ij}}$$

$$\geq \lceil \frac{B}{\gamma_i} \rceil \sum_{v_j \in \mathbb{N}_1(v_i)} \frac{\omega_i^* \times 1 + (1 - \omega_i^*) \times 0}{l_{ij}}$$

$$= \lceil \frac{B}{\gamma_i} \rceil \sum_{v_j \in \mathbb{N}_1(v_i)} \frac{\omega_i^*}{l_{ij}} \geq v^*.$$

Applying the Chernoff bound, we obtain the following bound:

$$Pr[\sum_{j:\widehat{\omega}_i l_{ij} > 0} \frac{1}{l_{ij}} \leq (1 - \sigma)\tau v^*] \leq e^{-\frac{\sigma^2}{2}\tau v^*}$$

---

[2]It is almost impossible that $\mathbb{M}$ is not a DS of $\mathbb{G}$. If not, we repeat the entire rounding process.

for arbitrary $0 < \sigma < 1$. To simplify this bound, let $\sigma = \sqrt{\frac{6}{7}}$, we get:

$$Pr\left[\sum_{j:\widehat{\omega}_i l_{ij}>0} \frac{1}{l_{ij}} \le (1-\sigma)\tau v^* \right] \le e^{-\frac{3}{7}\tau v^*} \le e^{-\frac{3}{7}\tau}$$

$$\le e^{-\frac{3\ln(n)}{\min\{\omega_i^* | v_i \in \mathbb{V}, \omega_i^* > 0\}}}$$

$$\le e^{-3\ln(n)} = \frac{1}{n^3}.$$

Applying the union bound, we get the probability that some independent node has the potential load $\widehat{v}$ less than $(1-\sigma)\tau v^*$,

$$Pr[\widehat{v} \le (1-\sigma)\tau v^*] \le n\frac{1}{n^3} = \frac{1}{n^2}.$$

Again, since $\sum_{n>0} \frac{1}{n^2}$ is a particular case of the Riemann zeta function, then $\sum_{n>0} \frac{1}{n^2}$ is bounded, i.e., $\sum_{n>0} \frac{1}{n^2} < \infty$ by the result of the Basel problem. Thus, according to the Borel-Cantelli lemma, $P[\widehat{v} \le (1-\sigma)\tau v^*] \sim 0$.

In summary, we get:

$$Pr[\text{a node is selected to be an independent node in one-hop}$$
$$\text{neighborhood} \bigcap \widehat{v} \ge (1-\sigma)\tau v^*]$$
$$= 1 \cdot (1 - \frac{1}{n^2}) \sim 1, \text{ when } n \sim \infty,$$

where $\sigma = \sqrt{\frac{6}{7}}$, and $\tau = \frac{7\ln(n)}{\min\{\omega_i^* | v_i \in \mathbb{V}, \omega_i^* > 0\}}$.

Furthermore, the minimum potential load on all the independent nodes produced by Algorithm 9 is upper bounded by $(1-\sigma)\tau v^*$ with probability 1, where v^* is the optimum solution of LP_{LBMIS}^*, and $\tau = \frac{7\ln(n)}{\min\{\omega_i^* | v_i \in \mathbb{V}, \omega_i^* > 0\}}$. Hence, Theorem 7.2 is proven. □

7.4.3 Connecting LBMIS

To solve the CMIS problem, one more step is needed after constructing an LBMIS, which is making the LBMIS connected. Next, we introduce how to find a minimum-sized set of LBMIS connectors to connect the constructed LBMIS by similar procedure as in [17].

We first divide the LBMIS $\mathbb{M}$ into disjoint node sets according to the following criterion: $\mathbb{M}_0 = \{v_0\}$ and $\mathbb{M}_l = \{v_i \mid v_i \in \mathbb{M}, \exists v_j \in \mathbb{M}_{l-1}, v_i \in \mathbb{N}_2(v_j), v_i \notin \bigcup_{k=0}^{l-1} \mathbb{M}_k\}$.

The sink node is put into $\mathbb{M}_0$. Clearly, $|\mathbb{M}_0| = 1$. All the independent nodes in the 2-hop neighborhood of the nodes in $\mathbb{M}_{l-1}$ are put into $\mathbb{M}_l$. Hence, l is called the *level* of an independent node. $\mathbb{M}_l$ represents the set of independent nodes of level l in $\mathbb{G}$ with respect to the node in $\mathbb{M}_0$. Additionally, suppose the maximum level

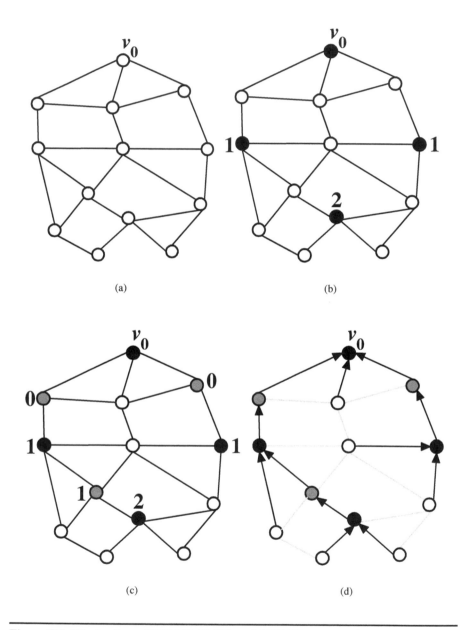

Figure 7.2: Illustration of LBMIS, CMIS, and LBDAT construction process.

of an independent node is L. For each $0 \leq i \leq L - 1$, let $\mathbb{S}_i$ be the set of the nodes adjacent to at least one node in $\mathbb{M}_i$ and at least one node in $\mathbb{M}_{i+1}$. Subsequently, compute a minimum-sized set of nodes $\mathbb{C}_i \subseteq \mathbb{S}_i$ to cover the nodes in set $\mathbb{M}_{i+1}$. Let $\mathbb{C} = \bigcup_{i=0}^{L-1} \mathbb{C}_i$ and therefore $\mathbb{D} = \mathbb{M} \bigcup \mathbb{C}$ is a CMIS of the original graph $\mathbb{G}$. We use the WSN shown in Figure 7.2 (a) as an example to explain the construction process of a CMIS. In Figure 7.2 (a), each circle represents a sensor node. As we mentioned early, the construction process consists of two phases. In the first phase, it solves the LBMIS problem by Algorithm 9 to obtain $\mathbb{M}$ which is shown in Figure 7.2 (b) by black circles. The number beside each independent node is the level of that node with respect to the sink node v_0. In the second phase, we choose the appropriate LBMIS connectors ($\mathbb{C}$), shown by gray nodes in Figure 7.2 (c), to connect all the nodes in $\mathbb{M}$ to form a CMIS ($\mathbb{D}$).

Next, we analyze the number of non-leaf nodes $|\mathbb{D}|$ produced by our proposed algorithms. The lemma below presents some additional properties of the constructed CMIS [17].

Lemma 7.3
The following statements are true.

1. *For each $0 \leq i < L$, each LBMIS connector in $\mathbb{C}_i$ is adjacent to at most four independent nodes in $\mathbb{M}_{i+1}$.*

2. *For each $1 \leq i < L - 1$, each independent node in $\mathbb{M}_i$ is adjacent to at most 11 LBMIS connectors in $\mathbb{C}_i$.*

3. *$|\mathbb{C}_0| \leq 12$.* □

Based on Lemma 7.3, we have the following theorem which indicates the number of non-leaf nodes produced by our proposed algorithm.

Theorem 7.3
The number of non-leaf nodes satisfies $|\mathbb{M}| + \lceil \frac{\mathbb{M}-1}{4} \rceil \leq |\mathbb{D}| \leq 2|\mathbb{M}|$.

Proof: According to the above proposed algorithm, each LBMIS connector connects the independent nodes in $\mathbb{M}_i$ and $\mathbb{M}_{i+1}$. Hence, $|\mathbb{C}| = |\bigcup_{i=0}^{L-1} \mathbb{C}_i| \leq \sum_{i=0}^{L-1} \max\{|\mathbb{M}_i|, |\mathbb{M}_{i+1}|\} \leq |\mathbb{M}|$. Moreover, based on Lemma 7.3, $|\mathbb{C}| \geq \lceil \frac{\mathbb{M}-1}{4} \rceil$. Finally, we get $|\mathbb{M}| + \lceil \frac{\mathbb{M}-1}{4} \rceil \leq |\mathbb{M} \bigcup \mathbb{C}| = |\mathbb{D}| \leq |\mathbb{M}| + |\mathbb{C}| \leq 2|\mathbb{M}|$. □

7.4.4 LBPNA for Non-leaf Nodes

After CMIS is constructed, we find an LBPNA for non-leaf nodes. The procedure is as follows:

1. $\forall v_i \in \mathbb{C}_0$, their parent is the sink node v_0.

2. According to the ID increasing order, for every $v_i \in \mathbb{C}_l$, and $l > 0$, its parent is assigned to be the adjacent node $v_j \in \mathbb{M}_{l-1}$ with the minimum traffic load.

3. According to the ID increasing order, for every $v_i \in \mathbb{M}_l$, and $l > 0$, its parent is assigned to be the adjacent node $v_j \in \mathbb{C}_{l-1}$ with the minimum traffic load.

We use the node ID to break the tie (small-ID with higher priority). After applying the above parent node assignment scheme to all the non-leaf nodes, $\forall v_i \in \mathbb{D}$, its parent node is decided. Furthermore, for each $v_i \in \mathbb{D}$, the traffic load of v_i introduced by its non-leaf child nodes is denoted by φ_i. Based on Lemma 7.3, $\forall v_i \in \mathbb{M}$, it has at most 12 non-leaf children, and $\forall v_i \in \mathbb{C}$, it has at most 4 non-leaf children. Considering that for $v_i \in \mathbb{D}$, it can has as many as $O(n)$ leaf children. We focus on studying the parent node assignment scheme for leaf nodes in the next section since any parent node assignment method for non-leaf nodes yields a constant approximation of at most 12.

7.5 Load-balanced Data Aggregation Tree

A tree structure is decided after the load-balanced parent node assignment (LBPNA) $\mathcal{A}^*$ is produced, which includes LBPNA for non-leaf nodes and leaf nodes. By assigning a direction of each link in the constructed tree from the children node to the parent node, we obtain an LBDAT. We already illustrate how to find an parent node assignment for non-leaf nodes. Hence, in this section, we first formulate the LBPNA for leaf nodes as an integer linear programming (ILP). Then, we present an approximation algorithm by applying the linear relaxation and random rounding technique. Finally, we exploit an example to illustrate how to build an LBDAT.

We already illustrate how to find a parent node assignment for non-leaf nodes. In this section, we study the LBPNA for leaf nodes.

7.5.1 ILP Formulation of LBPNA for Leaf Nodes

As we have already known, constructing an arbitrary aggregation tree with the maximum lifetime is NP-complete [112]. Through similar proving procedure, it can be shown that LBPNA is also an NP-complete problem. In this subsection, we first model LBPNA as an ILP.

We define a binary variable β_i to indicate whether the sensor v_i is a non-leaf node or not. β_i sets to be 1 *iff* the sensor v_i is a non-leaf node. Otherwise, β_i sets to be 0. Additionally, we assign a random variable ξ_{ij} for each link connecting a non-leaf node v_i and a leaf node v_j on the graph $\mathbb{G}$ modeled from a probabilistic WSN, i.e.,

$$\xi_{ij} = \begin{cases} 1, & \text{if non-leaf node } v_i \text{ is assigned to be} \\ & \quad \text{the parent of leaf node } v_j. \\ 0, & \text{otherwise.} \end{cases}$$

Consequently, LBPNA can be formulated as an integer linear programming ILP_{LBPNA} as follows:

$$\max \vartheta = \min\{\alpha_i = \sum_{v_j \in \mathbb{L}(v_i) \setminus \{v_i\}} \lceil \tfrac{B}{\gamma_i} \rceil \tfrac{1}{t_{ij}} + \varphi_i | \forall v_i \in \mathbb{D}\}$$

$$s.t. \quad \sum_{v_i \in \mathbb{N}_1(v_j)} \beta_i \xi_{ij} = 1, \; \forall v_j \notin \mathbb{D} \qquad (ILP_{LBPNA})$$

$$\xi_{ij} \in \{0,1\}.$$

The objective function ϑ is the minimum actual load (α_i) among all the non-leaf nodes. The first constraint states that each leaf node can be allocated to only one non-leaf node, whereas the second constraint indicates that ξ_{ij} is a binary variable. According to Definition 7.8, the number of *leaf child nodes* and the number of *non-leaf child nodes* both contribute to the actual load of a non-leaf node. The leaf child nodes of parent node v_i can be represented by $v_j : \beta_i \xi_{ij} > 0$. The traffic load introduced by non-leaf children to v_i is denoted by φ_i. Moreover, as stated in Lemma 7.3, the number of non-leaf children nodes of an independent parent node $v_i \in \mathbb{M}$ is no more than 12, whereas the number of non-leaf children nodes of an LBMIS connector parent node $v_i \in \mathbb{C}$ is no more than 4. Therefore, for simplicity, we assume that the total actual load of leaf children nodes is approximated to $12\lceil \tfrac{B}{\mathcal{R}} \rceil$ (i.e., $\sum_{v_j \in \{\mathbb{I}(v_i) \mid i \neq j\}} \lceil \tfrac{B}{\gamma_i} \rceil \tfrac{1}{t_{ij}} \simeq 12\lceil \tfrac{B}{\mathcal{R}} \rceil$).[3] By relaxing variable $\xi_{ij} \in \{0,1\}$ to $\xi_{ij} \in [0,1]$, we get the relaxed formulation which falls into a standard linear programming (LP) problem, denoted by LP^*_{LBPNA} as follows:

$$\max \quad \vartheta = \min\{\alpha_i \mid \forall v_i \in \mathbb{D}\}$$

$$s.t. \quad \sum_{v_i \in \mathbb{N}_1(v_j)} \beta_i \xi_{ij} = 1, \; \forall v_j \notin \mathbb{D}$$

$$0 \leq \varphi_i \leq 12\lceil \tfrac{B}{\mathcal{R}} \rceil \qquad (LP^*_{LBPNA})$$

$$\xi_{ij} \in [0,1].$$

In LP^*_{LBPNA}, $\alpha_i = \sum_{v_j \in \mathbb{L}(v_i) \setminus \{v_i\}, j:\beta_i \xi_{ij} > 0} \lceil \tfrac{B}{\gamma_i} \rceil \max\{1, \tfrac{\beta_i \xi_{ij}}{t_{ij}}\} + \varphi_i$. Using $\max\{1, \tfrac{\beta_i \xi_{ij}}{t_{ij}}\}$ is mainly because if v_j has some data ($\xi_{ij} > 0$ percent of v_j's data) been forwarded by v_i (v_i is the parent node of v_j), v_j must transmit at least one data packet to v_i since data packets are the basic communication units in a WSN. Due to the relaxation enlarging the optimization space, the solution of LP^*_{LBPNA} corresponds to an upper bound of the objective of ILP_{LBPNA}.

7.5.2 Randomized Approximation Algorithm

Given an instance of LBPNA modeled by the integer linear programming ILP_{LBPNA}, the sketch of the randomized approximation algorithm is shown in Algorithm 10. We summarize Algorithm 10 as follows: first, solve the relaxed linear programming LP^*_{LBPNA} to get an optimal fractional solution, denoted by $(\boldsymbol{\xi}^*, \vartheta^*)$, where

---

[3] It loses only a constant factor.

Algorithm 10: Approximation Algorithm for LBPNA

---

1 Solve LP^*_{LBPNA}. Let $(\boldsymbol{\xi}^*, \vartheta^*)$ be the optimum solution;

2 Sort the ξ^*_{ij} values in each row (for each i) of $\boldsymbol{\xi}^*$ in the decreasing order and
   then store the corresponding j (v_j's ID) in a two-dimensional array denoted by
   $A[n][m]$;

3 $\widehat{\xi}_{ij} = 0$;

4 **while** $k \leq \kappa = \dfrac{6\log(n)}{\delta^2 \min\{\xi^*_{ij}|1\leq i\leq n.1\leq j\leq m, \xi^*_{ij}>0\}}$ **do**

5 $\quad$ $k = 0, l = 0$;

6 $\quad$ **while** $k < n$ **do**

7 $\quad\quad$ $i = k$;

8 $\quad\quad$ **while** $l < m$ **do**

9 $\quad\quad\quad$ $j = A[k][l]$;

10 $\quad\quad\quad$ **if** $v_j \in \mathbb{N}_1(v_i)$ *and* $\widehat{\xi} = 0$ **then**

11 $\quad\quad\quad\quad$ $\widehat{\xi}_{ij} = 1$ with probability ξ^*_{ij};

12 $\quad\quad\quad\quad$ break;

13 $\quad\quad$ $k = k + 1$;

14 **return** $(\widehat{\boldsymbol{\xi}}, \widehat{\vartheta} = \min\{\alpha_i \mid \forall v_i \in \mathbb{D}\}$

$\boldsymbol{\xi}^* = << \xi^*_{11}, \cdots, \xi^*_{1m} >, < \xi^*_{21}, \cdots, \xi^*_{2m} >, \cdots, < \xi^*_{n1}, \cdots, \xi^*_{nm} >>$, and then round
ξ^*_{ij} to integers $\widehat{\xi}_{ij}$ by a random rounding procedure, which consists of five steps as
shown in lines 2-13 of Algorithm 10.

1. Sort the ξ^*_{ij} values in each row of $\boldsymbol{\xi}^*$ (for every $1 \leq i \leq n$) in the decreasing order and store the corresponding j (v_j's ID) in a two-dimensional array
   $A[n][m]$.

2. Set all $\widehat{\xi}_{ij}$ to be 0.

3. Start from the first row in the sorted array A. If there is no parent node assigned
   to v_i in its 1-hop neighborhood, then let $\widehat{\xi}_{ij} = 1$ with probability ξ^*_{ij}. Then, go
   to the next row in A.

4. Repeat step (3) till reach the end of array A.

5. Repeat steps (3) and (4) for κ times, where $\kappa = \dfrac{6\log(n)}{\delta^2 \min\{\xi^*_{ij} \mid 1\leq i\leq n.1\leq j\leq m, \xi^*_{ij}>0\}}$,
   and δ is any constant satisfying $0 < \delta < 1$.

Next, the correctness of the proposed approximation algorithm (Algorithm 10) is
proven in the following lemma. Due to space limitation, the proof of Lemma 7.4 is
omitted.

Lemma 7.4

$\forall v_i \in \mathbb{V} \setminus \mathbb{D}$ *is assigned a parent non-leaf node in its one-hop neighborhood almost surely after executing Algorithm 10.*

Proof: We first denote $\chi = \max\{\xi_{ij}^* \mid v_i \in \mathbb{N}_1(v_j), v_i \in \mathbb{D}, \xi_{ij}^* > 0\}$. Additionally, denote the probability that a leaf node $v_j \in \mathbb{V} \setminus \mathbb{D}$ is not assigned a parent non-leaf node in its one-hop neighborhood after executing κ times as $\mathbb{P}$. Then,

$$\mathbb{P} = \prod_{v_i \in \mathbb{N}_1(v_j),\, v_i \in \mathbb{D}} (1 - \xi_{ij}^*)^\kappa \leq (1 - \chi)^\kappa \leq e^{-\chi\kappa} = e^{-\chi \frac{6\log(n)}{\delta^2 \min\{\xi_{ij}^* \mid 1 \leq i \leq n. 1 \leq j \leq m, \xi_{ij}^* > 0\}}} \leq$$

$e^{-6\log(n)} \leq \frac{1}{n^6}$.

Now, the probability that a leaf node is not assigned a parent non-leaf node in its one-hop neighborhood after the random rounding is $\frac{1}{n^6}$, which implies $Pr[$a leaf node has no neighboring non-leaf node$] \leq n\frac{1}{n^6} = \frac{1}{n^5}$.

Similarly, according to the Borel-Cantelli lemma, the above probability is 0 almost surely, which implies that it is almost sure that every leaf node is assigned a parent non-leaf node in its one-hop neighborhood after executing Algorithm 10. □

Subsequently, we analyze the approximation factor of Algorithm 10 in Theorem 7.4.

Theorem 7.4

Let opt_{LBPNA} denote the optimal solution of LBPNA. Algorithm 10 yields an optimal fractional solution of $O(\log(n))opt_{LBPNA}$ with probability 1.

Proof: Considering any non-leaf node v_i and leaf node v_j, the expected actual load of v_i is as follows:

$$E[\sum_{v_j \in \mathbb{L}(v_i) \setminus \{v_i\}, j:\beta_i \widehat{\xi_{ij}} > 0} \lceil \frac{B}{\gamma_i} \rceil \max\{1, \frac{\beta_i \xi_{ij}}{l_{ij}}\} + \varphi_i] \tag{7.1}$$

$$= E[\sum_{v_j \in \mathbb{L}(v_i) \setminus \{v_i\}, j:\beta_i \widehat{\xi_{ij}} > 0} \lceil \frac{B}{\gamma_i} \rceil \max\{1, \frac{\beta_i \xi_{ij}}{l_{ij}}\}] + \varphi_i \tag{7.2}$$

$$= \lceil \frac{B}{\gamma_i} \rceil E[\sum_{v_j \in \mathbb{L}(v_i) \setminus \{v_i\}, j:\beta_i \widehat{\xi_{ij}} > 0} \max\{1, \frac{\beta_i \xi_{ij}}{l_{ij}}\}] + \varphi_i \tag{7.3}$$

$$= \lceil \frac{B}{\gamma_i} \rceil \sum_{v_j \in \mathbb{L}(v_i) \setminus \{v_i\}, j:\beta_i \widehat{\xi_{ij}} > 0} \max\{1, \frac{\beta_i}{l_{ij}} E[\widehat{\xi_{ij}}]\} + \varphi_i \tag{7.4}$$

$$\geq \lceil \frac{B}{\gamma_i} \rceil \sum_{v_j \in \mathbb{L}(v_i) \setminus \{v_i\}, j:\beta_i \widehat{\xi_{ij}} > 0} \max\{1, \frac{\beta_i}{l_{ij}} \xi_{ij}^*\} + \varphi_i \tag{7.5}$$

$$= \vartheta^*. \tag{7.6}$$

In the derivation, we exploit the facts that β_i and $\widehat{\xi}_{ij}$ are independent, and the procedure setting $\widehat{\xi}_{ij} = 1$ with probability ξ_{ij}^* is repeated κ times. Hence, $E[\widehat{\xi}_{ij}] \geq 1 \times \xi_{ij}^* + 0 \times (1 - \xi_{ij}^*) = \xi_{ij}^*$.

Applying the Chernoff bound, we obtain the following bound:

$$Pr[\sum_{v_j \in \mathbb{L}(v_i) \setminus \{v_i\}, j:\beta_i \widehat{\xi}_{ij} > 0} \lceil \frac{B}{\gamma_i} \rceil \max\{1, \frac{\beta_i \widehat{\xi}_{ij}}{l_{ij}}\} + \varphi_i \leq (1 - \delta)\kappa\vartheta^*] \leq e^{-\frac{\delta^2}{2}\kappa\vartheta^*}$$

for arbitrary $0 < \delta < 1$. Then, we have

$$Pr[\sum_{v_j \in \mathbb{L}(v_i) \setminus \{v_i\}, j:\beta_i \widehat{\xi}_{ij} > 0} \lceil \frac{B}{\gamma_i} \rceil \max\{1, \frac{b\beta_i \widehat{\xi}_{ij}}{l_{ij}}\} + \varphi_i]$$

$$\leq (1 - \delta)\kappa\vartheta^*] \leq e^{-\frac{\delta^2}{2}\kappa} \tag{7.7}$$

$$= e^{\frac{6\log(n)}{\delta^2 \min\{\xi_{ij}^* \mid 1 \leq i \leq n. 1 \leq j \leq m, \xi_{ij}^* > 0\}} \frac{\delta^2}{2}} \tag{7.8}$$

$$\leq e^{-3\log(n)} \leq \frac{1}{n^3}. \tag{7.9}$$

Inequality (7.7) holds, since

$$\vartheta^* = \min\{\sum_{v_j \in \mathbb{L}(v_i) \setminus \{v_i\}, j:\beta_i \xi_{ij}^* > 0} \lceil \frac{B}{\gamma_i} \rceil \max\{1, \frac{\beta_i \xi_{ij}^*}{l_{ij}}\} + \varphi_i \mid \forall v_i \in \mathbb{D}\} \geq 1.$$

we have

$$Pr[\sum_{j:\beta_i \widehat{\xi}_{ij} > 0} \lceil \frac{B}{\gamma_i} \rceil \frac{1}{l_{ij}} + \varphi_i + 1 \leq (1 - \lambda)\kappa\vartheta^*] \leq e^{-\lambda\kappa\vartheta^*} \leq e^{-\lambda\kappa}$$

$$= e^{-\frac{3\Delta\log(n)}{\Delta}} \leq \frac{1}{e^3 \ln(n)} \leq \frac{1}{n^3}.$$

Summing over all non-leaf nodes $v_i \in \mathbb{D}$, we obtain the probability that some non-leaf node has the actual load less than $(1 - \delta)\kappa\vartheta^*$ as follows: $Pr[\widehat{\vartheta} \leq (1 - \delta)\kappa\vartheta^*] \leq n\frac{1}{n^3} = \frac{1}{n^2}$. Again, according to Borel-Cantelli lemma, $Pr[\widehat{\vartheta} \leq (1 - \delta)\kappa\vartheta^*] \sim 0$.

Then, considering Lemma 7.4, we have $Pr[\text{each leaf node is assigned to a parent non-leaf node} \cap \widehat{y} \geq (1 - \delta)\kappa(\vartheta^*)] \geq (1 - \frac{1}{n^3})(1 - \frac{1}{n^2}) \sim 1$, when $n \sim \infty$, for $0 < \delta < 1$, and $\kappa = \frac{6\log(n)}{\delta^2 \min\{\xi_{ij}^* \mid 1 \leq i \leq n. 1 \leq j \leq m, \xi_{ij}^* > 0\}}$. □

Hence, Algorithm 10 yields an solution upper bounded by $O(\log(n))opt_{LBPNA}$. Moreover, this bound can be verified in polynomial time. After $\mathcal{A}^*$ is decided, a tree can be obtained by assigning each link a direction from the children to the parent.

7.6 Performance Evaluation

Since there are no existing works studying the LBDAT construction problem for probabilistic WSNs currently, in the simulations, LBDAT is compared with the recently published DS-based data aggregation algorithm denoted by DAT. We compare both algorithms in terms of the number of non-leaf nodes, network lifetime, which is defined as the time duration until the first non-leaf node runs out of energy, and the network residual energy.

7.6.1 Simulation Environment

We build our own simulator where all the nodes have the same transmission range 50 m and all the nodes are deployed uniformly and randomly in a square area of size 300 $m \times$ 300 m. For each specific setting, 100 instances are generated. The results are averaged over these 100 instances (all results are rounded to integers). Moreover, a random value between 0.5 and 0.98 is assigned to the transmission success ratio (ι_{ij}) value associated to a pair of nodes (v_i and v_j) inside the transmission range. Otherwise, a random value between 0 and 0.5 is assigned to ι_{ij} associated to a pair of nodes beyond the transmission range. Every sensor node produces a packet with size 1 during each report time interval. The data receiving rate γ_i of each node v_i is randomly generated from the value between 0 and 10. The energy consumption model is that every node has the same initial 1000 units of energy. Receiving a packet consumes 1 unit of energy, while transmitting a packet consumes 2 units of energy. In the simulation, we consider the following tunable parameters: the node transmission range and the total number of nodes deployed in the square area.

7.6.2 Scenario 1: Change side length of square area

In this scenario, all nodes have the same transmission range of 20 m and 100 nodes are deployed uniformly and randomly in a square area. The side length of the square area is incremented from 100 m to 150 m by 10 m. The impact of the area side length on the number of non-leaf nodes, the network lifetime, and the network residual energy of both algorithms is presented in Figure 7.3.

From Figure 7.3(a), we can see that, with the increase of the area of the network deployed region, the number of non-leaf nodes increases for both algorithms (DAT and LBDAT). This is because the probabilistic WSN becomes thinner and more non-leaf nodes are needed to maintain the connectivity of the constructed CMIS. There is no obvious trend showing which algorithm might produce more non-leaf nodes when constructing a DAT.

From Figure 7.3(b), we know that the network lifetime increases for both algorithms with the side length of the deployed area increasing. It is obvious that the density of the network becomes much thinner with the side length of the deployed area increasing. As to data aggregation, the thinner the network is, the fewer neighbors of each non-leaf node. In other words, the aggregated data are less on each non-leaf node when the network becomes thinner. Hence, network lifetime is increasing

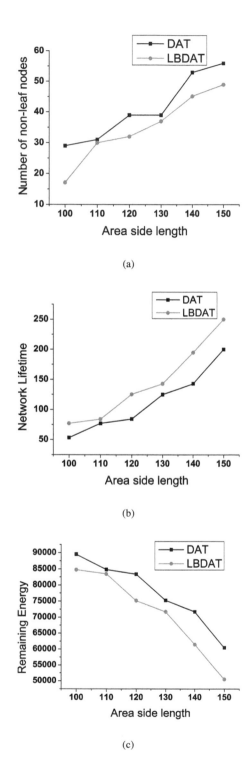

(a)

(b)

(c)

Figure 7.3: Simulation results of Scenario 1.

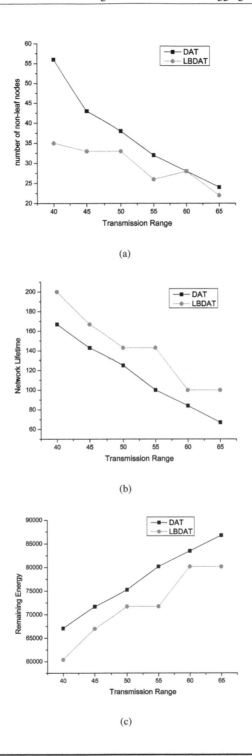

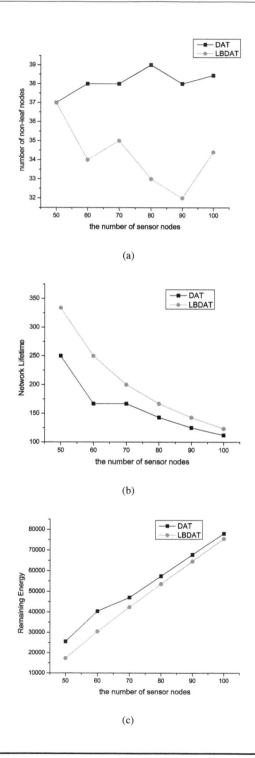

Figure 7.5: Simulation results for Scenario 3.

for both algorithms. Additionally, LBDAT prolongs network lifetime by 32% on average compared with DAT. The results demonstrate that load-balanced parent node assignment can improve network lifetime.

Figure 7.3(c) shows the remaining energy over the whole network of both algorithms. With the increase of the side length of the deployed area, the remaining energy decreases for both algorithms. As the WSN becomes thinner, more nodes are selected as non-leaf nodes to maintain the connectivity of the constructed CMIS. Additionally, the traffic load on a non-leaf node is less as mentioned earlier, hence, the remaining energy decreases with the deployed area increasing. From Figure 7.3(c), LBDAT has less remaining energy than DAT. This is because LBDAT considers the load-balance factor when building a CMIS and doing parent node assignment. Thus, the lifetime of the whole network is extended, which means the remaining energy of the network is less than DAT. In summary, Figure 7.3 indicates that constructing an LBDAT can balance the energy consumption on each non-leaf node, and make the lifetime of the whole network prolonged considerably.

7.6.3 Scenario 2: Change node transmission range

When 100 nodes are deployed uniformly and randomly in a square area of 300 $m \times$ 300 m, the impacts of the node transmission range on the number of non-leaf nodes, the network lifetime, and the network residual energy of LBDAT and DAT are shown in Figure 7.4.

From Figure 7.4(a), we can see that, with the increase of the node transmission range, the number of non-leaf nodes decreases for both algorithms (DAT, and LBDAT). This is because there are more nodes in the circle with the node transmission range increasing and for the network to become denser. Hence, the connectivity of the constructed CMIS can still be maintained even using fewer non-leaf nodes.

From Figure 7.4(b), we know that the network lifetime decreases for both algorithms with the node transmission range increasing. The fact is that the network becomes denser with the node transmission range increasing. The denser the network is, the more neighbors of each non-leaf node. Since we perform data aggregation, the aggregated data are increasing on each non-leaf node when the network becomes denser. Hence, network lifetime is decreasing for both algorithms. Moreover, LBDAT prolongs network lifetime by 28% on average compared to DAT.

Figure 7.4(c) shows the remaining energy over the whole network of both algorithms. With the increase of the node transmission range, the remaining energy increases for both algorithms. This is because a bunch of redundant sensors exist in the more and more crowded network. Thus the remaining energy increases for both algorithms with the network lifetime decreasing.

7.6.4 Scenario 3: Change total number of nodes

Suppose that all the nodes have the same transmission range of 50 m and all the nodes are deployed uniformly and randomly in a square area of 300 $m \times$ 300 m. The impacts of the number of nodes on the number of non-leaf nodes, the network lifetime, and the network residual energy of both algorithms are shown in Figure 7.5.

From Figure 7.5(a), we can see that, with the increase of the number of the sensor nodes, the number of non-leaf nodes almost keeps stable (from 32 to 39) for both algorithms (DAT, and LBDAT). This is because the area of the network deployed region and the transmission range of n nodes keep fixed.

From Figure 7.5(b), we can see that the network lifetime decreases for both algorithms with the number of nodes increasing. This is because we perform data aggregation in a more and more crowded network. Additionally, we can see LBDAT prolongs network lifetime by 42% on average compared to DAT. The results demonstrate that the LBPNA can improve network lifetime significantly.

Figure 7.5(c) shows the remaining energy over the whole network of both algorithms. With the increase of the number of nodes, the remaining energy increases for both algorithms. This is because the WSN becomes denser, and a lot of redundant sensor nodes exist in the WSN.

7.7   Conclusion

In this chapter, we address the fundamental problems of constructing a load-balanced DAT in probabilistic WSNs. We first solve the CMIS problem, which is NP-hard, in two phases. In the first phase, we aim to find the optimal MIS such that the minimum *potential load* of all the independent nodes is maximized. To this end, a near optimal approximation algorithm is proposed. In the second phase, the minimum-sized set of LBMIS connectors are found to make the LBMIS connected. The theoretical lower and upper bounds of the number of non-leaf nodes are analyzed as well. Subsequently, we study the LBDAT construction problem and propose an approximation algorithm by using the linear relaxing and random rounding techniques. After an LBPNA is decided by assigning a direction to each link, we obtain an LBDAT. The simulation results show that the proposed algorithms can extend network lifetime significantly.

Our next step is to come up with a sophisticated model to integrate the aforementioned three phases together and analyze the overall performance of the LBDAT construction problem. This is because three phases algorithm might lead to performance loss/improvement since we did not investigate the correlations among them. Another direction is to design distributed algorithms for the LBDAT construction problem under both DNM and PNM.

APPLICATIONS

Chapter 8

Reliable and Energy Efficient Target Coverage for Wireless Sensor Networks

CONTENTS

8.1 Introduction

Wireless sensor networks (WSNs) consist of a large number of ad hoc networked, low-power, short-lived and unreliable micro-sensors, which are limited in computation, memory capacity and radio range [120]. Many sensors are deployed in regions of interest to collect related information or report that some event has taken place in that area. Therefore, WSNs are widely applied to battlefield surveillance, health care applications, environment and habitat monitoring, home appliance, smart space, and inventory tracking [52]. A basic and important function of WSNs is to monitor areas or targets for a long period. Since sensors are often deployed in remote or inaccessible environments where replenishing energy is usually impossible, a critical issue in WSN applications is conserving sensors' energy and prolonging the network lifetime while guaranteeing the coverage of desired areas or targets, which is called the coverage problem.

Coverage is a fundamental problem in WSNs for environment monitoring and surveillance purposes. The coverage concept is subject to a wide range of interpretations due to a variety of sensors and applications. Generally, coverage which has direct effect on the network performance can be considered as the measure of quality of service in a WSN [121]. In general, coverage problems either deploy sensors to cover the sensing field completely [122, 123], or make sure that all the sensing field is covered by a certain number of sensors, such as one-coverage or k-coverage [123, 124], or select active sensors in a densely deployed WSN to cover all the sensing field [125, 126, 127, 128, 15]. The last case is known as an *activity scheduling problem* (ASP) [129]. Recently, a lot of researchers paid attention to ASP. Generally speaking, according to the requirement of sensing tasks, ASP may be divided into four classes: *target coverage* [126, 15], *area coverage* [125, 130], *barrier coverage* [127], and *patrol coverage* [128].

In this chapter, we address an ASP on *target coverage*. As we already know, the goal of ASP is to maximize the network lifetime on the premise of preserving the sensing coverage. Many algorithms propose to organize sensors into a number of subsets, such that each set completely covers all the sensing field, and scheduling the time to make these subsets activated successively, in which only one set is active at any time instant. Through avoiding redundant sensors that waste energy, the network lifetime is prolonged. The above problem is NP-hard [126], so many optimization techniques are applied, such as genetic algorithms, linear programming, greedy algorithms [15, 131, 132, 133, 134, 79, 135], etc. However, another prominent issue in the target coverage problem is how to improve reliability of the whole system. Few studies have been done on this aspect. Nevertheless, due to interference of dangerous environments, nodes may become unavailable (e.g., physical damage, lack of power resources), malfunction or totally missing. Improving reliability of target coverage in WSNs is worth putting more efforts into research in the future. In this work, we also consider the reliability of the target coverage problem.

Our approach differs from the aforementioned solutions by adding reliable restrictions to get non-disjoint sensor sets and by allowing the sets to operate for different time intervals. Our approach is to find the active non-disjoint sensor sets to fully cover all targets in a distributed manner to maximize the network lifetime within a preset reliability threshold. We introduce a new concept which is *failure probability* of sensor sets for reliability purposes. Only the failure probability of sensor sets greater than the preset threshold can be set to active. The reliability of a network is a measure of the quality of service (QoS) of the sensing function and is subject to a wide range of interpretations due to a large variety of sensors and applications. Thus in the study, we use failure probability as the metric. All the current works just use at least one node to cover targets for energy saving issues. However, one critical node's failure might cause the whole network fail. Hence we choose the lowest failure probability's sensor sets to monitor all targets, that will improve the network lifetime as well as the reliability of the whole system.

Consequently, our main contributions in this work are summarized as follows:

■ We introduce the *failure probability* concept into the target coverage problem. To the best of our knowledge, this is the first study addressing the problem in the literature. Almost all the related works make such assumption — only if the sensor is not out of energy, can it cover the targets in its sensing range. However, in reality, there is a failure probability associated with each sensor as well as sensor sets.

■ We formalize the target coverage problem with reliability concern as α-reliable maximum sensor cover (α-RMSC) problem.

■ We propose a distributed algorithm to solve the α-RMSC problem. The algorithm can precisely control the failure rate of the whole system, which is a critical fact in many applications, such as military surveillance systems and environment monitoring systems.

■ The simulation results validate that the proposed algorithm has perfect reliability control and also has similar network lifetime compared to the latest algorithm [126]. Thus the proposed algorithm does not sacrifice the energy consumption to achieve system reliability.

The rest of the chapter is organized as follows. In Section 8.2, we review previous related works. In Section 8.3, we first introduce some preliminaries, such as network model and related definitions, and then formalize the α-reliable maximum sensor cover problem (α-RMSC). In Section 8.4, a novel heuristic greedy algorithm is proposed to solve the α-RMSC problem. We show the simulation results in Section 8.5. Finally we concludes this chapter and give future work directions in Section 8.6.

8.2 Related Work

The coverage problem in WSNs has been intensively investigated. In this section, we summarize the related studies according to two groups: the target coverage problem and other coverage problems.

8.2.1 Target Coverage

There exists a wealth of literatures on the target coverage problem in WSNs. We can divide the target coverage problem into (1) target coverage problem with unique sensing range [126, 136], (2) target coverage problem with multiple sensing ranges [15, 137], (3) target coverage problem in directional WSNs [138], and (4) other variants, such as target area coverage [129] in which the sensors cover not only a set of targets, but also deal with a small area coverage, and target Q-coverage [139], in which each target needs to be covered by different numbers of sensors.

The authors in [126, 136] propose an efficient way to extend the network lifetime by organizing the sensors into a maximal number of sensor sets that are activated successively. Only the sensors from the current active set are responsible for monitoring all targets and for transmitting the collected data, while all other nodes are in a low-energy sleep mode. The authors in [15] considered the status scheduling of a sensor with multiple sensing ranges. A collaborative task scheduling algorithms was proposed to minimize the energy consumption. It employed a two-level scheduling approach to the execution of tasks collaboratively at a group and individual levels among neighboring sensor nodes. The authors in [137] propose an integer linear programming (ILP) based approximation to maximize the network lifetime, where the sensors have different sensing ranges. In [138], the authors organize the directions of sensors into a group of non-disjoint cover sets. One cover set which can cover all the targets satisfying their coverage quality requirement is activated at one time. In [129], the authors propose a geometric-based activity scheduling scheme to address the target area problem. By means of computational geometry, the sensors can self-determine when to sleep or wake up while preserving the coverage requirement. In [139], the authors prove the target Q-coverage problem is NP-Hard. A greedy algorithm is proposed to efficiently solve the target Q-coverage problem.

8.2.2 Other Coverage

Many studies have focused on characterizing the *area coverage* and designing algorithms to achieve desired area coverage. In [140], the authors studied the coverage of a grid-based WSNs. They derived the necessary and sufficient conditions on the sensing range and failure rate of sensors in order to ensure that the whole network is covered as well as connected. In [141], a probing-based density control algorithm is proposed to extend the network lifetime. Again, the basic idea is to turn off redundant sensors to save energy. A sleeping sensor wakes up occasionally to probe its neighborhood and starts working if there are no other working sensors in its probing range. The desired redundancy of working sensors can be achieved by adjusting the probing range of sensors.

Another line of work on coverage studies the path exposure of moving objects in WSNs, which is a quantitative measure of how well sensors can detect objects moving in the network. In [142], the authors proposed algorithms to find paths which are most or least likely to be detected by sensors in a WSN. The authors further defined and studied the path exposure of a moving object in a WSN [143], which is a quantitative measure of how well an object, moving on an arbitrary path, can be detected by a WSN. An algorithm is developed to find minimum exposure paths in WSNs, where the probability of a moving object being detected is minimized. Along this line, [144] investigates deployment strategies for WSNs performing target detection. The goal of sensor deployment is to maximize the exposure of the least exposed path in the network.

Different criteria can categorize the current coverage problem into different types. One category includes *partially coverage* and *complete coverage* problems. In [145], a fractional coverage scheme (FCS) is proposed to achieve the desired partially coverage with a minimum number of active sensors for energy conservation. Another one includes connected coverage and non-connected coverage problem. In [146], the authors propose a density control algorithm for WSNs to keep as few as possible sensors in active state to achieve a connected coverage of a specific area of interest.

8.2.3   Remarks

Most previous work focused on saving energy to prolong network lifetime. In this chapter, we approach the coverage problems from a different perspective by studying the reliability of the whole system. The reliability of the whole system is determined by the basic network parameters (e.g., deployment strategy, sensor density, physical environment, etc.), but it can fundamentally impact the performance in a WSN. For example, one critical node's failure might cause the whole network to fail. Hence, controlling and improving reliability of the target coverage problem in WSNs is worth more research.

8.3   Network Model and Related Definitions

Before diving into solving the target coverage problem, we give some fundamental definitions and notations used throughout this chapter.

8.3.1   Network Model

We consider a WSN consisting of a set S of n sensors that are also called nodes. Each $s \in S$ can sense m interested targets (denoted by T) in its *sensing range* and communicate with nodes in its transmission range. Each target t has a unique identification number denoted by t_j. We make the natural assumption that there are no two sensors at the same location. Also, each sensor $s \in S$ has a unique identification number, denoted by s_i. The sensors are distributed over a large two-dimensional area. We refer to the region as the *monitoring area*.

We assume that the sensing and transmission ranges of node s are open discs, centered at s_i, with radius r_{s_i} and R_{s_i} respectively, where $R_{s_i} > r_{s_i}$. We assume a sensor s_i covers a target t_j if the Euclidean distance between the sensor s_i and the target t_j is smaller than or equal to the sensing range r_{s_i}. Nodes are termed adjacent or neighbors if they are included in the transmission range of each other.

We assume each sensor s_i has a *failure probability* associated with each target t_j in the monitored area (denoted by fp_{ij}). The failure probability can be set by manufacturer initially. However many factors can effect the parameter, say weather in the monitoring area, interferences to the sensors, or unexpected accidents. If we know the environment in advance, we can predict or calculate the failure probability easily.

The time is divided into time slots and we assume that the sensors have synchronized clocks which notify them at the beginning of each time slot. Sensor $s_i \in S$ has an initial energy B_i and, as a normalization, we assume that every sensor consumes e unit of energy in each time slot in which it is *active*. For saving energy, a sensor may be in a *sleep* mode, in which it does not communicate with its neighbors nor sense its vicinity. A sensor in sleep mode consumes only negligible energy, which we assume to be zero. This is also called *duty-cycling*.

8.3.2 Related Definitions

All targets in the monitoring area are covered by at least one sensor and this is called a target coverage problem. In most cases, sensors are densely deployed to tolerat the failure. It is not reasonable to turn on all sensors in the monitoring area to cover all the targets, because more than one sensor can cover the same target. So it is necessary to divide the n sensors to a couple of subsets, and each subset can cover all relevant targets. In each time slot, only one subset is active. The duty to monitor targets is cycled over the whole WSN. The main purpose of duty cycling is save energy and prolong the lifetime of the WSN.

Definition 8.1   *Target set.* The target set is set of all targets in the monitoring area, which is denoted by $T = \{t_1, t_2, \cdots, t_j, \cdots, t_m\}$

Definition 8.2   *Sensor cover.* Assume a WSN consists of a set of n sensors, $S = \{s_1, s_2, \cdots, s_i, \cdots, s_n\}$. Any subset C_r of S that can completely cover the target set is termed a sensor cover, $C_r = \{s_1, s_2, \cdots, s_l\}$, and if we can find maximum k of such sensor covers, then $r \in [1, k]$.

If there does not exist a sensor cover C_r such that all the nodes in C_r have non-zero energy, then the network is said to have a *coverage hole*.

Definition 8.3   *Network lifetime.* The network lifetime is the time interval from the activation of the network until the first time a coverage hole appears.

Taking the reliability into consideration, the definition of the classic target coverage problem needs to be improved. In reality, the failure of sensors cannot be ignored, so we add the *failure probability* as a property to each sensor. And then we can calculate the failure probability for each *sensor cover*.

Definition 8.4 *Target failure probability*. Given a target t_j in the Target Set $T = \{t_1, t_2, \cdots, t_m\}$, $\forall j \in [1, m]$, we can find the sensor set $S_j = \{s_1, s_2, \cdots, s_l\}$, $\forall l \in [1, n]$, where S_j is a subset of the whole sensor set S. Every node s_i in the sensor set S_j can cover the given target t_j with the probability fp_{ij}. The target failure probability (denoted by tp_j) is defined as the probability that sensor s_i fails to cover target t_j.

$$tp_j = \prod_{i=1}^{l} fp_{ij} \tag{8.1}$$

As we have already mentioned more than one sensor can cover the same target. So we can divide the whole sensor set into non-disjoint sensor covers. Let some sensor covers be active and some sensor covers sleep, then the network lifetime can be prolonged. In this situation, how to calculate the target failure probability becomes complex. Let us assume there are l where $l \in [1, n]$ sensors can cover target t_j. Each sensor s_i has a failure probability fp_{ij} to cover the target t_j. So the failure probability to cover the target by all l sensors should be the product of each sensor's failure probability, i.e.,

$$fp_{1j} * fp_{2j} * \cdots * fp_{lj},$$

which is formulated in Equation (8.1). The purpose of calculating target failure probability is to calculate sensor cover failure probability. Next we introduce the definition of sensor cover failure probability.

Definition 8.5 *Sensor cover failure probability*. The probability that a sensor cover $C_r = \{s_1, s_2, \cdots, s_l\}$, $\forall l \in [1, n]$; $\forall r \in [1, k]$, where k is the maximum number of sensor covers we can find, fails to cover all the targets in the target set $T = \{t_1, t_2, \cdots, t_m\}$, which is denoted by cp_r.

$$cp_r = 1 - \prod_{j=1}^{m} (1 - tp_j) \tag{8.2}$$

$$= 1 - \prod_{j=1}^{m} (1 - \prod_{i=1}^{l} fp_{ij}). \tag{8.3}$$

We introduce the above definition to monitor the reliability of each sensor cover. The higher value of the sensor cover failure probability, the more reliable the sensor cover. Hence when we schedule the sensor covers, we know how to choose the desired sensor cover to be active in order to improve the whole network's reliability.

From Definition 8.2, we know every sensor cover must completely cover all targets in the monitoring area. Based on Equation (8.1), we can calculate each target's failure probability tp_j where $j \in [1, m]$. The failure probability is equal to (1

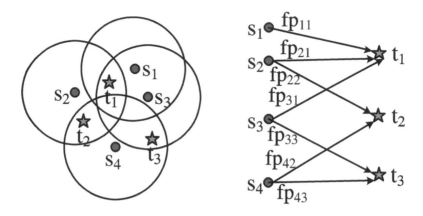

Figure 8.1: Example with three targets and four sensors.

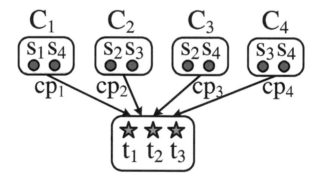

Figure 8.2: Illustration of four sensor covers.

— success probability). So we first calculate the success probability of covering each target, which is $1 - tp_j$. Then the success probability of covering the target set $T = \{t_1, t_2, \cdots, t_m\}$ is $(1 - tp_1) * (1 - tp_2) * \cdots * (1 - tp_m)$. Using 1 we deduct the above success probability of covering the target set, we finally get the Equation (8.2). If we apply Equation (8.1) to Equation (8.2), then we can get Equation (8.3).

Figure 8.1 shows an example with four sensor nodes $S = \{s_1, s_2, s_3, s_4\}$ and three targets $T = \{t_1, t_2, t_3\}$. The coverage relationship between sensors and targets is also illustrated in the figure. From the coverage information we can find maximum four sensor covers, i.e., $C_1 = \{s_1, s_4\}$, $C_2 = \{s_2, s_3\}$, $C_3 = \{s_2, s_4\}$, and $C_4 = \{s_3, s_4\}$, shown in Figure 8.2. For convenience, we assume 1 is the lifetime of each senor. Therefore, if we let set C_1 active for 0.33 time slot, C_2 active for 0.5 time slot, C_3 active for 0.33 time slot, C_4 active for 0.33 time slot, then the network lifetime is 1.49 time slots. Thus, duty cycling WSN can improve the network lifetime.

According to Equation (8.1), we can calculate each target's failure probability.

Let us use target 1 (t_1) as an example. From the coverage information in Figure 8.1, we know three nodes $S_1 = \{s_1, s_2, s_3\}$ can cover t_1, so the target failure probability of t_1 is:

$$tp_1 = fp_{11} * fp_{21} * fp_{31}. \tag{8.4}$$

Based on Equation (8.2), we can calculate each sensor cover's failure probability. If we use sensor cover $C_3 = \{s_2, s_4\}$ as an example, then

$$
\begin{aligned}
cp_3 &= 1 - \prod_{j=1}^{3}(1 - tp_j) \\
&= 1 - (1 - tp_1)(1 - tp_2)(1 - tp_3) \\
&= 1 - (1 - fp_{21})(1 - fp_{22} * fp_{42})(1 - fp_{43}).
\end{aligned}
\tag{8.5}
$$

Previous algorithms did not consider the network reliability. Hence, the network lifetime of our scheme can not be compared directly to the lifetime of previous algorithms. Thus, we proposed a new measurement matrix.

Definition 8.6 *Reliable Lifetime* is equal to current active sensor cover's failure probability times current network lifetime.

8.3.3 Problem Formulation

In this subsection, we proceed to define the α-reliable maximum sensor covers problem.

Definition 8.7 *α-reliable maximum sensor covers (α-RMSC) problem.* Given a collection S and a finite set T, find a family of sensor covers C with time weights $tw_1, tw_2, \cdots, tw_k$ in $[0, 1]$ and sensor cover failure probabilities $cp_1, \cdots, cp_k$, where k is the maximum number of sensor covers we can find

1. $\forall i \in [1, k], cp_i \leq \alpha$, where α is the user pre-defined maximum failure probability.

2. Maximize $\sum_{i=1}^{k} tw_i$, where each sensor s appears in C with a total weight of at most 1 where 1 is the lifetime of each sensor.

From Definition 8.7, we know we want to determine a number of sensor covers $C_1, C_2, \cdots, C_r, \cdots, C_k$, where each sensor cover $C_r, r \in [1, k]$ completely covers all the targets with $cp_r \leq \alpha$, to maximize the network lifetime $\sum_{i=1}^{k} tw_i$, where $\forall i \in [1, k]$, tw_i is the time interval while the sensor cover C_i is active. Note that if a sensor belongs to more than one sensor cover, then the sum of the time intervals of those sensor covers cannot be greater than 1. This is because each sensor cannot be active more than its lifetime 1.

The solutions of α-RMSC guarantee that the success probability of covering T is larger than $1 - \alpha$. So by setting different value of α, WSNs can gain different reliability levels.

8.4   Our Proposed Algorithm

We now describe the α-RMSC heuristic algorithm. The α-RMSC problem is closely related to the MSC problem, which is proposed to solve the traditional target coverage problem in [126]. The difference is we are trying to maximum the network's reliability as well as the network lifetime. The MSC problem is NP-hard. From the work in [126], we can claim that the α-RMSC problem is also NP-hard. Hence, we provide a heuristic approach to solve the α-RMSC problem.

Algorithm 11: : α-RMSC Heuristic Algorithm

Require: S: sensor set; T: target set; e: sensor lifetime granularity; α: the maximum failure probability;

Ensure: $C_1, \cdots, C_k$: α-reliable sensor covers:

1: **for** each s in S **do**
2:     $s.lifetime = 1$;
3: **end for**
4: SENSORS = S;
5: i = 0;
6: **while** each target is covered by at least one sensor in SENSORS, and the $SENSORS.cp \leq \alpha$ **do**
7:     $i = i + 1$;
8:     $C_i = \emptyset$;
9:     **while** $C_i.cp \leq \alpha$ **do**
10:        $s = $ **Greatest_Contribution_Sensor**(T, S, C_i);
11:        $C_i = C_i \bigcup s$;
12:    **end while**
13:    **for** each sensor $s \in C_i$ **do**
14:        $s.lifetime = s.lifetime - e$;
15:        **if** $s.lifetime == 0$ **then**
16:            $SENSORS = SENSORS - s$;
17:        **end if**
18:    **end for**
19: **end while**
20: **return** $C_1, \cdots, C_i$

8.4.1 α-RMSC Heuristic Algorithm Overview

Our heuristic algorithm (see Algorithm 11) takes S (sensor set), T (target set), e (sensor lifetime granularity), where $e \in (0,1]$, and α (the pre-set threshold) as the input parameters. Consequently, the algorithm returns all possible sensor covers $C_1, C_2, \cdots, C_k$.

Algorithm 11 recursively builds sensor covers from lines 6 to 19. The set SENSORS maintains the list of sensors that have residual energy greater than zero; thus these sensors can participate in additional sensor covers. At each step, a target is selected, in line 6, to be covered. Once the target has been selected, the algorithm selects the sensor with the greatest contribution that covers the selected target. Sensor contribution functions (i.e., greatest_contribution_sensor function shown in Line 10 in Algorithm 11) are defined in Algorithm 12. Once a sensor has been selected, it is added to the current sensor cover in line 11. When all targets are covered, the new sensor cover is formed.

8.4.2 Contribution Function

How to find the greatest contributed sensor is critical to solve the α-RMSC problem. The contribution of a sensor s_j to a sensor cover C_r is defined as follows:

Definition 8.8 *Contribution.* We assume sensor s_j can cover a set of targets $\{t_1, t_2, \cdots, t_l\}, l \in [1, m]$, the contribution of s_j to C_r is:

$$Con(C_r, s_j) = \sum_{i=1}^{k} tp_i * (1 - fp_{ji}) * w_i \qquad (8.6)$$

The assignment of weight values $w_1, w_2, \cdots, w_l$ is based on application specification. Generally, we can give evenly distributed weight values. However in some cases, some targets are sparsely covered, then their weight values should be a large number.

We still use the Figure 8.1 as the example to illustrate the contribution function. If Algorithm 11 puts s_1 and s_2 into current sensor cover, the algorithm cannot stop at this point, because the target t_3 has not been covered yet. The next step is to decide which sensor needs to be put into the current sensor cover — s_3 or s_4. At this time we need to calculate the contribution of s_3 and s_4 respectively according to Equation (8.6). From Figure 8.1, we know s_3 can cover $\{t_1, t_3\}$. s_4 can cover $\{t_2, t_3\}$. For simplifying the calculation, we assume all sensors' failure probabilities are the same, denoted by p. Then,

$$\begin{aligned} &Con(\{s1, s2\}, s3) \\ &= tp_1 * (1 - fp_{31}) * w_1 + tp_3 * (1 - fp_{33}) * w_3 \\ &= p^2 * (1 - p) + 1 * (1 - p)) = (p^2 + 1)(1 - p). \end{aligned} \qquad (8.7)$$

$$Con(\{s1,s2\},s4)$$

$$
\begin{aligned}
&= tp_2 * (1 - fp_{42}) * w_2 + tp_3 * (1 - fp_{43}) * w_3 \\
&= p * (1 - p) + 1 * (1 - p)) = (p + 1)(1 - p).
\end{aligned}
\tag{8.8}
$$

From the result of Equation (8.7) and (8.8), we know $Con(\{s_1,s_2\},s_3) < Con(\{s_1,s_2\},s_4)$, which means s_4 contributes more to sensor cover $\{s_1,s_2\}$ than s_3. Thus, s_4 will be chosen to add to the current sensor cover. The result is reasonable intuitively. $\{s_1,s_2\}$ covers t_1 twice and t_2 once. Both s_3 and s_4 can cover t_3. However, s_4 can cover t_2, which is only covered once by $\{s_1,s_2\}$. Hence, s_4 might improve the reliability to cover t_2 compared with s_3.

Algorithm 12: Greatest_Contribution_Sensor

Require: T; S; C_r;

Ensure: *candidate_sensor*: greatest contributed sensor;

1:   $greatest\_contribition = 0$;
2: **for** each sensor $s \in (S - C_r)$ **do**
3:     $contribution = 0$;
4:     **for** each target $t_j \in T$ **do**
5:       **if** s covers t_j **then**
6:         $current\_fp = 1$;
7:         **for** each sensor $cs_i \in C_r$ **do**
8:           **if** cs_i covers t_j **then**
9:             $current\_fp = current\_fp * cs.fp_{ij}$;
10:           **end if**
11:         **end for**
12:         $contribution = contribution + current\_fp * (1 - cs.fp_{ij}) * w_j$;
13:       **end if**
14:     **end for**
15:     **if** $contribution > greatest\_contribution$ **then**
16:       $grestest\_contribution = contribution$;
17:       $candidate\_sensor = s$;
18:     **end if**
19: **end for**
20: **return** *candidate_sensor*

The Algorithm 12 recursively calculates each target's failure probability *current_fp* from lines 7 to 11. The contribution (see detail in Definition 8.8) of each unchosen sensor to current sensor cover is calculated based on Equation (8.6) in line 12. The most contributed sensor *candidate_sensor* is outputted in line 20.

8.4.3 Relation between MSC and α-RMSC

Each solution of our proposed α-RMSC is also a solution of MSC [126]. Not like the situation in MSC in which only the lifetime of sensors is taken into consideration, in the α-RMSC, we also use the failure probability for each sensor cover as the selection criterion, which may be especially useful in the applications that are sensitive to failures. An important point is that lower failure probability means higher network reliability might lead to shorter network lifetime. Therefore we need to trade off based on the actual needs. Moreover, when we set α as 1, the α-RMSC problem converts to the MSC problem.

8.5 Performance Evaluation

In this section we evaluate the performance of the α-RMSC and MSC [126]. We simulate a stationary network with $n = 36$ sensor nodes (each with an initial energy 1, each time slot consume e energy), deployed uniformly at random in a 30 $m \times$ 20 m area. $m = 3$ target points randomly put in the monitoring area. We assume the sensing range, transmission range, and failure probability are equal for all the sensors in the network. We examine the lifetime attained by α-RMSC compared to the lifetime of MSC. Each time slot is 1 unit long. In the simulation, we consider the following tunable parameters:

- fp, the failure probability of each sensor node. We vary the number from 0.02 to 0.88 with an increment of 0.02.

- α, the user pre-defined maximum failure rate of the whole network. We vary the number from 0.015625 to 1 with an increment of timing the original number by 2.

- e, sensor lifetime granularity. We vary the number from 0.0125 to 1 with an increment of multiplying the original number by 2.

8.5.1 Simulation 1: Control Failure Probability

In this simulation, we show how well our algorithm controls the system's failure probability to satisfy different user pre-defined α values.

In Figure 8.3, we measure the reliable lifetime (see detail in Definition 8.6) and failure probability when the preset maximum failure probability α varies from 0.015625 to 1 by using our α-RMSC algorithm. From Figure 8.3(b) and Figure 8.3(c), we find the failure probability of the network is controlled very well. The circle-dotted line is completely below the rectangle-dotted line. The rectangle-dotted line is the user required maximum failure probability. And the circle-dotted line is the failure probability calculated by Equation (8.3). Based on our proposed α-RMSC algorithm, as the number of iterations increases, the failure probability must be less than or equal to the preset maximum failure probability α. In fact, the simulation

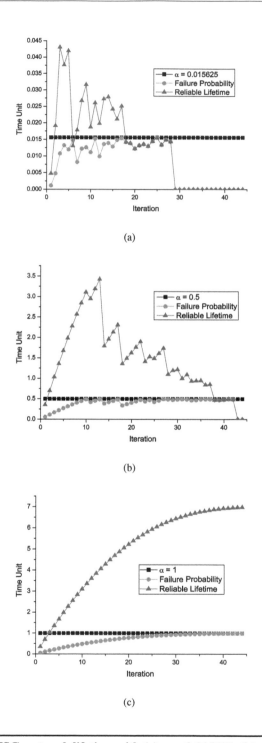

(a)

(b)

(c)

Figure 8.3: α-**RMSC, network lifetime with (a)** $\alpha = 0.015625$, **(b)** $\alpha = 0.5$, **(c)** $\alpha = 1$.

results verify our proposed algorithm. In Figure 8.3(a), since $\alpha = 0.015625$, which is a very small number, the limitation of the number of sensors deployed in a fixed area causes difficult algorithm converges. Thus, the network reliable lifetime is also unstable. However, in Figure 8.3(b) ($\alpha = 0.5$), the algorithm converges very quickly, roughly at iteration 20. The network reliable lifetime decreases as the iteration number increases. Since we use a greedy criterion to find α-reliable sensor covers, in Figure 8.3(c) ($\alpha = 1$), it converges in iteration 20. And there is a trade-off between the network lifetime and the network reliability. The less α, the higher network reliability, but the smaller network lifetime. When α is set to 1, the α-RMSC problem converts to the MSC problem.

8.5.2 Simulation 2: Comparison between α-RMSC and MSC

In this section, we first analyze the performance of MSC [126], and then compare the performance of MSC to our α-RMSC scheme.

From Figure 8.4, we measure the network reliable lifetime and failure probability when e varies from 0.125 to 1 using MSC [126] algorithm. The network reliable lifetime decreases when the sensor lifetime granularity e increases. If in each time slot, every sensor consumes more energy, then the network lifetime should be decreased. The network reliable lifetime increases with the number of iterations, as each node can now participate in more sensor covers and then the overall failure probability increases as well. As a conclusion from Figure 8.4, e values did not change the network reliability too much but did affect the network lifetime.

In Figure 8.5, we plot the reliable lifetime computed by α-RMSC and MSC [126], depending on the number of iterations. We vary the user pre-set threshold value α from $\frac{1}{2^6}$ to 1, with the increment of multiplying the original value by 2. We only show $\alpha = 0.5$ and $\alpha = 1$ in Figure 8.5. From the figure we can see, when $\alpha = 1$ the reliable lifetime of α-RMSC is close to the lifetime of MSC. It is reasonable, because our α-RMSC algorithm spent some energy to calculate sensor cover failure probability. When the α became smaller, the objective was to improve network reliability more than to improve the network lifetime. Thus we found the lifetime line goes down very quickly in Figure 8.5, where $\alpha = 0.5$.

The simulation results can be summarized as follows:

■ For a specific number of targets, the network lifetime output by our α-RMSC increases with the number of sensors.

■ For a specific number of sensors, the network lifetime increases as the number of targets to be monitored decrease.

■ Sensor lifetime granulation e only affects network lifetime. It does not affect network reliability too much.

■ For smaller maximum failure probability values α, the lifetime value decreases over time as the result of fewer sensor covers.

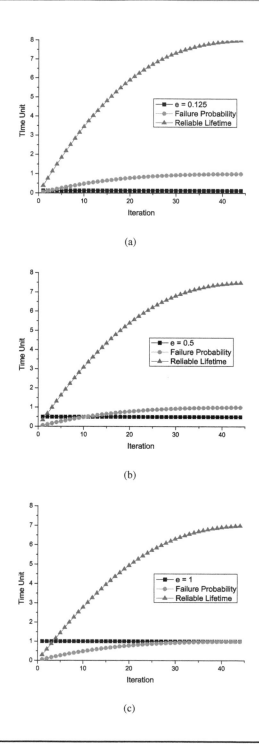

Figure 8.4: MSC, network lifetime with (a) $e = 0.0125$, **(b)** $e = 0.5$, **(c)** $e = 1$.

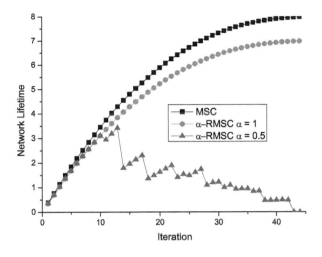

Figure 8.5: Comparison of α-RMSC and MSC.

- There is a trade-off between the higher lifetime value and the high reliability of the whole system.

- α-RMSC has a fast converge time, thus it is scalable to large sensor networks.

8.6   Conclusion

WSNs are battery powered, therefore prolonging the network lifetime through a power-aware node organization is highly desirable. Poor communication links and hazard unknown environment make the make improved reliability of WSNs a critical issue. An efficient method for energy saving and control system reliability is to schedule the sensor node activity such that every sensor alternates between sleep and active state to meet the user preset maximum failure probability. One solution is to organize the sensor nodes in sensor covers, so that every cover completely monitors all the targets. These covers are activated in turn. At a specific time only one sensor set is responsible for sensing the targets, while all other sensors are in sleep state. The failure probability of each sensor cover must be greater than the user preset maximum failure probability. This problem is modeled as the α-reliable maximum sensor covers problem, which is a NP-hard problem. Moreover, we propose an efficient heuristic greedy algorithm to solve the problem. Simulation results demonstrate that our proposed method can control the system's reliability easily without sacrificing the network lifetime too much.

Chapter 9

CDS-based Multi-regional Query Processing in Wireless Sensor Networks

CONTENTS

9.1 Introduction

Wireless sensor networks (WSNs) have variety of applications due to low cost and easy deployment, e.g., environmental monitoring, battlefield surveillance, and traffic control [120]. One major task of WSNs is to collect sensed data from sensor nodes deployed in the monitoring area. Hence, WSNs are also called *data-centric* networks where the sensed data of all the sensor nodes can be viewed as a database. Due to the fact that sensor nodes are usually randomly deployed in hostile or unreachable areas, the only way that users can interact with WSNs and obtain useful information is to disseminate queries from a special node called the sink (or base station). Therefore, query processing is one of the most important issues in WSNs and has been widely studied [147]-[148]. Early research of query processing mainly focuses on how to improve query efficiency [149][150] while minimizing the energy consumption. With the emergence of *real-time query processing* [151], e.g., fire monitoring in forests, *minimum latency* becomes another primary concern. Hence, *minimum latency query scheduling*, which aims to derive a collision-free query scheduling plan with minimum latency, shows its advantage in time-sensitive applications.

Ever since the query scheduling problem was introduced, efforts have been made to design different scheduling strategies for different query applications. Since limited power is still a severe constraint in WSNs, the energy efficient scheduling problem is investigated in [152][153][154][155]. After that, the authors in [156][157][115] investigated the minimum latency query processing problem. Additionally, the authors in [158][159][160] aimed to improve network performance such as throughput and QoS when designing a query scheduling.

SELECT $\mathcal{R}_1$.maxGasdensity, $\mathcal{R}_2$.maxTemperature
FROM $\mathcal{R}_1, \mathcal{R}_2$
WHERE $\mathcal{R}_1 = (11^{th}\text{-floor} \cup 12^{th}\text{-floor})$ and $\mathcal{R}_2 = (12^{th}\text{-floor} \cup 13^{th}\text{-floor})$

Figure 9.1: An MRQ example.

The previous research made great contributions to the development of query scheduling technologies. However, consider the following scenario: a WSN is deployed in a building, where the sensors at the 11th and 12th floors report gas leakages and the sensors at the 12th and 13th floors report fire alarms. In order to get the detailed information, the query shown in Figure 9.1 is disseminated to different regions (subareas) asking for the maximum gas density of the 11th and 12th floors and the maximum temperature of the 12th and 13th floors. The aforementioned techniques ([152][153][154]-[159][160][155][115]) are inefficient when dealing with the above query. The reasons are as follows. First, in most existing works, a user disseminates a query by broadcasting the query to the entire network even though the required data is from a particular small region. It wastes a large amount of energy, especially in large scale WSNs. Second, the scheduling strategies proposed in the existing works construct a single fixed scheduling tree consisting of all the sensor nodes. Avoiding collisions on a fixed global scheduling tree results in more latency. Hence, It is not suitable for scheduling multiple queries without correlations and targeting at multiple regions simultaneously. Last, the assumptions on disseminating time of queries with temporal and spatial considerations in the existing works are no longer reasonable when dealing with queries without those characteristics.

In order to address the aforementioned issues, we investigate the minimum latency multi-regional query scheduling (ML-MRQS) problem in this chapter. A multi-regional query (MRQ) is defined as a query that targets relevant data from multiple regions (overlapping may exist) of a WSN, where each region is a subarea of the WSN. Furthermore, for an MRQ, there is no limitation in query disseminating time interval or assumptions on temporal or spatial correlations. The traditional multi-query processing can be thought of as a special case of the MRQ processing if the query region is the entire network. Therefore, compared with the traditional query model, the MRQ model is more practical and general. Consequently, we investigate the ML-MRQS problem which aims to get a collision-free scheduling plan for an MRQ with minimum latency. The contributions of this chapter are summarized as follows:

1. In this chapter, we investigate the ML-MRQS problem for WSNs which is NP-hard, and propose a heuristic collision free scheduling algorithm, named multi-regional query scheduling algorithm (MRQSA). In MRQSA, we construct a CDS-based scheduling tree for each region. All those scheduling trees form a scheduling forest serving as the data transmission structure during the data transmission procedure. Furthermore, in order to improve the parallelism of MRQSA, we propose to assign higher priorities to the sensor nodes which may cause multiple collisions. During the scheduling procedure, transmissions on different trees are scheduled concurrently. In order to avoid collisions, priority and interference are considered in the entire forest.

2. We theoretically analyze the latency of MRQSA which is upper bounded by $23A + B + C$ for an MRQ with m query regions $\mathcal{R}_1, \mathcal{R}_2, ..., \mathcal{R}_m$, where $A = \max\limits_{i=1}^{m} D_i^{left}$, $B = \max\limits_{i=1}^{m}\{(23D_i + 5\Delta + 21)k_i\}$, $C = \sum\limits_{i=1}^{m} H_i + 5\Delta - m + 17$, m is the number of regions, Δ is the maximum node degree in the WSN, D_i is the diameter of $\mathcal{R}_i$, k_i is the maximum overlapped degree of sensor nodes in $\mathcal{R}_i$, H_i represents the distance of $\mathcal{R}_i$ with respect to the sink, and D_i^{left} is the diameter of the non-overlapped part of $\mathcal{R}_i$.

3. We also conduct extensive simulations to validate the performance of MRQSA. The simulation results indicate that the proposed MRQSA has a better performance compared with the most recently published multi-query scheduling algorithm C-DCQS [157]. MRQSA reduces latency by 49.3% on average for different number of query regions, 50.7% on average for different network densities, 42.7% on average for different region sizes, and 51.63%

on average for different interference/transmission ranges compared with C-DCQS.

The rest of this chapter is organized as follows: in Section 9.2, we summarize the related works. In Section 9.3, we introduce the terminologies and define the ML-MRQSA problem. In Section 9.4, we present the details of our scheduling algorithm followed by theoretical analysis. The simulation evaluations are presented in Section 9.5. Finally, we conclude this chapter and point out some future research directions in Section 11.9.

9.2   Related Work

Query scheduling is one of the most significant technologies in query processing [161]. It has attracted a large amount of research interest in the past few years. In this section, we summarize the existing works and the proposed solutions while providing a classification in terms of their query models.

For clarity, D is used to denote network diameter, Δ is used to represent the maximum node degree, R is the network radius, R' represents the inferior network radius, and n is the number of sensor nodes in a WSN (network size).

9.2.1   Periodic Query Scheduling

Periodic query processing is a widely used query model in WSNs. Under this query model, queries are disseminated into a WSN periodically. *Data collection* [73][74] and *data aggregation* [115] are two example periodic query processing methods.

In [10], the authors studied the periodic data collection scheduling problem. Two centralized algorithms are proposed to address the snapshot data collection scheduling problem and continuous data collection scheduling problem, respectively. Comprehensive theoretical analysis indicates that the latency bounds of the algorithms are order optimal. A centralized fit-greedy approximation scheduling algorithm for periodic data aggregation is proposed in [113]. The authors adopted a *connected dominating set* (CDS)-based tree as a scheduling tree. Two distributed data aggregation scheduling algorithms are proposed in [155] and [115], respectively. In [155], a CDS-based scheduling tree is constructed while in [115] a cluster-based network architecture is employed. The latency bounds for the scheduling algorithms proposed in [155] and [115] are $24D + 6\Delta + 16$ and $4R' + 2\Delta - 2$, respectively. In [114], the authors studied the minimum latency data aggregation scheduling problem as a TDMA scheduling problem above the MAC layer. A distributed algorithm is proposed with a latency bound of $16R + \Delta - 14$. Furthermore, The overall delay for data aggregation under any interference model is investigated, which is proven to be lower bounded by $max\{\log n, R\}$.

The authors in [162] proposed TDMA link scheduling algorithms for the purpose of maximizing network throughout. In this chapter, two interference models, the RTS/CTS (request-to-send and clear-to-send) model and protocol interference model, are considered. Both centralized and distributed algorithms are given. In [157], efforts are made to obtain conflict-free scheduling for periodic query processing which aims to balance the query latency and network throughput. The authors suggested a plan maker make a scheduling plan for queries, and then a scheduler is responsible for executing the scheduling plan without collisions. The tradeoffs between energy consumption and latency for data gathering in WSNs are investigated in [163]. In this chapter, the authors tried to reduce energy consumption, while guaranteeing that the data gathering procedure is finished within a specified latency constraint. The proposed algorithms include two off-line algorithms and one distributed on-line algorithm, which are based on the modulation scaling technique.

The authors in [152] and [153] proposed to schedule sensor nodes to go into sleep mode while no query requests are disseminated in periodic data query processing. The authors in [153] proposed to reduce the energy consumption of idle listening. However, the authors of [152] tried to find a scheduling strategy with minimum latency.

In [164], the periodic aggregation queries scheduling problem (PAQS) is investigated. The authors investigated the minimum delay scheduling problem for queries with different periods. A solution with routing, node and packet scheduling protocols is given to solve PAQS. The authors also proposed a scheme to test the schedulability of a given set of periodic queries, where the schedulability means whether the given queries can be scheduled within a finite delay.

9.2.2 Dynamic Query Scheduling

Instead of disseminating queries periodically, under a dynamic query model, queries are disseminated into a WSN on demand.

Due to the nature of resource limitations in WSNs, many researchers studied the query scheduling problem which can reduce energy consumption. In [165] and [166], the authors investigated the query optimization problem in cluster-based WSNs. The cluster-based hierarchy groups sensor nodes into clusters. A *cluster head* is elected from each cluster to take charge of the inter-cluster and intra-cluster data collection/aggregation. Hence, the cluster-based hierarchy can efficiently reduce data transmissions. In [167], the index-based query processing is studied. The authors proposed to divide the network into different squares with each square tagged with a particular index. With the help of the index, unnecessary data transmissions are avoided and the query processing time can be accelerated meanwhile. The query scheduling algorithm proposed in [154] solves the energy-efficient query scheduling problem for multiple queries with temporal and spatial correlations. This chapter is based on the assumption that multiple queries may have temporal and spatial overlapping in querying operations. Hence, the authors proposed to assign an optimal starting time to a newly arrived query based on the schedules of the existing queries.

In order to handle time sensitive queries, the authors of [156] investigated the real-time query scheduling problem in WSNs, which is defined as finding a scheduling plan that can return query result before a user defined deadline. The same components of plan maker and scheduler used in [157] are used in this chapter. However, an additional prediction procedure is carried out by the plan maker in [156] to see if the result can be returned before the deadline. Only queries that can meet the predefined deadline will be processed.

The authors of [159] investigated the optimal scheduling problem for queries whose arrival time obey the Poisson distribution and queries whose inter-arrival time obey the hyperexponential distribution, respectively. An optimal scheduling algorithm for the Poisson event arrival model is proposed while a suboptimal algorithm is proposed for the hyperexponential distribution inter-arrival model since the computation cost of the optimal solution under this model is very high.

In [160], the authors proposed to consider the quality of service (QoS) (response time and/or query lifetime) and quality of data (QoD) (the received data to be more evenly distributed) while dealing with query scheduling. Based on the QoS and QoD requirements which are specified by users, the execution order of queries are scheduled to maximize the ratio of award over cost.

The authors of [168] investigated the *area query* scheduling problem in WSNs. Besides querying for sensed data from a query region, in this chapter, the area information of the interested region can also be accommodated, e.g., find the area of the regions whose average temperature < 60 Fahrenheit. Instead of focusing on how to aggregate data in WSNs, in [163], the authors investigated how to aggregate queries to benefit query processing. A query aggregation algorithm is proposed in this chapter to aggregate queries with duplicate and overlapping.

9.2.3   Remarks

This chapter is different from all of the above mentioned works on some or all of the following aspects. They focus on either periodic query processing or dynamic query scheduling together with the network performance considerations, such as energy consumption, throughput, and latency. However, none of them focuses on the ML-MRQS problem. Therefore, in this chapter, we first give a formal definition of the ML-MRQS problem which is NP-Hard and then point out the limitations of using the existing methods to solve the problem. After that, a heuristic algorithm called MRQSA is proposed to solve the ML-MRQS problem.

9.3   Problem Formulation

9.3.1   Network Model

We consider a connected dense WSN consisting of n sensor nodes along with one *sink* (*base station*) denoted by s. Each node is equipped with a single, *half-duplex* radio. During a time slot, a node a can either send or receive data (not both) to or from all directions. The *transmission/interference range* of any sensor node is defined as a *unit disk* with radius 1 centered at the node. Hence, we can use an undirected unit disk graph (UDG) $G = (V, E)$ to represent the topology graph of a WSN, where V represents the set of all the sensor nodes, then $|V| = n + 1$; E denotes the *bidirectional* link among all the sensor nodes. $\forall a \in V$, a's one-hop neighbor set is denoted by $\mathcal{N}(a) = \{b \| a - b| \leq 1, b \in V, a \neq b\}$. An edge $(a,b) \in E$ exists if and only if $b \in \mathcal{N}(a)$.

We assume the network time is synchronized and slotted. Within a time slot, a data package can be sent from a sender and received by a receiver successfully as long as there is no interference. Additionally, we assume the global locations of all the sensor nodes are known.

9.3.2   *Multi-regional Query*

Let U represent the entire monitoring area of a WSN. A *region* is defined as a *consecutive* subarea within U, i.e., for a region $\mathcal{R}_i$ in a WSN, $\mathcal{R}_i \subseteq U$. Since the considered WSN is dense, the sensor nodes deployed in $\mathcal{R}_i$ is a connected component of G. For $\forall a \in V$, $\forall \mathcal{R}_i \subseteq U$, a is called a *regional node* (RN) of $\mathcal{R}_i$ if and only if a is deployed in $\mathcal{R}_i$. Furthermore, $\mathcal{S}(\mathcal{R}_i)$ is defined as the set of RNs in $\mathcal{R}_i$.

$$
\begin{aligned}
&\textbf{SELECT} \quad \mathcal{R}_1.\text{ds}, \ \mathcal{R}_2.\text{ds}, \ ..., \ \mathcal{R}_m.\text{ds} \\
&\textbf{FROM} \quad \ \ \ \mathcal{R}_1 \cup \mathcal{R}_2 \cup ... \cup \mathcal{R}_m \\
&\textbf{[WHERE} \quad \text{predicates]}
\end{aligned}
$$

Figure 9.2: The formal expression of an MRQ structure.

Definition 9.1   *Multi-regional Query* (MRQ). An MRQ is a query targeting at user interested data from multiple regions in a WSN. The formal expression of an MRQ is given in Figure 9.2 which shows an MRQ with m query regions. *SELECT* denotes the query results, where $\mathcal{R}_i.ds$ means the user interested data in region $\mathcal{R}_i$. *FROM* shows the multiple regions where the MRQ should be disseminated. The query conditions are specified in *WHERE*.

When an MRQ is disseminated to regions specified by the *FROM* clause, the RNs deployed in $\mathcal{R}_i$ only have to collect $\mathcal{R}_i.ds$. An example MRQ is shown in Figure 9.1. The MRQ targets at two regions with $\mathcal{R}_1 = \{11^{th} \bigcup 12^{th}\}$ floors and $\mathcal{R}_2 = \{12^{th} \bigcup 13^{th}\}$ floors. The purpose is to obtain the maximum gas density of $\mathcal{R}_1$ and the maximum temperature of $\mathcal{R}_2$. From this example, we can see that this MRQ targets data with no correlations from different regions simultaneously. Secondly, the target regions in an MRQ may overlap, such as $\mathcal{R}_1 \bigcap \mathcal{R}_2 = \{12^{th}\}$ floor. Thirdly, since the target regions of an MRQ are known, the MRQ can be directly disseminated to $\mathcal{R}_1 \bigcup \mathcal{R}_2 = \{11^{th} \bigcup 12^{th}\} \bigcup \{12^{th} \bigcup 13^{th}\} = \{11^{th} \bigcup 12^{th} \bigcup 13^{th}\}$ floors.

According to the above illustration, an MRQ only cares about data sets collected by sensor nodes deployed in the target query regions. Hence, MRQ has the following benefits. The query regions of an MRQ are known. Therefore, an MRQ can be directly disseminated to its corresponding query regions. Besides, sensor nodes that are deployed outside of the query regions, do not need to be involved in the MRQ unless they are relay nodes that help to transfer query results to the sink. In this way, energy consumption is reduced through reducing unnecessary data transmissions. Finally, since fewer sensor nodes are involved in data transmission, the possibility of reducing transmission conflicts is increased.

In this chapter, we investigate how to process an MRQ, which is dynamically disseminated into multiple regions of a WSN. In order to conserve network resources, query results are transmitted to the sink with aggregation. That is, on the routing tree, a parent node has to wait for all of its children to finish their data transmissions before starting its own transmission. We make no assumption on the correlations among data collected from different regions. Hence, the data can be aggregated during data transmission only if they are from the same region.

9.3.3 Problem Definition

Let $\mathcal{R}_i$ represent a region. A *regional query tree* (RQT) $\mathcal{T}_i$ is a tree constructed within $\mathcal{R}_i$ such that $\forall a \in \mathcal{S}(\mathcal{R}_i) \Leftrightarrow a \in \mathcal{T}_i$. The root of $\mathcal{T}_i$ is denoted by $r_{\mathcal{T}_i}$. Considering an MRQ with m query regions $\mathcal{R}_1, \mathcal{R}_2, ..., \mathcal{R}_m, \forall i, 1 \leq i \leq m, \mathcal{T}_i$ denotes the RQT constructed in $\mathcal{R}_i$ with $\mathcal{S}(\mathcal{R}_i)$. The *multi-regional query forest* (MRQF) denoted by $\mathcal{F}$ of an MRQ is the set of all the RQTs of this MRQ, i.e., $\mathcal{F} = \{\mathcal{T}_1, \mathcal{T}_2, ..., \mathcal{T}_m\}$. The set of all the sensor nodes in $\mathcal{F}$ is denoted by $\mathcal{S}(\mathcal{F})$, i.e., $\mathcal{S}(\mathcal{F}) = \mathcal{S}(\mathcal{R}_1) \bigcup \mathcal{S}(\mathcal{R}_2) \bigcup ... \bigcup \mathcal{S}(\mathcal{R}_m)$. The defined MRQF (RQTs) serves as the routing structure in our proposed algorithm.

The *overlapped degree* (OD) of an RN a, denoted by od_a, is defined as the number of RQTs it belongs to. For any region $\mathcal{R}_i$ with k RNs $n_1, n_2, ..., n_k$, let $\mathcal{O} = \{a|od_a = min(od_{n_1}, od_{n_2}, ..., od_{n_k})\}$, which is the set of RNs in $\mathcal{R}_i$ with the smallest OD. $\mathcal{A}$ is the set of RNs in $\mathcal{O}$ that are closest to the sink (the distance of each RN to the sink is measured by hop-distance). An *accessing node* (AN) is defined as the RN in $\mathcal{A}$ with the smallest ID.

Given an MRQF $\mathcal{F} = \{\mathcal{T}_1, \mathcal{T}_2, ..., \mathcal{T}_m\}$, let $\mathcal{I} = \overset{m}{\underset{i=1}{\bigcup}} \{a|a \in \mathcal{T}_i, \exists b, b \in \mathcal{T}_j, i \neq j, a \in \mathcal{N}(b), 1 \leq i$ and $j \leq m\} \bigcup \{a|od_a \geq 2\}$ represent the set of RNs that have at least one one-hop neighbor in another RQT and RNs whose ODs are no less than 2. $\forall a \in \mathcal{T}_i, \mathcal{D}(a)$ is defined as a's descendant set which contains the RNs in a subtree of $\mathcal{T}_i$ rooted at a. The *overlapped interference set* (OIS) of an MRQF is denoted by $\mathcal{S}(Ois) = \mathcal{I} \bigcup \mathcal{D}(\mathcal{I})$, where $\mathcal{D}(\mathcal{I})$ is the set of descendants of RNs in $\mathcal{I}$. The reason for including the descendants of $\mathcal{I}$ relies on the transmission constraint of aggregating data within a region during the data transmission.

Figure 9.3 shows an MRQF with two RQTs $\mathcal{T}_1$ and $\mathcal{T}_2$ which are constructed within query regions $\mathcal{R}_1$ and $\mathcal{R}_2$, respectively. An RN is represented by a black circle with its ID inside. The ANs are black and the RNs in the overlapped interference set $\mathcal{S}(Ois)$ are gray. From Figure 9.3, we can see that $od_3 = 1$ since it is only on $\mathcal{T}_1$, while $od_{14} = 2$ because it is on both $\mathcal{T}_1$ and $\mathcal{T}_2$. RN 0 is the AN of $\mathcal{T}_1$ since it has the smallest

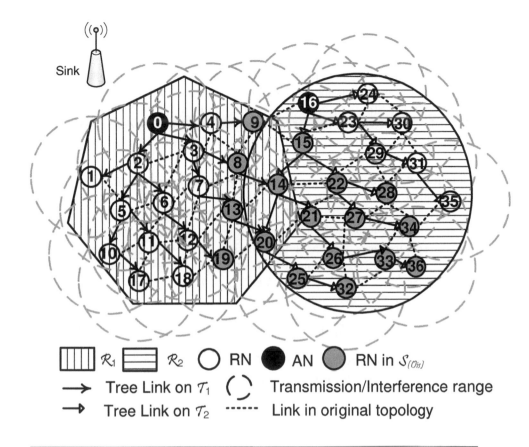

| | $\mathcal{R}_1$ | | $\mathcal{R}_2$ | ○ RN | ● AN | ○ RN in $\mathcal{S}_{(Ois)}$ |

→ Tree Link on $\mathcal{T}_1$ ( ) Transmission/Interference range

� Tree Link on $\mathcal{T}_2$ ------ Link in original topology

Figure 9.3: An example MRQF.

OD and is closest to the sink and the AN of $\mathcal{T}_2$ is RN 16. The OIS of the MRQF is $\mathcal{S}(Ois) = \mathcal{I} \bigcup \mathcal{D}(\mathcal{I}) = \{8, 9, 13, 14, 15, 19, 20, 21, 22, 25, 26, 27, 28, 32, 33, 34\}$, where each RN in $\mathcal{I}$ has at least one one-hop neighbor in another RQT, and the RNs in $\mathcal{D}(\mathcal{I})$ are descendants of RNs in $\mathcal{I}$.

Interference model: In this chapter, we employed the classical interference model which is widely used in the previous works [152]-[160][155][115]. Given two simultaneous transmissions $a \to b$ and $c \to d$, where a and c are senders, and c and d are receivers. Transmission $c \to d$ is said to conflict with transmission $a \to b$ if and only if $d \in \mathcal{N}(a)$ or $b \in \mathcal{N}(c)$. The the definition of a *non-concurrent set* is given below.

Definition 9.2 *Non-concurrent set* (NCS). Given an MRQ with m regions, its corresponding MRQF is $\mathcal{F} = \{\mathcal{T}_1, \mathcal{T}_2, ..., \mathcal{T}_m\}$. $\forall a \in \mathcal{T}_i$, $\mathcal{P}_i(a)$ is the parent node of a in $\mathcal{T}_i$. Then, transmission $a \to \mathcal{P}_i(a)$'s NCS is defined as $\mathcal{NCS}(a \to \mathcal{P}_i(a)) = \mathcal{N}_{\mathcal{F}}(\mathcal{P}_i(a)) \bigcup \{\bigcup_{b \in \mathcal{N}_{\mathcal{F}}(a)} Ch_{\mathcal{F}}(b)\} \bigcup \{a, \mathcal{P}_i(a)\} \bigcup \{Ch_{\mathcal{F}}(a)\}$, where $\mathcal{N}_{\mathcal{F}}(\mathcal{P}_i(a))$ is

$\mathcal{P}_i(a)$'s one-hop neighbor set in $\mathcal{F}$, $\mathcal{N}_{\mathcal{F}}(a)$ is a's one-hop neighbor set in $\mathcal{F}$, and $\mathcal{C}h(b)$ is b's children node set in $\mathcal{F}$.

RNs may be in multiple RQTs for a particular time slot or any RN can participate in only one transmission. Hence, for a, $\mathcal{C}h(a)$ and $\mathcal{P}_i(a)$ are also needed to be included in $\mathcal{NCS}(a \to \mathcal{P}_i(a))$, which is different from the NCS defined in the traditional case based on a single fixed scheduling tree. Then, we define the minimum latency multi-regional query scheduling problem (ML-MRQS) as follows.

Definition 9.3 *Minimum latency multi-regional query scheduling* (ML-MRQS). If an MRQ targets at m regions $\mathcal{R}_1, \mathcal{R}_2, ..., \mathcal{R}_m$, the RQT constructed in $\mathcal{R}_i$ is represented by $\mathcal{T}_i$. $\mathcal{F} = \{\mathcal{T}_1, \mathcal{T}_2, ..., \mathcal{T}_m\}$ is the MRQF. Let $Sch_{\mathcal{T}_i}$ be the set of all *child* $\to$ *parent* transmissions in $\mathcal{T}_i$, $Sch_{\mathcal{F}} = Sch_{\mathcal{T}_1} \bigcup Sch_{\mathcal{T}_2} \bigcup ... \bigcup Sch_{\mathcal{T}_m}$, and $Sch_{\mathcal{T}_i}^{t_j}$ represents the set of transmissions in $\mathcal{T}_i$ that are scheduled at time t_j. An ML-MRQS is a sequence of transmission sets denoted by $SCH = \{\{Sch_{\mathcal{T}_1}^{t_1}, Sch_{\mathcal{T}_2}^{t_1}, ..., Sch_{\mathcal{T}_m}^{t_1}\}, \{Sch_{\mathcal{T}_1}^{t_2}, Sch_{\mathcal{T}_2}^{t_2}, ..., Sch_{\mathcal{T}_m}^{t_2}\}, ..., \{Sch_{\mathcal{T}_1}^{t_L}, Sch_{\mathcal{T}_2}^{t_L}, ..., Sch_{\mathcal{T}_m}^{t_L}\}\}$ satisfying the following conditions:

(i). $Sch_{\mathcal{T}_k}^{t_i} \bigcap Sch_{\mathcal{T}_k}^{t_j} = \emptyset, \forall i \neq j, 1 \leq k \leq m, 1 \leq i, j \leq L$.

(ii). $\forall a \to \mathcal{P}(a) \in Sch_{\mathcal{T}_k}^{t_i}, \forall b \to \mathcal{P}(b) \in Sch_{\mathcal{T}_j}^{t_i}, a \notin \mathcal{NCS}(b \to \mathcal{P}(b))$ and $b \notin \mathcal{NCS}(a \to \mathcal{P}(a))$, where $\forall k \neq j, 1 \leq k, j \leq m, 1 \leq i \leq L$.

(iii). $\bigcup_{i=1}^{L} Sch_{\mathcal{T}_k}^{t_i} = Sch_{\mathcal{T}_k}$, where $1 \leq k \leq m$.

(iv). $\bigcup_{i=1}^{L} \bigcup_{k=1}^{m} Sch_{\mathcal{T}_k}^{t_i} = Sch_{\mathcal{F}}$.

(v). Data are aggregated from $\bigcup_{k=1}^{m} Sch_{\mathcal{T}_k}^{t_i}$ to $Sch_{\mathcal{F}} - \bigcup_{t=1}^{t_i} \bigcup_{k=1}^{m} Sch_{\mathcal{T}_k}^{t}$ at time slot t_i, for all $t = 1, 2, ...L$. Data are aggregated to $\bigcup_{k=1}^{m} r_{\mathcal{T}_k}$ in L time slots, where L is the latency of the MRQS.

(vi). L is minimized.

In Definition 9.3, constraint (*i*) specifies that a particular transmission in an RQT can only be scheduled once. Constraint (*ii*) guarantees that at a particular time, the transmissions scheduled in different RQTs should have no collisions. Constraint (*iii*) forces all the transmissions in an RQT getting their scheduling time slots, while constraint (*iv*) guarantees that all the transmissions in an MRQF are scheduled. Constraint (*v*) denotes the transmission set scheduled at time t_i, and it also guarantees that the query result of each region is collected to the AN. Constraint (*vi*) denotes that the latency should be minimized.

The ML-MRQS problem under the UDG model is NP-hard. This is based on the fact that the *minimum latency data aggregation scheduling* problem is NP-hard [109] under the UDG model, which can be considered as a special case of ML-MRQS when the target region of an MRQ is the entire network.

9.4   Multi-regional Query Scheduling

In this section, we propose a scheduling algorithm for ML-MRQS which consists of two phases. First, an MRQF is constructed which serves as the data transmission structure during the data transmission procedure. Subsequently, a collision-free scheduling plan is generated.

9.4.1   *Construction of MRQF*

In a WSN represented by $G = (V, E)$, a *dominating set* (DS) is defined as a subset of sensor nodes in the WSN with the property that for each sensor node in the WSN, it is either in the subset or adjacent to at least one node in the subset. If the induced subgraph of a DS is connected, we call this DS a connected dominating set (CDS). Nodes in a CDS are called dominators or connectors and others are called dominatees. Ever since the CDS was introduced, it has been widely used in many applications as a virtual backbone for efficient routing [28][169].

In order to obtain an efficient routing tree for the data transmission of an ML-MRQS, we use a CDS-based tree as the routing tree. There are many existing works studying how to construct a CDS [23][25][80][42]. In this chapter, the method proposed in [17] is adopted. Note that other algorithms can also be used. The RQTs constructed in all the regions form an MRQF, and are used as the data transmission structure of the ML-MRQS.

The construction of $\mathcal{T}_i$ in $\mathcal{R}_i$ can be briefly described as follows:

Step 1: Find the AN a of $\mathcal{R}_i$.

Step 2: Build a *breadth first searching* (BFS) tree rooted at a with all the RNs in $\mathcal{R}_i$. During the BFS tree construction procedure, obtain a *maximal independent set* (MIS) $\mathcal{M}_i$. The RNs in $\mathcal{M}_i$ colored black are dominators.

Step 3: Find a set of sensor nodes $\mathcal{C}_i$ to connect the nodes in $\mathcal{M}_i$ to form a CDS. The RNs in $\mathcal{C}_i$ colored gray are connectors and satisfy the fact that for any RN in $\mathcal{C}_i$, its parent node and children nodes are dominators.

Step 4: Color RNs in $\mathcal{S}(\mathcal{R}_i) \backslash (\mathcal{M}_i \bigcup \mathcal{C}_i)$, which are dominatees, as white, and pick each dominatee b a dominator from $\mathcal{N}(b) \bigcap \mathcal{M}_i$ as its parent. This step allocates each dominatee to a dominator, such that every dominatee can transmit data to its allocated dominator during the data transmission process.

Figure 9.4 shows the procedure of constructing a CDS-based routing tree using $\mathcal{T}_1$ in Figure 9.3, where the number inside each circle is the node's ID, the number beside the circle is the depth, and links are represented by lines or dashed lines. Figure 9.4(a) shows the construction of a BFS tree rooted at the AN (RN 0). In Figure 9.4(b), a dominating set is selected with which the RNs in the WSN is either in the dominating set or have a one-hop neighbor in the dominating set, where the selected RNs in this step are dominators. In order to guarantee data transmission, a connector set is chosen to make the dominating set connected as shown in Figure 9.4(c), where the RNs selected in this step are connectors. The next step is to allocate each dominatee (RNs except dominators and connectors) to a dominator, so that the dominatee only needs to forward data to its allocated dominator as shown in Figure 9.5(a), where the

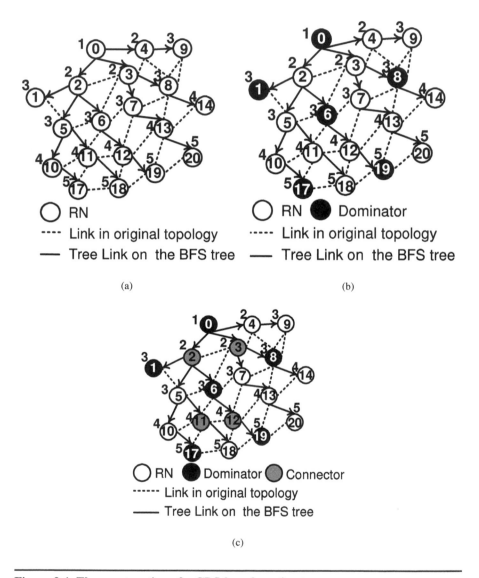

(a)

(b)

(c)

Figure 9.4: The construction of a CDS-based routing tree.

data structures of the RNs are updated based on the constructed CDS-based routing tree, e.g., the depth of a dominatee is updated to one more than the depth of its parent (dominator).

Particularly, due to the existence of overlapped regions, RQTs in an MRQF may overlap with each other (as shown in Figure 9.3), which is different from a traditional forest. Therefore, the ML-MRQS problem is much more challenging than the traditional scheduling problem because RNs in $\mathcal{S}(\mathcal{O}is)$ may cause collisions more than once.

9.4.2 MRQSA

9.4.2.1 Scheduling Initialization

At the beginning of the initialization, the RNs in $\mathcal{F}$ are divided into two parts: $\mathcal{S}(\mathcal{O}is)$ and $\mathcal{F}^{left}$, where $\mathcal{S}(\mathcal{O}is)$ is the OIS of an MRQF, and $\mathcal{F}^{left} = \mathcal{F}\backslash\mathcal{S}(\mathcal{O}is)$. Let $\mathcal{T}_i^{left} = \mathcal{T}_i\backslash\mathcal{S}(\mathcal{O}is)$, then $\mathcal{F}^{left}$ can be represented by $\{\mathcal{T}_1^{left}, \mathcal{T}_2^{left}, ..., \mathcal{T}_m^{left}\}$ satisfying $\forall i \neq j$, $\mathcal{T}_i^{left} \bigcap \mathcal{T}_j^{left} = \emptyset$. The behind-the-scene meaning of separating $\mathcal{S}(\mathcal{O}is)$ from $\mathcal{F}^{left}$ is to improve the parallelism of MRQS. The RNs in $\mathcal{S}(\mathcal{O}is)$ may influence the data transmissions of multiple RQTs because their interference range may overlap with multiple regions. Therefore, the RNs in $\mathcal{S}(\mathcal{O}is)$ are given a higher priority in MRQSA. It follows that the RNs in $\mathcal{S}(\mathcal{O}is)$ may finish their scheduling first. After that, the scheduling among the $\mathcal{T}^{left}$s can be paralleled without collision.

The $\mathcal{S}(\mathcal{O}is)$ can be specified by checking all the RNs in the MRQF: from $\mathcal{T}_1$ to $\mathcal{T}_m$. If one RN in $\mathcal{T}_i$ has a one-hop neighbor in $\mathcal{T}_j$ ($i \neq j$) or the OD of the RN is no less than 2, mark it as an RN in $\mathcal{S}(\mathcal{O}is)$, i.e., if $\forall a \in \mathcal{T}_i$, $\exists b, b \in \mathcal{T}_j, a \in \mathcal{N}(b), i \neq j$, or $od_a \geq 2$, put a in $\mathcal{S}(\mathcal{O}is)$. Subsequently, find the descendants of all the elements in $\mathcal{S}(\mathcal{O}is)$.

9.4.2.2 Scheduling Algorithm

After the initialization phase, all information needed for scheduling is collected. For simplicity, let $\mathcal{S}ch^l$ represent the schedule plan in the l-th iteration, which is a set of transmissions scheduled in the l-th time slot, $l = 1, 2, ..., L$. $\mathcal{NCS}(\mathcal{S}ch^l)$ represents the NCS of the transmissions in $\mathcal{S}ch^l$. Let $\mathcal{T}_i^{Ois} = \mathcal{T}_i \bigcap \mathcal{S}(\mathcal{O}is)$.

A transmission $a \to b$ on $\mathcal{T}_i$ is said **ready** to be scheduled if RN a is a leaf node or a's children nodes in $\mathcal{T}_i$ have already obtained their scheduling time slots and the priority of a is the highest compared with other scheduling candidates in $\mathcal{T}_i$. The priority is given to an RN in the order of "RN in the OIS", "color is white", "depth is larger", and "ID is smaller". The scheduling multi-regional query scheduling algorithm (MRQSA), is shown in *Algorithm 13*.

Algorithm 13 shows that MRQSA takes the MRQF as an input. It runs iteratively from line 3 to line 9 and finally outputs the scheduling plan $\mathcal{S}$ and the latency L of $\mathcal{S}$. During the l-th iteration ($l = 1, 2, ..., L$), the first step is to initialize $\mathcal{S}ch^l$ and $\mathcal{NC}(\mathcal{S}ch^l)$ to be empty sets (line 5-6), then schedule the transmissions whose senders are RNs in $\mathcal{S}(\mathcal{O}is)$ and can be scheduled in the l-th iteration as line 7 shows. Finally,

---

Algorithm 13: MRQSA

---

input : MRQF

output: schedule plan $S = \{Sch^1, Sch^2, ..., Sch^m\}$ and latency L

1   $S \leftarrow \emptyset$;

2   $l \leftarrow 0$;

3   **while** $\mathcal{F}^{left} \neq \emptyset$ *or* $S(\mathcal{O}is) \neq \emptyset$ **do**

4   l++;

5      $Sch^l \leftarrow \emptyset$;

6      $\mathcal{N}CS(Sch^l) \leftarrow \emptyset$;

7   Schedule $S(\mathcal{O}is)$ at the l-th time slot, find the conflict-free transmissions in $S(\mathcal{O}is)$ and add these transmissions to Sch^l followed by updating $\mathcal{N}CS(Sch^l)$ (as shown in *Algorithm 14*);

8   Schedule $\mathcal{F}^{left}$ at the l-th time slot, find the conflict-free transmissions in $\mathcal{F}^{left}$ and add these transmissions to Sch^l (as shown in *Algorithm 14*);

9   Add Sch^l to S;

10   return $S, L = l$

---

the transmissions whose senders are RNs in $\mathcal{F}^{left}$ can be scheduled in the l-th iteration as shown in line 8. MRQSA generates a scheduling plan iteratively until all the *child* → *parent* transmissions in each RQT are scheduled. The method of scheduling $S(\mathcal{O}is)$ or $\mathcal{F}^{left}$ is presented in *Algorithm 14*.

Algorithm 14 takes the $\mathcal{O}S$ (objective set) as an input, where $\mathcal{O}S = S(\mathcal{O}is)$ when scheduling $S(\mathcal{O}is)$ and $\mathcal{O}S = \mathcal{F}^{left}$ when scheduling $\mathcal{F}^{left}$, respectively. The outputs of *Algorithm 14* are the transmissions whose senders are in the considered $\mathcal{O}S$ that can be scheduled in the l-th iteration. From $\mathcal{T}_1$ to $\mathcal{T}_m$ (line 4), *Algorithm 14* runs iteratively to find a ready transmission $a \to \mathcal{P}_i(a)$ with the highest priority and $a \notin \mathcal{N}CS(Sch^l)$ as line 5 shows. If such a transmission is found, add it to Sch^l (line 7). Subsequently, the $\mathcal{N}CS(Sch^l)$ is updated to be $\mathcal{N}CS(Sch^l) \bigcup \mathcal{N}CS(a \to \mathcal{P}_i(a))$ (line 8). If all the transmissions in the MRQF involving a are scheduled, a should be removed from the $\mathcal{O}S$ (line 10-11). The procedure will continue until no more ready transmission that can be scheduled without collision is found in this iteration.

Figure 9.5(a) shows the CDS-based RQTs derived from Figure 9.3, where the numbers beside the RNs represent the depth while the numbers within the circles represent their IDs. At the beginning of the first slot, Sch^1 and $\mathcal{N}CS(Sch^1)$ are initialized to be empty sets. Starting from $\mathcal{T}_1$, we first search for ready transmissions $a \to \mathcal{P}_i(a)$ where $a \in \mathcal{T}_1^{\mathcal{O}is}$. RN 14 is selected since it is ready and its priority is the highest. Then, transmission 20 → 19 is added to Sch^1 and $\mathcal{N}CS(Sch^1) = \emptyset \bigcup \mathcal{N}CS(20 \to 19) = \{12, 13, 18, 20\} \bigcup \{20, 21, 22, 26, 32\} \bigcup \{19, 20\} \bigcup \emptyset = \{12, 13, 18, 19, 20, 21, 22, 26, 32\}$, where $\{12, 13, 18, 20\}$ is 19's one-hop neighbor set, the second set $\{20, 21, 22, 26, 32\}$ is 20's one-hop neighbors' children set in the MRQF, the third set $\{19, 20\}$ is the set of sender and receiver of the transmission, and $\emptyset$ is 20's children in $\mathcal{F}$. Similarly, the next ready transmission 9 → 8 is found and added to Sch^1.

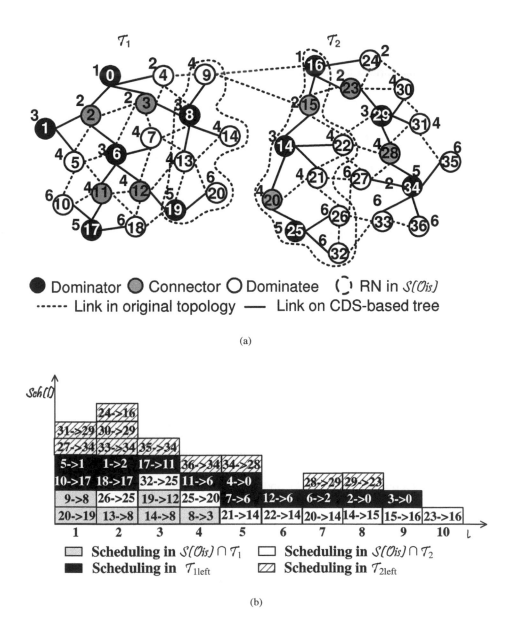

(a)

(b)

Figure 9.5: (a) RQTs. (b) Scheduling plan.

Algorithm 14: Schedule $\mathcal{S}(\mathcal{O}is)$ or $\mathcal{F}^{left}$ in the l-th slot

---

input : $\mathcal{OS}$(Objective Set)-$\mathcal{S}(\mathcal{O}is)$ or $\mathcal{F}^{left}$, $\mathcal{NCS}(Sch^l)$
output: schedule plan Sch^l for time slot l

1   $flag = 1$;
2   **while** $flag \neq 0$ **do**
3     $flag \leftarrow 0$;
4   **for** $i : 1 \rightarrow m$ **do**
5   **if** $a \in \mathcal{OS}$ and $a \rightarrow P_i(a))$ is ready and $a \notin \mathcal{NCS}(Sch^l)$ **then**
6         $flag \leftarrow 1$;
7         $Sch^l \leftarrow Sch^l \bigcup a \rightarrow P_i(a)$;
8         $\mathcal{NCS}(Sch^l) \leftarrow \mathcal{NCS}(Sch^l) \bigcup \mathcal{NCS}(a \rightarrow P_i(a))$;
9         $od_a - -$;
10   **if** $od_a == 0$ **then**
11          $\lfloor$ $\mathcal{OS} \leftarrow \mathcal{OS} \backslash \{a\}$;

12   return Sch^l, $\mathcal{NCS}(Sch^l)$

Then $\mathcal{NCS}(9 \rightarrow 8) = \{3,4,7,8,9,13,14,15,22,23,24\}$ is added to $\mathcal{NCS}(Sch^1)$, i.e., $\mathcal{NCS}(Sch^1) = \{12,13,18,19,20,21,22,26,32\} \bigcup \{3,4,7,8,9,13,14,15,22,23,24\} = \{3,4,7,8,9,12,13,14,15,18,19,20,21,22,23,24,26,32\}$. Then we check the remaining transmissions whose senders are in $\mathcal{T}_1^{Ois}$. It turns out that no more transmissions can be scheduled. Then MRQSA moves to $\mathcal{T}_2^{Ois}$. Unfortunately, there are no transmissions in $\mathcal{T}_2^{Ois}$ that can be scheduled without collision. Subsequently, MRQSA moves back to check $\mathcal{T}_i^{left}$, and continues from $\mathcal{T}_1^{left}$ to $\mathcal{T}_2^{left}$ to check the transmissions whose senders are in $\mathcal{T}_i^{left}$. Transmissions $10 \rightarrow 17$ and $5 \rightarrow 1$ are found in $\mathcal{T}_1$ and transmissions $27 \rightarrow 34$ and $31 \rightarrow 29$ are found in $\mathcal{T}_2$. The first iteration stops here since no more transmissions can be scheduled without collision. MRQSA repeats this procedure until all the transmissions in the MRQF are scheduled. Figure 9.5(b) shows the entire scheduling plan generated by MRQSA, where the x-axis represents the time slot and the y-axis represents transmissions scheduled in each corresponding time slot.

After obtaining the scheduling plan, every transmission can be scheduled in its assigned time slot.

In spite of the fact that we divide MRQF into two parts logically, the scheduling of $\mathcal{S}(\mathcal{O}is)$ and $\mathcal{F}^{left}$ may not necessarily be implemented sequentially. When generating a scheduling plan in each iteration, transmissions with senders in $\mathcal{S}(\mathcal{O}is)$ have a higher priority, but transmissions whose senders are in $\mathcal{F}_{left}$ can also be scheduled as long as they are ready and will not cause collisions with the current plan. This strategy aims to improve the parallelism of the MRQSA to minimize the latency. Particularly, the priority strategy guarantees the new added transmissions whose senders are dominators or connectors will not conflict with the scheduled transmissions whose senders are dominatees, and it also guarantees the scheduling of the transmission whose sender has a smaller depth will not conflict with a scheduled transmission whose sender has a higher depth.

9.4.3 Performance Analysis

In this subsection, we analyze the latency performance of MRQSA. Compared with traditional query scheduling methods, an MRQ can be directly disseminated to the query regions instead of the entire network. Intuitively, this procedure will reduce latency and energy consumption. Therefore, we only focus on the latency performance of MRQSA consisting of the latency of scheduling $\mathcal{F}^{left}$, the latency of scheduling $\mathcal{S}(\mathcal{O}is)$, and the latency of scheduling ANs to transmit the final aggregated values to the sink.

For convenience, The *diameter* of a region is defined as the maximum distance between any two RNs in this region. Let D_i^{left} represent the diameter of the non-overlapped part of region $\mathcal{R}_i$ ($\mathcal{T}_i^{left}$), D_i be the diameter of $\mathcal{R}_i$, k_i be the maximum overlapped degree of $\mathcal{T}_i^{Ois}$ which is defined as the maximum OD of the RNs in $\mathcal{T}_i^{Ois}$, and H_i represent the distance of $\mathcal{R}_i$ with respect to the sink, which is the hop distance from the AN of $\mathcal{T}_i$ to the sink.

The following lemma shows some properties of the constructed CDS.

Lemma 9.1
The following statements are true.

(i). Each connector has at most 5 one-hop neighbors which are dominators [113].

(ii). Each dominator has at most 20 two-hop neighbors which are dominators [113].

(iii). Considering the CDS-based tree excluding the dominatees, the nodes with depth $2i + 1$ (odd) are dominators and the ones with depth $2i$ (even) are connectors [170].

(iv). The depth of a CDS-based tree constructed within a region is no more than two times of the diameter of the region, where the diameter of the region is the maximum distance between any two sensor nodes in this region [155].

Lemma 9.2
The latency of scheduling $\mathcal{F}^{left}$ is no more than $23 \max\limits_{i=1}^{m} D_i^{left} + 5\Delta + 17$.

Proof: Based on the aforementioned illustration, $\mathcal{F}^{left} = \{\mathcal{T}_1^{left}, \mathcal{T}_2^{left}, ..., \mathcal{T}_m^{left}\}$ satisfies $\forall i \neq j$, $\mathcal{T}_i^{left} \cap \mathcal{T}_j^{left} = \emptyset$. Since the m subsets of $\mathcal{F}^{left}$ are pairwise disjoint, the scheduling of each subset is independent of others. As a consequence, the latency of scheduling $\mathcal{F}^{left}$ actually depends on the latency of $\mathcal{T}_i^{left}$ whose latency is the largest among all the subsets. Additionally, since the RNs in $\mathcal{T}_i^{Ois}$ are leaves or subtrees, each $\mathcal{T}_i^{left}$ is still a CDS-based tree based on the construction procedure of an MRQF in Section 9.4.1. According to MRQSA, the scheduling of T_i^{left} consists of two phases:

(1) Scheduling from dominatees to dominators. For any transmission $a \rightarrow \mathcal{P}_i(a)$, where a is a dominatee and $\mathcal{P}_i(a)$ is a dominator, $\mathcal{NCS}(a \rightarrow \mathcal{P}_i(a)) = \mathcal{N}_{T_i}(\mathcal{P}_i(a)) \bigcup \{ \bigcup_{b \in \mathcal{N}_{T_i}(a) - Ch(a)} Ch(b) \} - \{a, \mathcal{P}_i(a)\}$. Intuitively, $\|\mathcal{N}_{T_i}(\mathcal{P}_i(a))\| \le \Delta - 1$, where $\| \cdot \|$ is the cardinality. Meanwhile, among a's one-hop neighbors, only dominators or connectors may have dominatee children. Based on Lemma 1, the number of children of a's one-hop neighbors which are connectors is bounded by 20, and the number of children of a's one-hop neighbors which are dominators is bounded by 4Δ (a's parent is excluded). Hence, the latency of this phase is at most $20 + 4\Delta + \Delta - 1 + 1 = 5\Delta + 20$.

(2) Scheduling between dominators and connectors. After the first phase, all the transmissions from dominatees to their parents are scheduled. Let $k = 1, 2, ..., ly_{imax}$ represent the depth of the unscheduled RNs on T_i^{left}, where ly_{imax} is the largest depth. $ly_{imax} \le 2D_i^{left}$ based on Lemma 1 (iv).

Case 1: $ly_{imax} = 2j + 1$ is odd.

In this case, the transmissions from the dominators with depth $2j + 1$ to their parents which are connectors with depth $2j$ are scheduled first. For any transmission $a \rightarrow \mathcal{P}_i(a)$, it cannot be scheduled together with transmissions whose senders are dominators with depth $2j + 1$ in $\mathcal{NCS}(a \rightarrow \mathcal{P}_i(a))$. Since $\mathcal{P}_i(a)$ is a connector, it has at most five neighbors which are dominators including a. However, $\mathcal{P}_i(a)$'s parent will not be scheduled together with a. Thus, $\|\mathcal{N}_{T_i}(\mathcal{P}_i(a))\| = 3$. Moreover, a's one-hop neighbors which are connectors or dominators do not have any extra dominator children with depth $2j + 1$. Hence, $\|\mathcal{NCS}(a \rightarrow \mathcal{P}_i(a))\| = 3$, that is, at most $3 + 1 = 4$ time slots are needed to finish the transmission scheduling from depth $2j + 1$ to depth $2j$.

After that, we consider the scheduling from depth $2j$ to depth $2j - 1$, where for any transmission $a \rightarrow \mathcal{P}_i(a)$, a is a connector with depth $2j$ and $\mathcal{P}_i(a)$ is a dominator with depth $2j - 1$. Only the transmissions whose senders are connectors with depth $2j$ in $\mathcal{NCS}(a \rightarrow \mathcal{P}_i(a))$ are considered in this turn. Based on Lemma 1 (ii), $\mathcal{P}_i(a)$ has at most 20 one-hop neighbors which are connectors including a and $\mathcal{P}_i(a)$'s parent. Hence, $\|\mathcal{NCS}(a \rightarrow \mathcal{P}_i(a))\| = 18$. Similarly, a's one-hop neighbors which are connectors or dominators do not have any extra connector children in depth $2j$. Therefore, at most 19 time slots are needed to finish the transmission scheduling from the $(2j)$-th to the $(2j - 1)$-th depth. Specially, the scheduling from depth 2 to depth 1 needs at most 20 time slots due to the fact that the ANs have no parent in T_i.

The scheduling between dominators and connectors rotates until all the data are collected by the AN in depth 1. The latency from depth $2j + 1$ to depth 1 is represented by

$$l_{odd} \le 20 + 4 + (19 + 4)(j - 1) \tag{9.1}$$

$$= 23(ly_{imax} - 1)/2 + 1 \le 23D_i^{left} - 10. \tag{9.2}$$

Case 2: $ly_{imax} = 2j$ is even.

The analysis of case 2 is similar to case 1. Note that the transmission latency from depth $2j+1$ to depth $2j$ is removed. Hence, we denote the latency as

$$l_{even} \leq 20 + (19+4)(j-1) \tag{9.3}$$

$$= 23(l_{yimax})/2 - 3 \leq 23D_i^{left} - 3. \tag{9.4}$$

In summary, the latency of scheduling $\mathcal{F}^{left}$ denoted by

$$l_{\mathcal{F}^{left}} = \max\{l_{T_1^{left}}, l_{T_2^{left}}, \cdots, l_{T_m^{left}}\} \tag{9.5}$$

$$\leq 20 + 5\Delta + 23 \max_{i=1}^{m} D_i^{left} - 3 \tag{9.6}$$

$$= 23 \max_{i=1}^{m} D_i^{left} + 5\Delta + 17. \tag{9.7}$$

$\square$

Lemma 9.3
The latency of scheduling $\mathcal{S}(\mathcal{O}is)$ is no more than $\max_{i=1}^{m}\{(23D_i + 5\Delta + 21)k_i\}$.

Proof: According to the definition of $\mathcal{S}(\mathcal{O}is)$ (Section 9.3.3), $\mathcal{S}(\mathcal{O}is)$ is a set of RNs whose interference range overlapped with other RN (or RNs) in other region (or regions) and their descendants. $\mathcal{S}(\mathcal{O}is)$ can be represented by $\{T_1^{Ois}, T_2^{Ois}, ..., T_m^{Ois}\}$. Let $\{k_1, k_2, ..., k_m\}$ represent the maximum OD of each T_i^{Ois}. For each T_i^{Ois}, let l_{iomin} represent the smallest depth of the RNs in T_i^{Ois} and l_{iomax} represent the largest depth of the RNs in T_i^{Ois}. Since T_i^{Ois} may not be consecutive, we extend T_i^{Ois} to a subset consisting of the RNs in T_i from depth l_{iomin} to depth l_{iomax}. Evidently, the latency of T_i^{Ois} is no more than the latency of extended T_i^{Ois}. It follows that $l_{iomax} - l_{iomin} \leq 2D_i$.

In order to get the latency of scheduling T_i^{Ois}, we first consider the scheduling of T_i^{Ois} without considering the overlapping. Then, the scheduling of T_i^{Ois} contains two parts. The first part is the same as the first phase of T_i^{left} and the second part can be divided into the following four cases. The analysis of each case is similar to that of T_i^{left}.

Case 1: Both l_{iomin} and l_{iomax} are odd. Then, the latency is

$$l_{oo} = (4+19)(l_{iomax} - l_{iomin})/2 \tag{9.8}$$

$$\leq 23D_i. \tag{9.9}$$

Particularly, if the AN is included, then $l_{oo} = 23D_i + 1$.

Case 2: l_{iomin} is odd while l_{iomax} is even. Then, the latency is

$$l_{oe} = (19+4)((l_{iomax} - l_{iomin})/2 - 1) + 19 \tag{9.10}$$

$$\leq 23D_i - 4. \tag{9.11}$$

Particularly, if the AN is included, then $l_{oe} = 23D_i - 3$.

Case 3: l_{iomin} is even while l_{iomax} is odd. Then, the latency is

$$l_{eo} = (4+19)((l_{iomax} - l_{iomin})/2 - 1) + 4 \tag{9.12}$$
$$\leq 23D_i - 19. \tag{9.13}$$

Case 4: Both l_{iomin} and l_{iomax} are even. Then, the latency is

$$l_{ee} = (19+4)(l_{iomax} - l_{iomin})/2 \tag{9.14}$$
$$\leq 23D_i. \tag{9.15}$$

In summary, the latency of scheduling the extended T_i^{Ois} when the k_i overlappings are considered is

$$l_{T_i^{Ois}} \leq k_i * (20 + 5\Delta + 23D_i + 1) \tag{9.16}$$
$$= (23D_i + 5\Delta + 21)k_i. \tag{9.17}$$

Therefore, the latency of scheduling $\mathcal{S}(\mathcal{O}is)$ is upper bounded by

$$l_{\mathcal{S}(Ois)} = \max_{i=1}^{m}\{(23D_i + 5\Delta + 21)k_i\}. \tag{9.18}$$

$\square$

Lemma 9.4

The latency of scheduling the ANs is no more than $\sum_{i=1}^{m}(H_i - 1)$.

Proof: Since there are m RQ instances, the number of ANs is at most m. In the worst case, no two ANs can be scheduled concurrently. Hence, the latency is at most $l_{ANs} = \sum_{i=1}^{m}(H_i - 1)$. One time slot is deducted since the scheduling of the transmission from an AN to its parent has already been considered in the first two phases. $\square$

In practice, the time for ANs to finish their data collecting may vary. The actual latency may be much less than the bound shown in Lemma 4.

Theorem 9.1

The upper bound of the latency of MRQSA is no more than $23A + B + C$, where $A = \max_{i}^{m} D_i^{left}$, $B = \max_{i=1}^{m}\{(23D_i + 5\Delta + 21)k_i\}$, and $C = \sum_{i=1}^{m} H_i + 5\Delta - m + 17$.

Proof: According to the illustration of MRQSA in Section 9.4.2.2, the scheduling of $\mathcal{S}(\mathcal{O}is)$ and $\mathcal{F}^{left}$ can overlap in some extent. In the worst case, the scheduling of $\mathcal{S}(\mathcal{O}is)$ and $\mathcal{F}^{left}$ is sequential. Then, the latency is upper bounded by

$$L = l_{\mathcal{F}^{left}} + l_{\mathcal{S}(Ois)} + l_{ANs} \tag{9.19}$$

$$= 23\max_{i=1}^{m} D_i^{left} + \max_{i=1}^{m}\{(23D_i + 5\Delta + 21)k_i\} + \sum_{i=1}^{m} H_i + 5\Delta - m + 17. \tag{9.20}$$

$\square$

9.5 Performance Evaluation

In this section, we compare the performance of MRQSA with the centralized version of DCQS denoted by C-DCQS proposed in [157]. C-DCQS is the most recently published algorithm that solves the multi-query scheduling problem. We compare the two algorithms in terms of latency under three different scenarios. Since energy consumption is still one of the primary concerns in WSNs, the performance of our proposed MRQSA in terms of energy consumption is evaluated as well.

9.5.1 Simulation Environment

In our simulations, we deploy sensor nodes in a monitoring area of $1000\,m \times 1000\,m$, where all the sensor nodes have the same interference/transmission range of $50\,m$. The sink is deployed in the upper left corner. Moreover, the regions in an MRQ are set as squares. During the result collecting procedure of an MRQ, the results from a particular region will be aggregated and the results from different regions cannot be aggregated. The energy consumption model is that receiving and transmitting a packet both consume 1 unit of energy and 1000 units of energy are assigned to each sensor node during the initialization process. Besides, the latency is defined as the time interval between the first RN beginning to transmit data and the last packet of the results received by the sink. The sensor nodes involved in MRQSA contain RNs and the sensor nodes on the shortest paths from ANs to the sink. To be fair, we make the following improvements to C-DCQS. Firstly, we use the same algorithm in [17] to construct a CDS-based tree as the routing tree for C-DCQS. Secondly, the sensor node involved in an MRQ under C-DCQS is defined as a subtree rooted at the sink which has the least number of sensor nodes but contains all the RNs. In this way, the number of sensor nodes involved in an MRQ is minimized under C-DCQS. The simulation is implemented using C++ in Microsoft Visual Studio 2010 development environment. The results shown in the following are the averaged values by executing the same procedure 100 times.

9.5.2 Simulation Results

We examine the latency of C-DCQS and MRQSA when the number of query regions of an MRQ varies as shown in Figure 9.6(a). We fix each query region as a square of $600\,m \times 600\,m$ and the regions are randomly generated in the WSN. The number of query regions increases from 2 to 10 by 1. From Figure 9.6(a), we can see that, with the size of an MRQ increasing (the size of an MRQ means the number of regions in this MRQ), the latency of both C-DCQS and MRQSA increases. This is because more data transmissions are needed when collecting the results from new added regions to the sink. It follows that more time is needed to guarantee collision-free scheduling. MRQSA has a better performance because MRQSA tries to maximize the parallelism of scheduling by concurrently scheduling transmissions of different regions. C-DCQS calculates the minimum time interval for each query in different regions and schedules transmissions in different regions sequentially according to

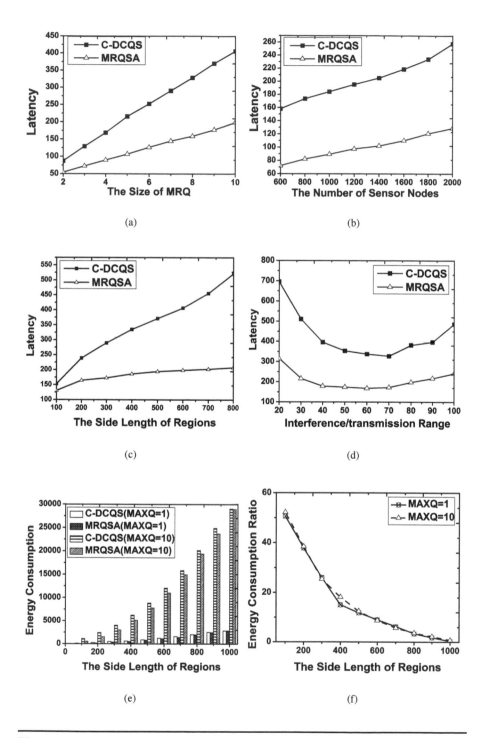

Figure 9.6: Simulation results.

the single fixed scheduling tree. Thus, more time is needed for C-DCQS. MRQSA reduces latency by 49.3% on average compared with C-DCQS.

We study the influence of network density on the latency of C-DCQS and MRQSA as shown in Figure 9.6(b). In this case, the query region is fixed to be a square of 600 m × 600 m and the number of query regions is 5. We increase the number of sensor nodes in the WSN from 600 to 2000 by 200 each time. Figure 9.6(b) shows that the latencies of both C-DCQS and MRQSA increase with the increasing of node density. This is because the number of RNs in each query region increases in a denser network. Therefore, more time is required to finish the scheduling. However, the increasing of MRQSA is more stable than C-DCQS, especially, when the number of the sensor nodes increases from 1400 to 2000, the latency of C-DCQS increases sharply. This is because for C-DCQS, with the increasing of the network density, the intermediate nodes in a fixed single routing tree have to wait longer to collect results from their children. Hence, the minimum interval time required to avoid collisions between the scheduling of different regions increases. MRQSA reduces latency by 50.7% on average when compared with C-DCQS under different network densities.

We study the latency of C-DCQS and MRQSA when changing the size of regions as shown in Figure 9.6(c). In this case, the number of sensor nodes is 1500, the number of query regions is 10, and the size of regions changes from 100 m × 100 m to 800 m × 800 m. We can see from Figure 9.6(c) that with the regions' size increases, the latency of both C-DCQS and MRQSA increase. Especially, the incremental ratio of C-DCQS is higher than MRQSA. This is because with the increasing of region size, the overlapped degree of sensor nodes increases, i.e., more sensor nodes or transmissions are involved in more than one query regions. Hence, in order to guarantee a collision free scheduling for an MRQ, C-DCQS requires longer interval time for scheduling of different regions. By contrast, MRQSA shows its advantage of dealing with an MRQ with overlapping, since it concurrently schedules transmissions in all the regions. Furthermore, the transmissions whose senders are in the overlapped regions are scheduled first which reduces the potential collision possibility. MRQSA reduces the latency by 42.7% on average compared with C-DCQS.

We evaluate the impact of interference/transmission range on the latency of C-DCQS and MRQSA as shown in Figure 9.6(d). In this case, the number of the sensor nodes is 1500, the number of query regions is 10, the size of the query regions is 400 m × 400 m and the interference/transmission range changes from 20 m to 100 m. From Figure 9.6(d) we can see that the latency of C-DCQS decreases when the interference/transmission range changes from 20 m to 70 m followed by an increase when the interference/transmission range changes from 70 m to 100 m. The latency of MRQSA shows a similar variation tendency, i.e., the latency of MRQSA decreases when the interference/transmission range changes from 20 m to 60 m followed by an increase when the interference/transmission range changes from 60 m to 100 m. For simplicity, we use the decreasing phase and increasing phase to describe the two phases.

Since the in-network data transmission relies on the multi-hop communication among sensor nodes, when the interference/transmission range is small, the delay caused by multi-hop communication plays a main role in the latency. With the increasing of interference/transmission range, although there are more interferences, the delay caused by multi-hop latency drops with a higher ratio. Hence, the total transmission latency decrease as shown in the decreasing phase in Figure 9.6(d). However, when the interference/transmission range exceeds some threshold, such as 70 m in C-DCQS and 60 m in MRQSA, the in-network data transmission suffers from interference, which results in the increasing phase. In summary, Figure 9.6(d) shows that MRQSA reduces the latency by 51.63% on average compared with C-DCQS.

In order to study the energy performance of MRQSA, we evaluate the energy consumption of C-DCQS and MRQSA with respect to different sizes of query regions as shown in Figure 9.6(e) and Figure 9.6(f). Figure 9.6(e) shows the total energy cost for all the sensor nodes that participate in a data aggregation process. Figure 9.6(f) shows the energy performance ratio which is evaluated as $\xi = \frac{\overline{CEC} - \overline{CEM}}{\overline{CEC}}$, where $\overline{CEC}$ is the average energy cost of C-DCQS and $\overline{CEM}$ is the average energy cost of MRQSA. To be fair, we only consider the energy consumption during the data aggregation process and ignore the cost of the query dissemination process. In this case, the number of the sensor nodes is 1500, the interference/transmission range is 50 m, 1 and 10 query regions are considered, and the size of the query regions changes from 100 $m \times 100\ m$ to 1000 $m \times 1000\ m$. From Figure 9.6(e) and Figure 9.6(f) we can see that the performance of MRQSA in terms of energy consumption is better than C-DCQS. Since C-DCQS is based on a global aggregation tree, when the size of query regions is smaller than the monitored target area, more sensor nodes participate in the data aggregation process. Hence, C-DCQS consumes more energy compared with MRQSA. However, with the increasing of the size of query regions, the difference between global aggregation tree and local aggregation tree decreases. Therefore, the difference of energy consumptions of those two methods decreases.

9.6 Conclusion

In this chapter, we investigate the ML-MRQS problem in WSNs. We first introduce the concept of an MRQ, and then show the challenge and importance of studying the ML-MRQS problem. As the ML-MRQS problem is NP-hard, a heuristic scheduling algorithm, called MRQSA, is proposed. The theoretical analysis shows that MRQSA has a latency bounded by $23A + B + C$, where $A = \max\limits_{i=1}^{m} D_i^{left}$ is the maximum latency for non-overlapped regions, $B = \max\limits_{i=1}^{m}\{(23D_i + 5\Delta + 21)k_i\}$ is the maximum latency for overlapped regions, $C = \sum\limits_{i=1}^{m} H_i + 5\Delta - m + 17$ is the accumulated latency for data transmission from the accessing nodes to the sink, m is the number of regions, Δ is the maximum node degree in the WSN, D_i is the diameter of $\mathcal{R}_i$, k_i is the maximum overlapped degree of sensor nodes in $\mathcal{R}_i$, H_i represents the distance of $\mathcal{R}_i$ with respect to the sink, and D_i^{left} is the diameter of the non-overlapped part of $\mathcal{R}_i$. The simulation results show that MRQSA reduces latency by 49.3% on average when the number of query regions increases, 50.7% on average when the network density increases,

42.7% on average with the increasing of the region size, and 51.63% on average with the interference/transmission range changed compared with C-DCQS, while guaranteeing energy efficiency.

Chapter 10

CDS-based Snapshot and Continuous Data Collection in Dual-radio Multi-channel Wireless Sensor Networks

CONTENTS

10.1 Introduction

Wireless sensor networks (WSNs) are mainly used for collecting data from the physical world. Data gathering can be categorized as *data aggregation* [17]-[113], which obtains aggregated values from WSNs, e.g., maximum, minimum or/and average value of all the data, and *data collection* [1]-[10], which gathers all the data from a network without any data aggregation. For data collection, the union of all the sensing values from all the sensors at a particular time instance is called a *snapshot* [171, 11, 10]. The problem of collecting all the data of one snapshot is called *snapshot data collection* (SDC). Similarly, the problem of collecting multiple continuous snapshots is called *continuous data collection* (CDC). Different from wired networks, WSNs suffer from the interference problem, which degrades the network performance. Consequently, *network capacity*, which can reflect the achievable data transmission rate, is usually used as an important measurement to evaluate network performance. Particularly, for a data collection WSN, we use the average data receiving rate at the sink during the data collection process, referred to as *data collection capacity* [171, 11, 10], to measure its achievable network capacity, i.e., data collection capacity reflects how fast data are collected to the sink. In this chapter, we study the snapshot and continuous data collection problems, as well as their achievable capacities for WSNs.

After the first work [172], extensive works emerged to study the network capacity issue for variety of network scenarios, e.g., multicast capacity [173]-[174], unicast capacity [175], [176], broadcast capacity [177], SDC capacity [1]-[178], [171] *etc.* Most of the previous studies on network capacity are for single-radio single-channel WSNs [1]-[178], [171], [172]-[179], where a network consists of a number of nodes, each with only one radio, and all the nodes communicate over a common single channel. Because of the inherent limitations of such networks, transmissions suffer from the *radio confliction* problem [180]-[181] and the *channel interference* problem [182]-[183], [181] seriously. This degrades network performance significantly. The radio confliction problem is caused by the fact that each node is equipped with only one radio, which means a node can only work on a *half-duplex* mode, i.e., this node cannot receive and transmit data simultaneously. The channel interference problem is caused by all the nodes working over a common channel. When one node transmits data, all the other nodes within its interference radius cannot receive any other data and all the other transmissions interfering with this transmission cannot be carried out simultaneously. Fortunately, many current off-the-shelf sensor nodes are capable of working over multiple orthogonal channels, e.g., IEEE 802.11 b/g standard supports 3 orthogonal channels and IEEE 802.11a standard supports 13 orthogonal channels [184], [185] respectively, which can greatly mitigate the channel interference problem. Furthermore, with the development of hardware technologies and the decreasing hardware cost, a sensor node can be equipped with multiple radios. This helps with solving the radio confliction problem. Therefore, multi-radio multi-channel WSNs are currently becoming more attractive [180]-[181].

Different from the previous works which investigate the capacity issues for single-radio single-channel WSNs, we study the network capacity problem for both CDC and SDC in dual-radio multi-channel WSNs under the *protocol interference model*. Similarly as [171], we define *capacity* as the data rate at the sink to continuously receive data from sensor nodes. We propose two channel scheduling algorithms for both CDC and SDC respectively. The motivation of this chapter lies in the fact that dual-radio multi-channel WSNs can make nodes work in a *full-duplex* manner without incurring high hardware cost, while the channel interference problem can be mitigated significantly. To the best of our knowledge, most of the previous works focus on addressing the SDC capacity problem, while our work is investigating the CDC capacity problem in detail under the protocol interference model. Besides, this chapter is suitable for dual-radio multi-channel WSNs. The main contributions of this chapter are as follows:

1. For the SDC problem in single-radio multi-channel WSNs, we propose a new *multi-path scheduling* algorithm. We prove that this algorithm can achieve the order-optimal network capacity $\Theta(W)$ and has a tighter lower bound $\frac{W}{2\lceil(1.81\rho^2+c_1\rho+c_2)/H\rceil}$ compared with the previously best result in [171], which is $\frac{W}{8\rho^2}$, where W is the channel bandwidth, H is the number of orthogonal channels, ρ is the ratio of the interference radius over the transmission radius of a node, $c_1 = \frac{2\pi}{\sqrt{3}} + \frac{\pi}{2} + 1$, and $c_2 = \frac{\pi}{\sqrt{3}} + \frac{\pi}{2} + 2$.

2. We propose a novel *pipeline scheduling* algorithm that combines *compressive data gathering* (CDG) [1] and *pipeline* together, which significantly improves the CDC capacity for dual-radio multi-channel WSNs. We also prove that the achievable asymptotic network capacity of this algorithm in a long run is $\frac{nW}{12M\lceil(3.63\rho^2+c_3\rho+c_4)/H\rceil}$ when $\Delta_e \leq 12$ or $\frac{nW}{M\Delta_e\lceil(3.63\rho^2+c_3\rho+c_4)/H\rceil}$ when $\Delta_e > 12$, where n is the number of the sensors, M is a constant value and usually $M \ll n$, and Δ_e is the maximum number of the leaf nodes having a same parent in the routing tree (i.e., data collection tree), $c_3 = \frac{8\pi}{\sqrt{3}} + \pi + 2$, and $c_4 = \frac{8\pi}{\sqrt{3}} + 2\pi + 6$. A straightforward upper bound of data collection of a dual-radio WSN is $2W$, since a dual-radio sink can simultaneously receive two packets at most. Due to the benefit brought by the pipeline technique and CDG, analysis shows that our pipeline scheduling algorithm can even achieve a capacity higher than $2W$.

3. For completeness, we also examine the performance of the proposed pipeline

scheduling algorithm in single-radio multi-channel WSNs, denoted by the *single-radio-based pipeline scheduling* algorithm. Theoretical analysis shows that for a long run CDC, the lower bound of the achievable asymptotic network capacity of the single-radio-based pipeline scheduling algorithm for single-radio multi-channel WSNs is $\frac{nW}{16M\lceil(3.63\rho^2+c_3\rho+c_4)/H\rceil}$ when $\Delta_e \leq 12$ or $\frac{nW}{M(\Delta_e+4)\lceil(3.63\rho^2+c_3\rho+c_4)/H\rceil}$ when $\Delta_e > 12$.

4. The simulation results indicate that the proposed algorithms have a better SDC capacity compared with the previously best works. Particularly, when $\rho = 2$ and $H = 3$, for SDC in a WSN with 4000 nodes, the improvements of the capacity of our multi-path scheduling algorithm are 74.3% and 29% compared with BFS [171] and SLR [182] respectively. For CDC in a WSN with 10,000 nodes, our pipeline scheduling algorithm achieves a capacity 7.6 times of that of CDG [1], 22.8 times of that of BFS [171] and 19.4 times of that of SLR [182], respectively.

The rest of this chapter is organized as follows: Section 10.3 introduces the network model and preliminaries. Section 10.2 summarizes the related works. The multi-channel scheduling algorithm for SDC in single-radio multi-channel WSNs is proposed and analyzed in Section 10.4. Section 10.5 presents a novel multi-channel scheduling algorithm for CDC and its theoretical achievable asymptotic network capacity. The simulations to validate the performance of the proposed algorithms are shown in Section 10.6. We conclude this chapter and point out possible future research directions in Section 10.7.

10.2   Related Work

10.2.1   *Capacity for Single-radio Single-channel Wireless Networks*

Following the seminal work [172] by Gupta and Kumar, extensive works emerged to study the network capacity issue. The works in [186] and [187] focus more on the MAC layer to improve the network capacity. In [186], the network capacity with random-access scheduling is investigated. In this work, each link is assigned a channel access probability. Based on which some simple and distributed channel access strategies are proposed. Another similar work is [188], in which the authors studied the capacity of CSMA wireless networks. The authors formulated the models of a series of CSMA protocols and study the capacity of CSMA scheduling versus TDMA scheduling. They also proposed a CSMA scheme which combines a backbone-peripheral routing scheme and a dual carrier-sensing and dual channel scheme. In [189], the authors considered the scheduling problem where all the communication requests are single-hop and all the nodes transmit at a fixed power level. They proposed an algorithm to maximize the number of links in one time slot. Unlike [189], the authors in [187] considered the power-control problem. A family of approximation algorithms were presented to maximize the capacity of an arbitrary wireless networks.

The works in [173], [190], [174], [175], [176], and [191] study the multicast and/or unicast capacity of wireless networks. The multicast capacity for wireless ad hoc networks under the protocol interference model and the Gaussian channel model are investigated in [173] and [190] respectively. In [173], the authors showed that the network multicast capacity is $\Theta(\sqrt{\frac{n}{\log n}} \cdot \frac{W}{k})$ when $k = O(\frac{n}{\log n})$ and is $\Theta(W)$ when $k = \Omega(\frac{n}{\log n})$, where W is the bandwidth of a wireless channel, n is the number of the nodes in a network, and k is the number of the nodes involved in one multicast session. In [190], the authors showed that when $k \leq \theta_1 \frac{n}{(\log n)^{2\alpha+6}}$ and $n_s \geq \theta_2 n^{1/2+\beta}$, the capacity that each multicast session can achieve is at least $c_8 \frac{\sqrt{n}}{n_s \sqrt{k}}$, where k is the number of the receivers in one multicast session, n is the number of the nodes in the network, n_s is the number of the multicast sessions, θ_1, θ_2 and c_8 are constants and β is any positive real number. Another similar work [174] studies the upper and lower bounds of multicast capacity for hybrid wireless networks consisting of ordinary wireless nodes and multiple base stations connected by a high-bandwidth wired network. Considering the problem of characterizing the unicast capacity scaling in arbitrary wireless networks, the authors proposed a general cooperative communication scheme in [175]. The authors also presented a family of schemes that address the issues between multi-hop and cooperative communication when the path loss exponent is greater than 3. In [176], the authors studied the balanced unicast and multicast capacity of a wireless network consisting of n randomly placed nodes, and obtained the characterization of the scaling of the n^2-dimensional balanced unicast and $n2^n$-dimensional balanced multicast capacity regions under the Gaussian fading channel model. A more general (n, m, k)-casting capacity problem was investigated in [191], where n, m and k denote the total number of the nodes in the network, the number of destinations for each communication group, and the actual number of communication-group members that receive information respectively. In [191], the upper and lower bounds for the (n, m, k)-cast capacity were obtained for random wireless networks.

In [192], the authors investigated the network capacity scaling in mobile wireless ad hoc networks under the protocol interference model with infrastructure support. In [193], the authors studied the network capacity of hybrid wireless networks with directional antenna and delay constraints. Unlike previous works, the authors in [194] studied the capacity of multi-unicast for wireless networks from the algorithmic aspects, and they designed provably good algorithms for arbitrary instances. The broadcast capacity of wireless networks under the protocol interference model is investigated in [195], where the authors derived the upper and lower bounds of the broadcast capacity in arbitrary connected networks. When the authors in [196] studied the data gathering capacity of wireless networks under the protocol interference model, they concerned the per source node throughput in a network where a subset of nodes send data to some designated destinations while other nodes serve as relays. To gather data from WSNs, a multi-query processing technology is proposed in [157]. In that work, the authors considered how to obtain data efficiently with data aggregation and query scheduling. Under different communication organizations, the authors in [197] derived the many-to-one capacity bound under the protocol interference model. Another work studied the many-to-one capacity issue for WSNs [198], where the authors used data compression to improve the data gathering efficiency. They also studied the relation between a data compression scheme and the data gathering quality. In [199], the authors studied the scaling laws of WSNs based on an antenna sharing idea. In that work, the authors derived the many-to-one capacity bounds under different power constraints. In [200], the authors studied the multicast capacity of MANETs under the physical interference model, called motioncast. They considered the network capacity of MANETs in two particular situations, which are the LSRM (local-based speed-restricted) model and the GSRM (global-based speed-restricted) model. The multi-unicast capacity of wireless networks is studied in [201] via percolation theory. By applying percolation theory, the authors obtained a tighter capacity bound for arbitrary wireless networks.

The SDC capacity of WSNs is studied in [202], [171], [1], [203], [178], and [204]. In [202], the authors considered the collision-free delay-efficient data gathering problem. Furthermore, they proposed a family of path scheduling algorithms to collect all the data to the sink and obtained the network capacity through theoretical analysis. The authors of [171] extended the work of [202]. They derived tighter upper and lower bounds of the capacity of data collection for arbitrary WSNs. [1] is a work studying how to distribute the data collection task to the entire network to achieve load balancing. In this work, all the sensors transmit the same number of data packets during the data collection process. In [203] and [178], the authors investigated the capacity of data collection for WSNs under protocol interference model and physical interference model, respectively. They proposed a grid partition method which divides the network into small grids to collect data and then derived the network capacity. The worst-case capacity of data collection of a WSN is studied in [204] under the physical and protocol interference models.

The capacity and energy efficiency of wireless ad hoc networks with multi-packet reception under the physical interference model is investigated in [205]. With the multi-packet reception scheme, a tight bound of the network capacity is obtained. Furthermore, the authors showed that a tradeoff can be made between increasing the transport capacity and decreasing the energy efficiency. In [206], a scheduling partition method for large-scale wireless networks is proposed. This method decomposes a large network into many small zones, and then localized scheduling algorithms which can achieve the order optimal capacity as a global scheduling strategy are executed in each zone independently. A general framework to characterize the capacity of wireless ad hoc networks with arbitrary mobility patterns is studied in [207]. By relaxing the "homogeneous mixing" assumption in most existing works, the capacity of a heterogeneous network is analyzed. Another work [179] studies the relationship between the capacity and the delay of mobile wireless ad hoc networks, where the authors studied how much delay must be tolerated under a certain mobile pattern to achieve an improvement of the network capacity.

10.2.2 Capacity for Multi-channel Wireless Networks

Since wireless nodes can be equipped with multiple radios, and each radio can work over multiple orthogonal channels, multi-radio multi-channel wireless networks attract much research interest [208][209][210][211]. In [208], the authors studied the data aggregation issue in multi-channel WSNs under the protocol interference model. Particularly, they designed a constant factor approximation scheme for data aggregation in multi-channel WSNs modeled by unit disk graphs (UDGs). Unlike [208], we study the data collection capacity issue for WSNs. In [209], [210], and [211] the authors investigated the joint channel assignment and routing problem for multi-radio wireless mesh networks, software-defined radio networks, and multi-channel ad hoc wireless networks, respectively. They focused on the channel assignment and routing issues, while in data collection, especially continuous data collection, we focus on how to solve the data accumulation problem at the sensors near the sink to improve the achievable network capacity.

The issue of the capacity of multi-channel wireless networks also attracts a lot of attention [182][212][213] [183][181]. In [182] and [212], the authors studied the connectivity and capacity problem of multi-channel wireless networks. They considered a multi-channel wireless network under constraints on channel switching, proposed some routing and channel assignment strategies for multiple unicast communications and derived the per-flow capacity. The multicast capacity of multi-channel wireless networks is studied in [213]. In this work, the authors represented the upper bound capacity of per multicast as a function of the number of the sources, the number of the destinations per multicast, the number of the interfaces per node, and the number of the available channels. Subsequently, an order-optimal scheduling method is proposed under certain circumstances. In [183], the authors first proposed a multi-channel network architecture, called MC-MDA, where each node is equipped with multiple directional antennas, and then obtained the capacity of multiple unicast communications under arbitrary and random network models. The impact of the number of the channels, the number of the interfaces and the interface switching delay on the capacity of multi-channel wireless networks is investigated in [181]. In this work, the authors derived the network capacity under different situations for arbitrary and random networks.

10.2.3 Remarks

Unlike the above mentioned works, this chapter has the following two main characteristics. First, most of the mentioned works are specifically for single-radio single-channel wireless networks, while this chapter considers the network capacity for dual-radio multi-channel WSNs. Second, this chapter investigates the network capacity for CDC in detail under the protocol interference model, whereas most of the previous works study the network capacity for multicast or/and unicast, etc. The works that study the data collection capacity of wireless networks focus on the SDC problem which is a special case of CDC. Compared with them, the results proposed in this chapter are more universal.

Table 10.1: Notations Used in Chapter.

Notation	Description
$G(V,E)$	The network topology graph, V is the set of all the nodes, E is the set of all the possible links
n	The number of sensors in a WSN
ρ	The interference radius
$\lambda_1, \cdots, \lambda_H$	The H available orthogonal channels
W	The bandwidth of a channel
b	The size of a data packet
t	A time slot
τ	The time consumption of snapshot/continuous data collection
N	The number of snapshots in a continuous data collection
Υ	The snapshot/continuous data collection capacity
D/C	The set of dominators/connectors
G'	The graph constructed by nodes in D
L'	The radius of G'
T	The data collection tree
$\mathfrak{R}(A)$	The conflicting graph of A
$\Delta(\cdot)/\delta(\cdot)$	The maximum/minimum degree of a graph
Δ_e	The maximum number of leaf nodes having a same parent in T
$\delta^*(\cdot)$	The inductivity of a graph
β_r	The number of the dominators within a half-disk with radius r

10.3 Network Model and Preliminaries

In this section, we describe the network model and assumptions, construct the routing tree used for data collection, and introduce some necessary preliminaries. The frequently used notations in this chapter are listed in Table 10.1.

10.3.1 Network Model

We consider a WSN consisting of n sensors and one sink, represented by a connected undirected graph $G = (V,E)$, where V is the set of all the nodes in the network and E is the set of all the possible links among the nodes in V. Every sensor in the WSN produces one packet in a snapshot (defined in the subsequent paragraph). Each sensor has two radios and each radio has a fixed transmission radius normalized to one and a fixed interference radius, denoted by ρ, $\rho \geq 1$. Since we use the protocol interference model, for any receiving node v, v can receive a packet successfully from a transmitting node u if $\|u - v\| \leq 1$ and there is no other node s satisfying $\|s - v\| \leq \rho$ and trying to transmit a packet simultaneously over the same channel with u. Here

$\|-\|$ is the Euclidean distance. Furthermore, we say two links are *interfering* if at least one transmission over them will fail if they transmit data simultaneously. Each radio can work over H orthogonal channels, denoted by $\lambda_1, \lambda_2, \ldots, \lambda_H$ respectively. A fixed data-rate channel model [171] is adopted in this chapter, which means each sensor can transmit at a rate of W bits/second over a wireless channel. The size of all the packets transmitted in the network is set to be b bits. We also assume that all the transmissions are synchronized and the size of a time slot is $t = b/W$ seconds.

We formally define the problem as follows. For a WSN consisting of n sensors and one sink, every sensor produces a data packet with b bits at a particular time instant. The union of all the n data packets produced by the n sensors at a particular time instant is called a *snapshot*. The process to collect all the data of a snapshot to the sink is called *snapshot data collection* (SDC). The *SDC capacity* is defined as $\Upsilon = \frac{nb}{\tau}$, where τ is the time used to collect all the data of a snapshot to the sink, i.e., SDC capacity reflects the average data receiving rate at the sink during SDC. Similarly, the process to collect all the data of N continuous snapshots is called *continuous data collection* (CDC). The *CDC capacity* is defined as $\Upsilon = \frac{Nnb}{\tau}$, where τ now is the time required to collect all the data of these N snapshots to the sink, i.e., CDC capacity reflects the average data receiving rate at the sink during CDC. In this chapter, we study the SDC and CDC problems for WSNs, as well as their achievable network capacities.[1]

10.3.2 Routing Tree

Let $G(V,E)$ be a unit disk graph representing a WSN. We define the sink s_0 as the *center* of G. The *radius* of G with respect to s_0 is the maximum depth of the *breadth-first-search* (BFS) tree rooted at s_0. For a subset U of V, U is a *Dominating Set* (DS) of G if every node in V is either an element of U or adjacent[2] to at least one node in U. If the subgraph of G induced by U is connected, then U is called a *connected dominating set* (CDS) of G. Since CDS can serve as a virtual backbone of a WSN, it received a lot of attention [214, 169, 28, 17] recently.

Taking the WSN shown in Figure 10.1(a) as an example, we build a CDS-based routing tree T (shown in Figure 10.1(d)) using the method proposed in [17]. Let G represent the network in Figure 10.1(a). T is rooted at sink s_0 and can be built according to the following steps. First, construct a BFS tree on G beginning at the sink and obtain a *maximal independent set* (MIS) D according to the search sequence. As shown in Figure 10.1(b), the set of all the black nodes $\{s_0, s_5, s_7, s_9, s_{11}\}$ is a MIS of the network shown in Figure 10.1(a). Note that D is also a DS of G and an element in D is called a *dominator*. Clearly, every dominator is out of the communication range of any other dominators. Let G' be a graph on D in which two nodes in D linked by an edge if and only if these two nodes have a common neighbor in G, e.g., s_0 and s_7. Obviously, sink s_0 is in G' and we also denote s_0 as the center of G'. Suppose

---

[1] In this chapter, we use snapshot/continuous data collection capacity and network capacity interchangeably.

[2] In this chapter, if we say two nodes u and v are adjacent/connected, we mean u and v are within the communication range of each other, i.e. $\|u - v\| \leq 1$.

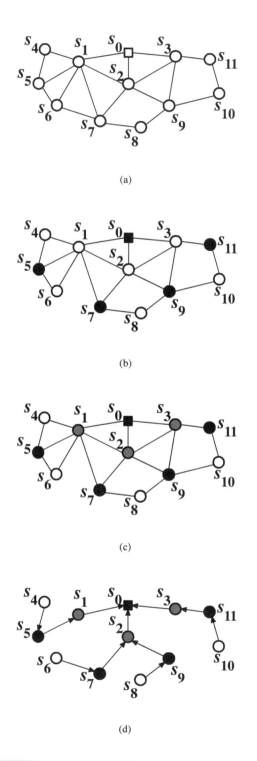

Figure 10.1: The construction of a CDS-based routing tree. s_0 is the sink. The black nodes in (b) are *dominators*, and the gray nodes in (c) are *connectors*.

that the radius of G' with respect to s_0 is L' and we denote the union of dominators at level l $(0 \leq l \leq L')$ as set D_l. Note that $D_0 = \{s_0\}$. Second, we choose nodes, also called *connectors*, to connect all the nodes in D to form a CDS. Let S_l $(0 \leq l \leq L')$ be the set of the nodes adjacent to at least one node in D_l and at least one node in D_{l+1} and compute a minimal cover $C_l \subseteq S_l$ for D_{l+1}. Let $C = \cup_0^{L'-1} C_l$ and therefore $D \cup C$ is a CDS of G. As shown in Figure 10.1(c), the blue nodes $\{s_1, s_2, s_3\}$ are connectors chosen to connect the dominators in $D_0 = \{s_0\}$ and $D_1 = \{s_5, s_7, s_9, s_{11}\}$. Meanwhile, the union of the dominators and connectors in Figure 10.1(c) forms a CDS of the network shown in Figure 10.1(a). Finally, for any other node u, also called a *dominatee*, not belonging to $D \cup C$, choose the nearest dominator as u's parent node. In this way, the routing tree T of G is obtained as shown in Figure 10.1(d).

For each link in T, we assign it a direction from the child node to the parent node along the data transmission flow to the sink as shown in Figure 10.1(d). Furthermore, the receiving (respectively, transmitting) node, i.e., parent (respectively, child) node, of a link is called a *head* (respectively, *tail*). Suppose that A is a set of links of T. The corresponding *conflicting graph* of A is denoted by $\Re(A) = (V_A, E_A)$, where each link in A is abstracted to a node in V_A and two nodes in V_A form an edge in E_A if the corresponding two links of these two nodes are interfering links.

Lemma 10.1 in [17] can be used to derive some useful results of the routing tree T.

Lemma 10.1

[17] *Suppose that O (respectively, O') is a disk (respectively, half-disk) with radius r, and U is a set of points with mutual distances of at least one. Then the number of the points α_r in a disk and the number of the points β_r in a half-disk are*

$$\alpha_r = |U \cap O| \leq \frac{2\pi}{\sqrt{3}} r^2 + \pi r + 1 \tag{10.1}$$

$$\beta_r = |U \cap O'| \leq \frac{\pi}{\sqrt{3}} r^2 + (\frac{\pi}{2} + 1) r + 1. \tag{10.2}$$

From Lemma 10.1, the authors in [17] derived the following properties of the routing tree T. First, for each $0 \leq l \leq L' - 1$, each connector in C_l is adjacent to at most four dominators in D_{l+1}. Second, for each $1 \leq l \leq L' - 1$, each dominator in D_l is adjacent to at most 11 connectors in C_l. Third, $|C_0| \leq 12$.

10.3.3 Vertex Coloring Problem

For a graph $G = (V, E)$, the *maximum degree* (respectively, *minimum degree*) of G is denoted by $\Delta(G)$ (respectively, $\delta(G)$). A subgraph of G on $U \subseteq V$ is denoted by $G(U)$. The *inductivity* of G is defined as $\delta^*(G) = \max_{U \subseteq V} \delta(G(U))$. A *vertex coloring* of G is a scheme of coloring all the vertices in G such that no two adjacent vertices share the same color. The *chromatic number* $\chi(G)$ of G is the least number of colors used to color G. Deciding the lower bound of $\chi(G)$ is a well-known NPC problem.

However, the upper bound of $\chi(G)$ has been derived in *graph theory* [215][17]. The following lemma was proven in [215] and [17].

Lemma 10.2
$\chi(G) \leq 1 + \delta^*(G)$ *and a vertex coloring scheme, called* first-fit coloring, *for G using at most* $1 + \delta^*(G)$ *colors can be found in polynomial time.*

Given a link set A of T, the channel assignment problem for A can be abstracted to the vertex coloring problem for its corresponding conflicting graph $\Re(A)$. If the tail (respectively, head) of every link in A is a dominator, then Lemma 10.3 in [17] gives the upper bound of $\delta^*(A)$.

Lemma 10.3
[17] $\delta^*(A) \leq \beta_{\rho+1} - 1$.

Lemma 10.3 implies that in the worst case, at most $\beta_{\rho+1}$ channels may be assigned to all the links in A without channel interference by a first-fit coloring method.

10.4   Capacity of SDC

In this section, we investigate the traditional SDC problem, propose a scheduling algorithm for this problem in single-radio multi-channel WSNs and analyze the achievable capacity of the proposed algorithm. Subsequently, we point out that the proposed algorithm and most existing works cannot improve the capacity of a network by the pipeline technology.

Since at any time slot, the sink can receive data from at most one neighboring sensor, the upper bound of the SDC capacity is W [171, 11, 10]. Aiming at this upper bound, we design a scheduling algorithm for SDC which is order-optimal and has a tighter lower bound than the previously best result [171].

10.4.1   Scheduling Algorithm for SDC

The idea of *single-path scheduling* has been employed in [202] and [171] to collect data for a WSN. However, their methods have a looser bound of the SDC capacity. In this subsection, we design a new *multi-path scheduling* algorithm based on the routing tree T built in Section 10.3, which is proven to have a better performance. We first study how to schedule a single path and then extend it to the scheduling of multi-path in the routing tree T.

For simplicity, we introduce the concept of *round*. A *round* is a period of time which consists of multiple continuous time slots. We take the path shown in Figure 10.2(a) as an example to explain the idea of the single path scheduling scheme. In Figure 10.2(a), the path, denoted by P, consists of one sink s_0 and three sensors s_1,

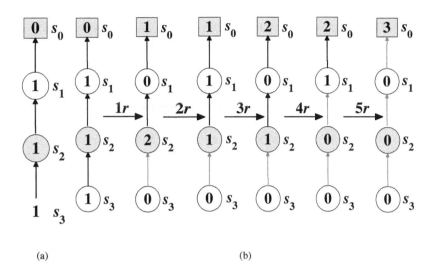

Figure 10.2: (a) A single path and (b) its scheduling (*r = round*).

s_2, and s_3, where s_0 and s_2 are dominators, s_1 is a connector, and s_3 is a dominatee. The value marked in each node is the number of the packets at this node to be transmitted during a time slot. Initially, every sensor on P has one packet and there is no packet at s_0. P_o (respectively, P_e) denotes the set of links on P whose heads (respectively, tails) are dominators and whose tails have at least one packet to be transmitted. For the path shown in Figure 10.2(a), $P_o = \{(s_3, s_2), (s_1, s_0)\}$ and $P_e = \{(s_2, s_1)\}$. We schedule P according to the following two steps and repeat them until all the packets have been collected by s_0.

Step 1: In an odd round, schedule every link in P_o once, i.e. *assign a dedicated channel and one dedicated time slot to each link in P_o.*

Step 2: In an even round, schedule every link in P_e once, i.e. *assign a dedicated channel and one dedicated time slot to each link in P_e.*

The detailed scheduling in Step 1 can be conducted in the following way: first, for any link $\iota_i \in P_o$, let $\mathbb{I}_{P_o}(\iota_i) = \{\iota_j | \iota_j \in P_o, \iota_i \text{ and } \iota_j \text{ are interfering links}\}$; second, sort the links in P_o according to $|\mathbb{I}_{P_o}(\iota_i)|$ ($1 \le i \le |P_o|$) in a non-decreasing order, where $|\cdot|$ denotes the cardinality of a set, and denote the resulting link sequence as $\{\iota_1', \iota_2', \dots, \iota_{|P_o|}'\}$; finally, during the i-th ($1 \le i \le \lceil \frac{|P_o|}{H} \rceil$) time slot of a round, let the j-th ($(i-1)H < j \le iH$) link in P_o work on channel $\lambda_{j\%H+1}$. Here, we sort the links in P_o first and subsequently assign channels based on the *first-fit coloring* scheme in Lemma 10.2. Furthermore, according to Lemma 10.2 and Lemma 10.3, the channel assignment plan for the links in P_o is interference/collision-free. The detailed scheduling in Step 2 is similar to that of Step 1.

The scheduling process of P in Figure 10.2(a) is shown in Figure 10.2(b). During the first (odd) round, links (s_3, s_2) and (s_1, s_0) are scheduled and the packets at s_3

and s_1 are transmitted to their parent nodes. After the first round, s_3 has no packet to transmit. During the second (even) schedule, link (s_2, s_1) is scheduled and s_2 transmits one packet to its parent node. This process continues until all the packets on path P has been transmitted to s_0.

We now consider the scheduling of the routing tree T built in Section 10.3. Suppose that there are m leaf nodes in T denoted by $s_1^l, s_2^l, \cdots, s_m^l$ respectively. The path from leaf node s_i^l ($1 \leq i \leq m$) to the sink s_0 is denoted by P_i. Two paths P_i and P_j are said *intersecting* if they have at least one common node besides the sink node. Assume path P_i and P_j are intersecting, the lowest common ancestor of s_i^l (P_i) and s_j^l (P_j), i.e., the common node of P_i and P_j having the largest number of hops from the sink, is called an *intersecting point* of P_i and P_j. If path P_i intersects with other paths, the route from s_i^l to the nearest intersecting point of P_i is called a *sub-path*, denoted by F_i. Otherwise, F_i is actually P_i.

Taking the routing tree $\widehat{T}$ shown in Figure 10.3(a) as an example, $\widehat{T}$ consists of one sink s_0 and 10 sensor nodes denoted by s_i ($1 \leq i \leq 10$). $\widehat{T}$ has three leaf nodes s_1, s_2, and s_3, which correspond to paths P_1, P_2, and P_3, respectively. In $\widehat{T}$, P_1 and P_2 are intersecting and their intersecting point is s_5. Nevertheless, P_1 and P_3, as well as P_2 and P_3, are not intersecting since they have no common node beside s_0. For P_1, the route from s_1 to s_5 is the sub-path of P_1, denoted by F_1. For P_3, since it is not intersecting with any path, F_3 is P_3 itself.

To schedule multiple paths on the routing tree T, we propose a *multi-path scheduling* algorithm as shown in Algorithm 15. In Algorithm 1, $\mathcal{P}$ is the set of paths been scheduled simultaneously in a round, $\mathcal{S}$ is the set of links from multiple paths that can be scheduled in a round, rd_i/rd_j indicates the number of available rounds that has been assigned to path P_i/P_j, and P_o^i/P_o^j (respectively, P_e^i/P_e^j) is the set of links on P_i/P_j whose heads (respectively, tails) are dominators and whose tails have at least one packet to be transmitted. From Algorithm 1, we can see that lines 2-10 are used to schedule path P_i according to the single-path scheduling algorithm. Lines 11-20 are used to find other paths that can be scheduled simultaneously with P_i according to the single-path scheduling algorithm at the same round.

We further explain the multi-path scheduling algorithm through the routing tree $\widehat{T}$ shown in Figure 10.3(a). Assume the interference radius $\rho = 1$, i.e. the interference radius is equal to the transmission radius, which implies each *round* consists of two time slots. Furthermore, we use $\mathcal{I}(s_i)$ ($1 \leq i \leq n$) to denote the set of sensor nodes that cannot be transmitted data simultaneously with s_i. For the nodes in $\widehat{T}$, we assume $\mathcal{I}(s_1) = \{s_4, s_5, s_6, s_7, s_8\}$, $\mathcal{I}(s_2) = \{s_6, s_7\}$, $\mathcal{I}(s_3) = \{s_9, s_{10}\}$, $\mathcal{I}(s_4) = \{s_1, s_5, s_7, s_8\}$, $\mathcal{I}(s_5) = \{s_1, s_4, s_7, s_8, s_{10}\}$, $\mathcal{I}(s_6) = \{s_1, s_2, s_7, s_8\}$, $\mathcal{I}(s_7) = \{s_1, s_2, s_4, s_5, s_6, s_8\}$, $\mathcal{I}(s_8) = \{s_1, s_4, s_5, s_6, s_7\}$, $\mathcal{I}(s_9) = \{s_3, s_{10}\}$, and $\mathcal{I}(s_{10}) = \{s_3, s_5, s_9\}$. Additionally, for path P_1, $P_o^1 = \{(s_1, s_4), (s_5, s_0)\}$ and $P_e^1 = \{(s_4, s_5)\}$, for path P_2, $P_o^2 = \{(s_2, s_6), (s_7, s_8), (s_5, s_0)\}$ and $P_e^2 = \{(s_6, s_7), (s_8, s_5)\}$, and for path P_3, $P_o^3 = \{(s_3, s_9), (s_{10}, s_0)\}$ and $P_e^3 = \{(s_9, s_{10})\}$. At the beginning of Algorithm 1, the network is shown in Figure 10.3(b) with the number inside each node denoting the number of the data packets at this node. According to the algorithm, during the first round, P_o^1 is scheduled, and path P_2 will not be scheduled since it is intersecting with P_1. P_3 also will not be scheduled since the link (s_{10}, s_0) in P_o^3 and the link (s_5, s_0)

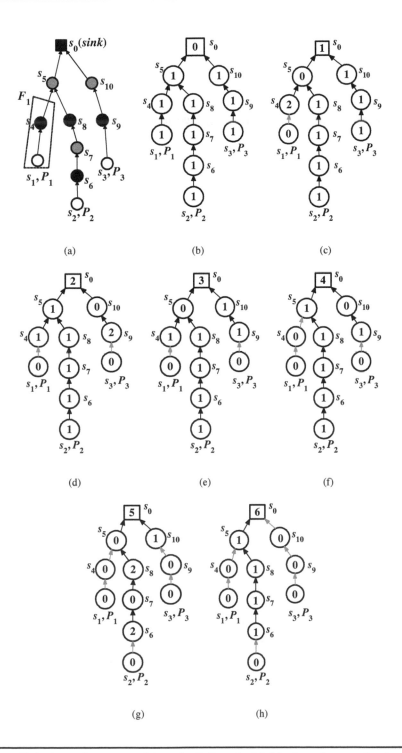

Figure 10.3: A routing tree and its scheduling. In (a), black nodes are dominators, gray nodes are connectors, and the other nodes are dominatees.

Algorithm 15: Multi-path Scheduling Algorithm

---

input : a routing tree T with m leaf nodes
output: a schedule plan for the routing tree T

1 **for** $i = 1; i \leq m; i++$ **do**
2   **while** *there is some data for transmission on F_i* **do**
3         $\mathcal{P} \leftarrow \{P_i\}$;
4         $\mathcal{S} \leftarrow \emptyset$;
5   **if** $rd_i \% 2 == 1$ **then**
6             $\mathcal{S} \leftarrow P_o^i$;
7   rd_i++;
8   **else if** $rd_i \% 2 == 0$ **then**
9             $\mathcal{S} \leftarrow P_e^i$;
10   rd_i++;
11   **for** $j = i+1; j \leq m; j++$ **do**
12   **if** P_j *is not intersecting with any path in $\mathcal{P}$ && there is some data for transmission on F_j* **then**
13   **if** $rd_j \% 2 == 1$ *&& all the transmissions in P_o^j and all the transmissions in $\mathcal{S}$ are interference/collision-free* **then**
14                     $\mathcal{P} \leftarrow \mathcal{P} \cup \{P_j\}$;
15                     $\mathcal{S} \leftarrow \mathcal{S} \cup P_o^j$;
16   rd_j++;
17   **if** $rd_j \% 2 == 0$ *&& all the transmissions in P_e^j and all the transmissions in $\mathcal{S}$ are interference/collision-free* **then**
18                     $\mathcal{P} \leftarrow \mathcal{P} \cup \{P_j\}$;
19                     $\mathcal{S} \leftarrow \mathcal{S} \cup P_e^j$;
20   rd_j++;
21   schedule the links in $\mathcal{S}$ in a round as in the single-path scheduling algorithm;
22   **if** *there is no data for transmission on F_i* **then**
23   remove all the links on F_i from T

---

in P_o^1 are not interference/confliction-free. Thus, after the first round, the network situation is shown in Figure 10.3(c). During the second round, P_e^1 will be scheduled. Now, all the links in P_e^1 and all the links in P_o^3 are conflict/interference-free (Here, we consider P_o^3 instead of P_e^3 is because for path P_3, the current round is the first available round.) Hence, P_1 and P_3 can be scheduled simultaneously at the second round. After the second round, the network situation is shown in Figure 10.3(d). Similarly, according to Algorithm 1, the network after the third, the fourth, the fifth, and the sixth rounds is shown in Figure 10.3(e), (f), (g), and (h), respectively. Finally, for $\widehat{T}$ shown in Figure 10.3(a), it will take 13 rounds to collect all the data packets to the sink by the multi-path scheduling algorithm. By contrast, it will take 18 rounds to collect all the data packets to the sink by the single-path scheduling algorithm.

10.4.2 Capacity Analysis

In this subsection, we analyze the achievable network capacity of the proposed multi-path scheduling algorithm. The upper bound of the SDC capacity is W which has been explained. Consequently, we focus on the lower bound of the SDC capacity. In the worst case, all the paths in the routing tree T are intersecting, i.e. they have a common intersecting point, which means only one path can be scheduled at any time. In order to derive the lower bound of the multi-path scheduling algorithm, we first investigate the number of the rounds needed to finish the scheduling of one single path and then study the number of the time slots in each round. Lemma 10.4 gives the maximum number of the rounds used for the scheduling of one single path.

Lemma 10.4
For a single path P of length L in T, it takes at most $2L - 1$ rounds to collect all the packets on P at the sink node.

Proof: Suppose that the node sequence on P is $s_1, s_2, \ldots, s_L, s_0$, where s_1 is the leaf node (dominatee), and s_0 is the sink node. Considering the building process of T, each link in P has either a dominator head or a dominator tail. According to the scheduling scheme of a single path, during the first (odd) round, the links in P_o are scheduled, which implies each non-dominator with at least one packet transmits this packet to its parent node. After the first round, the sink, receives one packet and all the other dominators of the links in P_o have two packets to be transmitted. During the second (even) round, the links in P_e are scheduled, which implies that every dominator in P_e transmits one packet to its parent node. As a result, the sensor $s_i \ (2 \leq i \leq L)$ has exactly one packet to be transmitted and a new odd-even scheduling round begins. In summary, after every two rounds, the sink receives one packet and the length of the data collection path decreases by 1. Since the length of P is L and s_0 is the destination of all the packets which does not have to transmit any data, it takes at most $2L - 1$ rounds to collect all the packets on P. □

From Lemma 10.4, it is straightforward to obtain the number of the rounds used to collect all the data on the sub-path F of P as shown in Corollary 10.1.

Corollary 10.1

For the sub-path F of length L_s in P , it takes at most $2L_s$ rounds to collect all the packets on F.

Proof. The proof of Corollary 10.1 is similar to that of Lemma 10.4. Note that the intersecting point is not a sink node in this case and thus it needs one round to transmit its packet. □

By Lemma 10.4 and Corollary 10.1, we can obtain the number of the rounds used to collect the packets on a path. The maximum number of the time slots in a round is as follows.

Lemma 10.5

In the single-path scheduling algorithm, a round has at most $\left\lceil \frac{\beta_{\rho+1}}{H} \right\rceil$ time slots, where $\beta_{\rho+1}$ is the number of the dominators in a half-disk with radius $\rho + 1$ and H is the number of available orthogonal channels.

Proof. During every odd (respectively, even) round, the scheduled links are links in P_o (respectively, P_e). Since the heads (respectively, tails) of links in P_o (respectively, P_e) are dominators, we can schedule all the links in P_o (respectively, P_e) in one time slot with at most $\beta_{\rho+1}$ channels in polynomial time by Lemma 10.3 and Lemma 10.2. Now, we have H available channels, which means we can finish the scheduling within $\left\lceil \frac{\beta_{\rho+1}}{H} \right\rceil$ time slots. Therefore, the lemma holds. □

Now we can obtain the lower bound of the achievable capacity of the multi-path scheduling algorithm as shown in Theorem 10.1.

Theorem 10.1

The capacity Υ at the sink of T of the multi-path scheduling algorithm is at least $\frac{W}{2\left\lceil (1.81\rho^2 + c_1\rho + c_2)/H \right\rceil}$, where $c_1 = \frac{2\pi}{\sqrt{3}} + \frac{\pi}{2} + 1$ and $c_2 = \frac{\pi}{\sqrt{3}} + \frac{\pi}{2} + 2$, which is order-optimal.

Proof. Suppose that T has m paths and the length of each path is L_i ($1 \leq i \leq m$). In the worst case, all the m paths cannot be scheduled concurrently. Then by Lemma 10.4, Corollary 10.1 and Lemma 10.5, the total time τ used to collect all the packets of T at the sink is at most $t \cdot \sum_{i=1}^{m} 2L_i \left\lceil \frac{\beta_{\rho+1}}{H} \right\rceil$. According to the multi-path scheduling algorithm, for any path P_i, the time used to collect packets on P_i is equal to the time used to collect packets on the corresponding sub-path F_i of P_i[3]. Therefore,

$$\tau \leq t \cdot \sum_{i=1}^{m} 2L_i \left\lceil \frac{\beta_{\rho+1}}{H} \right\rceil \tag{10.3}$$

---

[3] From lines 2–10 in Algorithm 1, the scheduling of path P_i is stopped when all the data packets on the sub-path F_i have been collected by the sink. As shown in Figures 10.3(f) and (g), after all the data packets on F_1 (the sub-path of P_1) have been collected by the sink, we begin to schedule path P_2. Additionally, based on the definition of a sub-path, F_3 in Figure 10.3 is P_3 itself since P_3 does not intersect with any path. Therefore, the time used to collect packets on P_i is equal to the time used to collect packets on the corresponding sub-path F_i of P_i.

$$= t \cdot \sum_{i=1}^{m} 2 |F_i| \left\lceil \frac{\beta_{\rho+1}}{H} \right\rceil \tag{10.4}$$

$$= 2t \left\lceil \frac{\beta_{\rho+1}}{H} \right\rceil \sum_{i=1}^{m} |F_i|. \tag{10.5}$$

Since the number of the links in T is equal to the number of the sensors in T, $\sum_{i=1}^{m} |F_i| = n$. Then, $\tau \le 2nt \left\lceil \frac{\beta_{\rho+1}}{H} \right\rceil$. Therefore, the capacity

$$\Upsilon = \frac{nb}{\tau} \tag{10.6}$$

$$\ge \frac{nb}{2nt \left\lceil \frac{\beta_{\rho+1}}{H} \right\rceil} \tag{10.7}$$

$$= \frac{b}{2t \left\lceil \frac{\beta_{\rho+1}}{H} \right\rceil} \tag{10.8}$$

$$= \frac{W}{2 \left\lceil \frac{\beta_{\rho+1}}{H} \right\rceil}. \tag{10.9}$$

From, Lemma 10.1, we have

$$\beta_{\rho+1} \le \frac{\pi}{\sqrt{3}} (\rho + 1)^2 + (\frac{\pi}{2} + 1)(\rho + 1) + 1 \tag{10.10}$$

$$= \frac{\pi}{\sqrt{3}} \rho^2 + (\frac{2\pi}{\sqrt{3}} + \frac{\pi}{2} + 1)\rho + \frac{\pi}{\sqrt{3}} + \frac{\pi}{2} + 2 \tag{10.11}$$

$$\approx 1.81\rho^2 + c_1\rho + c_2, \tag{10.12}$$

where $c_1 = \frac{2\pi}{\sqrt{3}} + \frac{\pi}{2} + 1$ and $c_2 = \frac{\pi}{\sqrt{3}} + \frac{\pi}{2} + 2$. This implies

$$\Upsilon \ge \frac{W}{2 \left\lceil \frac{\beta_{\rho+1}}{H} \right\rceil} \tag{10.13}$$

$$\ge \frac{W}{2 \left\lceil \frac{1.81\rho^2 + c_1\rho + c_2}{H} \right\rceil}. \tag{10.14}$$

Since H is a constant and the upper bound of Υ is W, Υ is order-optimal. □

From Theorem 10.1, we know that the achievable capacity of the multi-path scheduling algorithm is order-optimal, and it also has a tighter lower bound compared with the previously best result in [171], which has a lower bound of $\frac{W}{8\rho^2}$.

10.4.3 Discussion

When we address the CDC problem, an intuitive idea is to pipeline the existing SDC operations [171]. Nevertheless, such an idea cannot achieve a better performance.

This is because the sink can receive at most one data packet at a time slot. By pipeline, data transmissions at the nodes far from the sink are really accelerated. However, the fact that a sink can receive at most one packet at each time slot makes the data accumulate at the nodes near the sink. Finally, the network capacity cannot be improved even with pipeline. This motivates us to investigate new methods for CDC.

10.5 Capacity of CDC

Since multi-path scheduling algorithm and existing works with pipeline cannot improve the capacity of CDC, we propose a novel *pipeline scheduling* algorithm based on *compressive data gathering* (CDG) [1] in dual-radio multi-channel WSNs, which augments the CDC capacity significantly. Here we consider dual-radio multi-channel WSNs because dual radios can make a *half-duplex* single-radio node work in a *full-duplex* mode, i.e., a dual-radio node can receive and transmit data simultaneously with the two radios over different channels. Furthermore, the full-duplex working mode is in favor of pipeline. For completeness, we also analyze the achievable network capacity of the pipeline scheduling algorithm (a little modification is needed) in single-radio multi-channel WSNs.

10.5.1 Compressive Data Gathering (CDG)

CDG is first proposed in [1] for SDC in single-radio single-channel WSNs. The basic idea of CDG is to distribute the data collection load uniformly to all the nodes in the entire network. We take the data collection on a path consisting of L sensors $s_1, s_2, \ldots, s_L$ and one sink s_0 as shown in Figure 10.4 [1] as an example to explain CDG. In Figure 10.4, the packet produced at sensor s_j $(1 \leq j \leq L)$ is d_j. In the basic data collection shown in Figure 10.4(a), s_1 transmits one packet d_1 to s_2, s_2 transmits two packets d_1 and d_2 to s_3, and finally all the packets on the path are transmitted to s_0 by s_L. Obviously, nodes near the sink have more transmission load compared with nodes far from the sink in the basic data collection. To balance the transmission load, the authors in [1] proposed the CDG method as shown in Figure 10.4(b). Instead of transmitting the original data directly, s_1 multiplies its data with a random coefficient ϕ_{i1} $(1 \leq i \leq M)$, and sends the M results $\phi_{i1}d_1$ to s_2. Upon receiving $\phi_{i1}d_1$ $(1 \leq i \leq M)$ from s_1, s_2 multiplies its data d_2 with a random coefficient ϕ_{i2} $(1 \leq i \leq M)$, adds it to $\phi_{i1}d_1$, and then sends $\phi_{i1}d_1 + \phi_{i2}d_2$ as one data packet to s_3. Finally, s_L does the similar multiplication and addition and sends the result $\sum_{j=1}^{L} \phi_{ij}d_j$ $(1 \leq i \leq M)$ to s_0. After s_0 receives all the M packets, s_0 can restore the original packets based on the compressive sampling theory [1]. By CDG, all the sensors send M packets to their parent nodes, which achieves the goal to uniformly distribute the data collection task to the entire network. The number of the transmitted packets is $O(n^2)$ in Figure 10.4 (a) and is $O(NM)$ in Figure 10.4 (b), and usually $M \ll n$ for large scale WSNs. Therefore, CDG reduces the number of the transmitted packets.

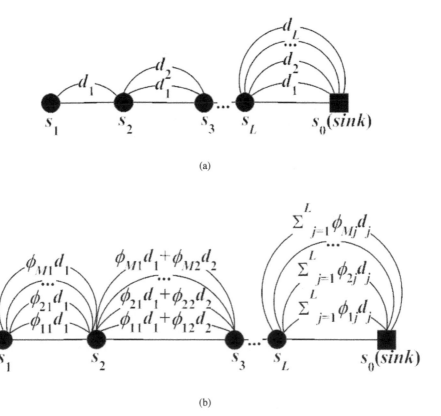

(a)

(b)

Figure 10.4: Comparing (a) basic data collection and (b) CDG [1].

10.5.2 Pipeline Scheduling

Thanks to the benefit brought by CDG, we can address the CDC problem with the pipeline technique. From the building process of the routing tree T, we know that the nodes in T can be divided into sets by levels $D_e, D_{L'}, C_{L'-1}, D_{L'-1}, C_{L'-2}, \ldots, D_1, C_0, D_0 = \{s_0 | s_0 \text{ is the sink}\}$ in a bottom-up way, where D_e is the set of all the dominatees, i.e., leaf nodes, D_i $(0 \leq i \leq L')$ is the set of the dominators at the i-th level, and C_i $(0 \leq i \leq L'-1)$ is the set of the connectors at the i-th level. Since every node has two radios, one radio can be dedicated to receive data and the other dedicated to transmit data. Therefore, the nodes at every level can receive and transmit data simultaneously over different channels. Consequently, for a CDC task consisting of N snapshots, we propose a *pipeline scheduling* algorithm as follows.

Step 1: The nodes at the dominatee level transit data packets to their parent nodes snapshot by snapshot in the CDG way. All the nodes in D_e transmit the packets of the j-th $(1 \leq j \leq N-1)$ snapshot to their parent nodes in the CDG way, i.e., for every node $s \in D_e$, s multiplies its data with M random coefficients respectively, and sends the M products to its parent node. After all the packets of the j-th snapshot have been transmitted successfully, the nodes in D_e immediately transmit the packets of the $(j+1)$-th snapshot in the CDG way.

Step 2: After the nodes at each dominator level receive all the data packets of the j-th snapshot, they transmit the data of the j-th snapshot to their parent nodes in the CDG way. After all the nodes in D_l $(1 \leq l \leq L')$ receive all the packets of the j-th snapshot from their child-level, they send the packets of the j-th snapshot to their parent nodes in the CDG way, i.e., every node $s \in D_l$ combines its packet of the j-th snapshot with the received packets of the j-th snapshot, and sends the M new packets to its parent node. After all the packets of the j-th snapshot have been transmitted successfully, the nodes in D_l immediately transmit the packets of the $(j+1)$-th snapshot to their parent nodes in the CDG way, if they have received all the packets of the $(j+1)$-th snapshot from their child-level.

Step 3: After the nodes at each connector level receive all the data packets of the j-th snapshot, they transmit the data of the j-th snapshot to their parent nodes in the CDG way. After all the nodes in C_l $(0 \leq l \leq L'-1)$ receive all the packets of the j-th snapshot from their child-level, they send the packets of the j-th snapshot to their parent nodes in the CDG way, i.e., every node $s \in C_l$ combines its packet of the j-th snapshot with the received packets of the j-th snapshot, and sends the M new packets to its parent node. After all the packets of the j-th snapshot have been transmitted successfully, the nodes in C_l immediately transmit the packets of the $(j+1)$-th snapshot in the CDG way if they have received all the packets of the $(j+1)$-th snapshot from their child-level.

Step 4: The sink restores the data of a snapshot in the CDG way after it receives all the packets of this snapshot.

Steps 1-4 provide the general frame of our pipeline scheduling scheme. Now, we discuss how to prevent radio confliction and channel interference in Steps 1-3. If two or more nodes have the same parent node, we call them *sibling* nodes. In Steps 1-3, radio confliction may arise if two or more sibling nodes send data to their parent node simultaneously even over different orthogonal channels. This is because every sensor only has one radio dedicated to receiving data. Suppose that there are at most Δ_e (respectively, Δ_d and Δ_c) nodes in D_e (respectively, D_l $(1 \leq l \leq L')$ and C_l $(1 \leq l \leq L' - 1)$) which have the same parent node. Usually, $\Delta_e < \Delta(T)$ except in one-hop WSNs, where any sensor is just one hop away from the sink, $\Delta_e = \Delta(T)$. Then, $\Delta_d \leq 4$ and $\Delta_c \leq 11$ (Note that $|C_0| \leq 12$.) (see Section 10.3.2). To avoid confliction, we divide the nodes in D_e (respectively, D_l $(1 \leq l \leq L')$ and C_l $(1 \leq l \leq L' - 1)$) into Δ_e (respectively, Δ_d and Δ_c) subsets to guarantee that each node belongs to one subset and no sibling nodes belong to the same subset. Then, when we schedule the nodes of each level, we schedule these subsets in a certain order. For the nodes in C_0, we schedule them in a certain order, e.g., the nodes with small IDs are scheduled with high priority.

Different from the multi-path scheduling algorithm, in which a sensor sends one packet over a link in one time slot, we employ the CDG way, where a sensor sends M packets for a snapshot. We now introduce the concept of a *super time slot* (STS) which consists of M time slots. In a STS, a sensor can send M packets over a channel for a snapshot. For the links working simultaneously, we assign channels and STSs in the similar way of the multi-path scheduling algorithm.

10.5.3 Capacity Analysis

In this subsection, we analyze the achievable network capacity of the proposed *pipeline scheduling* algorithm. For completeness, we also analyze the achievable network capacity of the pipeline scheduling algorithm (a little modification is needed since each sensor has one radio now) in single-radio multi-channel WSNs at the end of this subsection.

Lemma 10.6 indicates the inductivity (defined in Section 10.3.3) of the corresponding conflicting graph of the links scheduled simultaneously in the pipeline scheduling algorithm, which is used to obtain the upper bound of the number of the necessary channels to schedule these links.

Lemma 10.6

Suppose that A is the set of the links in T scheduled simultaneously in the pipeline scheduling algorithm, and $\Re(A)$ is the corresponding conflicting graph of A, then, $\delta^*(\Re(A)) \leq 2\beta_{\rho+2} - 1$, where $\delta^*(\Re(A))$ is the inductivity of $\Re(A)$ and $\beta_{\rho+2}$ is the number of the dominators within a half-disk of radius $\rho + 2$.

Proof: Since the sibling nodes at every level have been divided into different subsets and different subsets are scheduled in a certain order, there is no radio confliction among the links in A. Furthermore, for any link in A, either the tail or the head of this link is a dominator according to the building process of the routing tree T. Suppose that A' is a subset of A and e is the link in A' whose tail, denoted by $t(e)$, or head, denoted by $h(e)$, is the bottom most dominator among all the dominators in A'. Then, we prove the number of the links interfered with e, i.e., $\delta(\Re(A'))$, is at most $2\beta_{\rho+2} - 1$ case by case as follows.

Case 1: $t(e)$ is a dominator. In this case, assume that e' is another link in A' interfering with e and $t(e')$ is a dominator. Since $t(e)$ is the bottom most dominator, the necessary condition for e and e' to be interfering links is that $t(e')$ locates at the upper half-disk centered at $t(e)$ with radius $\rho + 1$. On the other hand, if $h(e')$ is a dominator, then the necessary condition for e and e' to be interfering links is that $h(e')$ locates at the upper half-disk centered at $t(e)$ with radius $\rho + 2$. By Lemma 10.1, the number of the dominators within a half-disk of radius $\rho + 2$ is at most $\beta_{\rho+2}$. Since every dominator in A' is associated with at most two links, there are at most $2\beta_{\rho+2}$ links within the half-disk of radius $\rho + 2$ centered at $t(e)$. Therefore, $\delta(\Re(A')) \leq 2\beta_{\rho+2} - 1$, where minus 1 means e is also in the half-disk. As a result, $\delta^*(\Re(A)) = \max\limits_{A' \subseteq A} \delta(\Re(A')) \leq 2\beta_{\rho+2} - 1$.

Case 2: $h(e)$ is a dominator. By the similar method as in Case 1, it can be proven that the conclusion also holds in this case. $\quad\square$

Based on the result of Lemma 10.6, we can determine the number of the STSs used to schedule all the links in A as follows.

Lemma 10.7

For the links of set A in Lemma 10.6, we can use $\left\lceil \frac{2\beta_{\rho+2}}{H} \right\rceil$ STSs to schedule them without channel interference.

Proof: By Lemma 10.6, $\delta^*(\Re(A)) \leq 2\beta_{\rho+2} - 1$. By Lemma 10.2, we can use $1 + \delta^*(\Re(A)) \leq 2\beta_{\rho+2}$ channels to schedule all the links in A in one STS simultaneously. Now, we have H channels, which implies we can schedule all the links in A in $\left\lceil \frac{2\beta_{\rho+2}}{H} \right\rceil$ STSs of H links in each STS. $\quad\square$

From the pipeline scheduling algorithm we know that the transport of subsequent snapshots has some time overlap with the transport of preceding snapshots. Therefore, we first analyze the time used to collect the packets of the first snapshot since it is the base of the pipeline, and then analyze the achievable capacity of the entire pipeline.

Theorem 10.2

The number of the time slots used to collect the packets of the first snapshot by the pipeline scheduling algorithm is at most $M \left\lceil \frac{2\beta_{\rho+2}}{H} \right\rceil (\Delta_e + 15L' + 1)$.

Proof: In Step 1 of the pipeline scheduling algorithm, we divide the nodes in D_e into Δ_e subsets and schedule them in a certain order. By Lemma 10.7, each scheduling uses at most $\left\lceil \frac{2\beta_{\rho+2}}{H} \right\rceil$ STSs. Consequently, Step 1 needs at most $\Delta_e \left\lceil \frac{2\beta_{\rho+2}}{H} \right\rceil$ STSs to finish the scheduling for the first snapshot. In Step 2 (respectively, Step 3), we divide the nodes in D_l ($1 \leq l \leq L'$) (respectively, C_l ($1 \leq l \leq L' - 1$)) into Δ_d (respectively, Δ_c) subsets and schedule them in a certain order. Since, $\Delta_d \leq 4$ (respectively, $\Delta_c \leq 11$), Step 2 (respectively, Step 3) needs at most $4L' \left\lceil \frac{2\beta_\rho + 2}{H} \right\rceil$ (respectively, $11(L'-1) \left\lceil \frac{2\beta_\rho + 2}{H} \right\rceil$) STSs to finish the scheduling for the first snapshot. Furthermore, it needs at most $12 \left\lceil \frac{2\beta_\rho + 2}{H} \right\rceil$ STSs for C_0 to transmit the packets for the first snapshot to the sink. In summary, the total number of the STSs used for the first snapshot is at most

$$\Delta_e \left\lceil \frac{2\beta_\rho + 2}{H} \right\rceil + 4L' \left\lceil \frac{2\beta_\rho + 2}{H} \right\rceil + 11(L'-1) \left\lceil \frac{2\beta_\rho + 2}{H} \right\rceil + 12 \left\lceil \frac{2\beta_\rho + 2}{H} \right\rceil \qquad (10.15)$$

$$= \left\lceil \frac{2\beta_\rho + 2}{H} \right\rceil (\Delta_e + 4L' + 11L' - 11 + 12) \qquad (10.16)$$

$$= \left\lceil \frac{2\beta_\rho + 2}{H} \right\rceil (\Delta_e + 15L' + 1). \qquad (10.17)$$

Since every STS has M time slots, then the number of the time slots used for the first snapshot is at most $M \left\lceil \frac{2\beta_\rho + 2}{H} \right\rceil (\Delta_e + 15L' + 1)$. □

On the basis of the result in Theorem 10.2, we obtain the time slots used to collect all the packets of N continuous snapshots for the pipeline scheduling algorithm as shown in Theorem 10.3.

Theorem 10.3

The time slots used for the pipeline scheduling algorithm to collect N continuous snapshots are at most $M \left\lceil \frac{2\beta_\rho + 2}{H} \right\rceil (\Delta_e + 15L' + 12N - 11)$ when $\Delta_e \leq 12$ or $M \left\lceil \frac{2\beta_\rho + 2}{H} \right\rceil (N\Delta_e + 15L' + 1)$ when $\Delta_e > 12$.

Proof: From the proof of Theorem 10.2 we know, it takes the nodes in D_e (respectively, D_l ($1 \leq l \leq L'$), C_l ($1 \leq l \leq L' - 1$) and C_0) at most $\Delta_e \left\lceil \frac{2\beta_\rho + 2}{H} \right\rceil$ (respectively, $4 \left\lceil \frac{2\beta_\rho + 2}{H} \right\rceil$, $11 \left\lceil \frac{2\beta_\rho + 2}{H} \right\rceil$ and $12 \left\lceil \frac{2\beta_\rho + 2}{H} \right\rceil$) STSs to transmit packets for a snapshot. In order to obtain the upper bound of the number of the time slots used, we assume the STSs used by nodes in D_e (respectively, D_l ($1 \leq l \leq L'$), C_l ($1 \leq l \leq L' - 1$) and C_0) are $\Delta_e \left\lceil \frac{2\beta_\rho + 2}{H} \right\rceil$ (respectively, $4 \left\lceil \frac{2\beta_\rho + 2}{H} \right\rceil$, $11 \left\lceil \frac{2\beta_\rho + 2}{H} \right\rceil$ and $12 \left\lceil \frac{2\beta_\rho + 2}{H} \right\rceil$) in the following proof. Then, we prove Theorem 10.3 by cases.

Case 1: $\Delta_e \leq 4$. For clearness, we use the transmission of two snapshots S-1 and S-2 shown in Figure 10.5(a) as an example for explanation. In Figure 10.5(a), the vertical axis denotes the levels in the routing tree T and the horizontal axis denotes time slots. $t_e = \Delta_e \left\lceil \frac{2\beta_\rho + 2}{H} \right\rceil$, $t_d = 4 \left\lceil \frac{2\beta_\rho + 2}{H} \right\rceil$, $t_c = 11 \left\lceil \frac{2\beta_\rho + 2}{H} \right\rceil$, and $t_0 = 12 \left\lceil \frac{2\beta_\rho + 2}{H} \right\rceil$, respectively. From Figure 10.5(a) we know, the nodes at the D_e-level begin to send packets of S-2 immediately after they send out the packets of S-1. Since $\Delta_e \leq 4$, after the nodes at the $D_{L'}$-level receive all the packets of S-2, they may still be busy with the transmission of the packets of S-1. Nevertheless, from the $C_{L'-1}$-level to the D_1-level, the pipeline can be utilized in a maximum degree, which implies whatever the packets of S-1 or the packets of S-2, they can be sent immediately. After the packets of S-2 are sent from the nodes at the D_0-level to the nodes in C_0, they may have to wait for a while at the nodes of the C_0-level, since the transmission for the packets of S-1 may last as long as $12 \left\lceil \frac{2\beta_\rho + 2}{H} \right\rceil$ STSs. This implies the sink will receive all the packets of S-2 in $12 \left\lceil \frac{2\beta_\rho + 2}{H} \right\rceil$ STSs after it receives all the packets of S-1. According to the description of the pipeline scheduling algorithm, the subsequent snapshots will be transmitted in the same way, which implies the sink will receive all the packets of a snapshot within at most every $12 \left\lceil \frac{2\beta_\rho + 2}{H} \right\rceil$ STSs, after it receives the packets of the first snapshot which takes at most $M \left\lceil \frac{2\beta_\rho + 2}{H} \right\rceil (\Delta_e + 15L' + 1)$ time slots by Theorem 10.2. As a result, the number of time slots used to collect the packets of N continuous snapshots by the pipeline scheduling algorithm is at most

$$M \left\lceil \frac{2\beta_\rho + 2}{H} \right\rceil (\Delta_e + 15L' + 1) + (N - 1) \cdot 12M \left\lceil \frac{2\beta_\rho + 2}{H} \right\rceil \tag{10.18}$$

$$= M \left\lceil \frac{2\beta_\rho + 2}{H} \right\rceil (\Delta_e + 15L' + 1 + 12N - 12) \tag{10.19}$$

$$= M \left\lceil \frac{2\beta_\rho + 2}{H} \right\rceil (\Delta_e + 15L' + 12N - 11). \tag{10.20}$$

Case 2: $4 < \Delta_e \le 11$ and *Case 3:* $\Delta_e = 12$. These two cases can be proven by the similar method used in Case 1. The number of the time slots used to collect the packets of N continuous snapshots is also at most $M \left\lceil \frac{2\beta_\rho + 2}{H} \right\rceil (\Delta_e + 15L' + 12N - 11)$.

Case 4: $\Delta_e > 12$. We use the data transmission of two snapshots shown in Figure 10.5(b) as an example to show the proof. The notations in Figure 10.5(b) are the same as those in Figure 10.5(a). Since $\Delta_e > 12$, the pipeline can be utilized in a maximum degree at the $D_{L'}$-level and continue to the C_0-level. Then, the sink can receive all the packets of a subsequent snapshot every $\Delta_e \left\lceil \frac{2\beta_\rho + 2}{H} \right\rceil$ STSs after it receives the packets of the first snapshot. Therefore, the number of the time slots used to collect the packets of N continuous snapshots is at most

$$M \left\lceil \frac{2\beta_\rho + 2}{H} \right\rceil (\Delta_e + 15L' + 1) + (N - 1) \cdot \Delta_e M \left\lceil \frac{2\beta_\rho + 2}{H} \right\rceil \tag{10.21}$$

$$= M \left\lceil \frac{2\beta_\rho + 2}{H} \right\rceil (\Delta_e + 15L' + 1 + (N - 1)\Delta_e) \tag{10.22}$$

$$= M \left\lceil \frac{2\beta_\rho + 2}{H} \right\rceil (N\Delta_e + 15L' + 1). \tag{10.23}$$

As a conclusion, Theorem 10.3 is true. □

Theorem 10.3 shows the number of the time slots used to collect N continuous snapshots. This prepares us to derive the achievable capacity of the pipeline scheduling algorithm. The lower bound of the achievable CDC capacity in a long-run is given in Theorem 10.4.

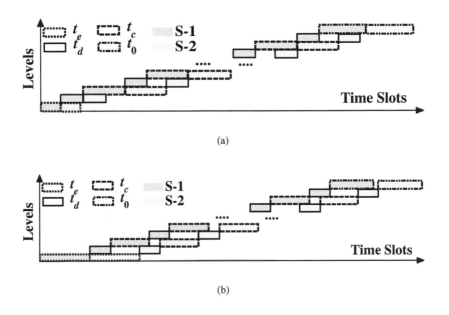

(a)

(b)

Figure 10.5: Data transport in (a) Case 1 and (b) Case 4.

Theorem 10.4

For a long-run CDC, the lower bound of the achievable asymptotic network capacity of the pipeline scheduling algorithm is $\dfrac{nW}{12M\lceil(3.63\rho^2+c_3\rho+c_4)/H\rceil}$ *when* $\Delta_e \leq 12$ *or* $\dfrac{nW}{M\Delta_e\lceil(3.63\rho^2+c_3\rho+c_4)/H\rceil}$ *when* $\Delta_e > 12$, *where* $c_3 = \frac{8\pi}{\sqrt{3}} + \pi + 2$ *and* $c_4 = \frac{8\pi}{\sqrt{3}} + 2\pi + 6$.

Proof. We prove Theorem 10.4 in two cases.

Case 1: $\Delta_e \leq 12$. In this case the number of the time slots used to collect N continuous snapshots is at most $M\left\lceil\frac{2\beta_\rho+2}{H}\right\rceil(\Delta_e+15L'+12N-11)$ as proven in Theorem 10.3. Therefore, the lower bound of the capacity of the pipeline scheduling algorithm is

$$\frac{N \cdot nb}{tM\left\lceil\frac{2\beta_\rho+2}{H}\right\rceil(\Delta_e+15L'+12N-11)} \tag{10.24}$$

$$= \frac{nb}{tM\left\lceil\frac{2\beta_\rho+2}{H}\right\rceil\left(\frac{\Delta_e}{N}+\frac{15L'}{N}+12-\frac{11}{N}\right)} \tag{10.25}$$

$$= \frac{nW}{M\left\lceil\frac{2\beta_\rho+2}{H}\right\rceil\left(\frac{\Delta_e}{N}+\frac{15L'}{N}+12-\frac{11}{N}\right)}. \tag{10.26}$$

When $N \to \infty$, the above equation approaches to $\frac{nW}{12M\left\lceil \frac{2\beta_{\rho+2}}{H} \right\rceil}$. From Lemma 10.1, we have

$$2\beta_{\rho+2} \leq 2[\frac{\pi}{\sqrt{3}}(\rho+2)^2 + (\frac{\pi}{2}+1)(\rho+2)+1] \tag{10.27}$$

$$= \frac{2\pi}{\sqrt{3}}\rho^2 + (\frac{8\pi}{\sqrt{3}}+\pi+2)\rho + \frac{8\pi}{\sqrt{3}}+2\pi+6 \tag{10.28}$$

$$\approx 3.63\rho^2 + c_3\rho + c_4, \tag{10.29}$$

where $c_3 = \frac{8\pi}{\sqrt{3}} + \pi + 2$ and $c_4 = \frac{8\pi}{\sqrt{3}} + 2\pi + 6$. This implies the asymptotic network capacity in this case is $\frac{nW}{12M\lceil(3.63\rho^2+c_3\rho+c_4)/H\rceil}$.

Case 2: $\Delta_e > 12$. The lower bound of the asymptotic network capacity in this case is $\frac{nW}{M\Delta_e\lceil(3.63\rho^2+c_3\rho+c_4)/H\rceil}$, which can be proven similarly as in Case 1. □

In a dual-radio multi-channel WSN, since every node has two radios, the upper bound of the network capacity is $2W$. This is because the sink can receive at most two packets in one time slot. From Theorem 10.4, when $\Delta_e \leq 12$ and $M \leq \frac{n}{24\lceil(3.63\rho^2+c_3\rho+c_4)/H\rceil}$, or $\Delta_e > 12$ and $M \leq \frac{n}{2\Delta_e\lceil(3.63\rho^2+c_3\rho+c_4)/H\rceil}$, the achievable CDC capacity of the pipeline scheduling algorithm is greater than $2W$. By checking the reasons carefully, we find the pipeline scheduling and CDG are responsible for this improvement. By forming a CDG based pipeline, the overlap of gathering multiple continuous snapshots conserves a lot of time, which accelerates the data collection process directly and significantly. These two reasons are also validated by the simulation results in Section 10.6 titled *Simulations and Results Analysis*.

Furthermore, we find that the pipeline scheduling algorithm is more effective for large scale WSNs, since large scale WSNs incur large size routing trees, which are more suitable for pipeline. The pipeline scheduling algorithm is also more effective for a long time CDC, which can also be seen from Theorem 10.4.

For completeness, we also analyze the achievable network capacity of the pipeline scheduling algorithm in single-radio multi-channel WSNs. Now, since each sensor node has one radio, we make some modifications of the pipeline scheduling algorithm as follows. For the nodes in D_e, D_l $(1 \leq l \leq L')$, and C_l $(1 \leq l \leq L'-1)$, instead of transmitting the packets of the $(j+1)$-th $(j \geq 1)$ snapshot immediately after transmitting the packets of the j-th snapshot, they wait until their parent nodes have transmitted all the data packets of the j-th snapshot successfully (Note that the transmission of the data from the first snapshot does not have the waiting process.). For convenience, we refer to the modified pipeline scheduling algorithm as the *single-radio-based pipeline scheduling* algorithm. Then, by the similar proof technique shown in Theorem 10.2, the following lemma can be proven.

Lemma 10.8

The number of the time slots used to collect the packets of the first snapshot by the single-radio-based pipeline scheduling algorithm for single-radio multi-channel WSNs is at most $M \left\lceil \frac{2\beta_{\rho+2}}{H} \right\rceil (\Delta_e + 15L' + 1)$.

On the basis of Lemma 10.8, the number of time slots used to collect all the packets of N continuous snapshots by the single-radio-based pipeline scheduling algorithm is shown in Theorem 10.5.

Theorem 10.5

The time slots used by the single-radio-based pipeline scheduling algorithm to collect N continuous snapshots for single-radio multi-channel WSNs are at most $M \left\lceil \frac{2\beta_\rho + 2}{H} \right\rceil (\Delta_e + 15L' + 16N - 15)$ when $\Delta_e \le 12$ or $M \left\lceil \frac{2\beta_\rho + 2}{H} \right\rceil (N\Delta_e + 15L' + 4N - 3)$ when $\Delta_e > 12$.

Proof: Similar as the analysis in the proof of Theorem 10.3, when $\Delta_e \le 12$, the sink will receive all the packets of a snapshot within every $12 \left\lceil \frac{2\beta_\rho + 2}{H} \right\rceil + 4 \left\lceil \frac{2\beta_\rho + 2}{H} \right\rceil = 16 \left\lceil \frac{2\beta_\rho + 2}{H} \right\rceil$ STSs after it receives the packets of the first snapshot, which implies the number of time slots used to collect the packets of N continuous snapshots by the single-radio-based pipeline scheduling algorithm is at most

$$M \left\lceil \frac{2\beta_\rho + 2}{H} \right\rceil (\Delta_e + 15L' + 1) + (N - 1) \cdot 16 \left\lceil \frac{2\beta_\rho + 2}{H} \right\rceil \tag{10.30}$$

$$= M \left\lceil \frac{2\beta_\rho + 2}{H} \right\rceil (\Delta_e + 15L' + 16N - 15). \tag{10.31}$$

Similarly, when when $\Delta_e > 12$, the sink will receive all the packets of a snapshot within every $\Delta_e \left\lceil \frac{2\beta_\rho + 2}{H} \right\rceil + 4 \left\lceil \frac{2\beta_\rho + 2}{H} \right\rceil = (\Delta_e + 4) \left\lceil \frac{2\beta_\rho + 2}{H} \right\rceil$ STSs after it receives the packets of the first snapshot, which implies the number of time slots used to collect all the packets of N continuous snapshots by the single-radio-based pipeline scheduling algorithm is at most

$$M \left\lceil \frac{2\beta_\rho + 2}{H} \right\rceil (\Delta_e + 15L' + 1) + (N - 1) \cdot (\Delta_e + 4) \left\lceil \frac{2\beta_\rho + 2}{H} \right\rceil \tag{10.32}$$

$$= M \left\lceil \frac{2\beta_\rho + 2}{H} \right\rceil (N\Delta_e + 15L' + 4N - 3). \tag{10.33}$$

□

Therefore, based on Theorem 10.5, the lower bound of the achievable CDC capacity of the single-radio-based pipeline scheduling algorithm in a long-run is shown in Theorem 10.6.

Theorem 10.6

For a long-run CDC, the lower bound of the achievable asymptotic network capacity of the single-radio-based pipeline scheduling algorithm for single-radio multi-channel WSNs is $\frac{nW}{16M \lceil (3.63\rho^2 + c_3\rho + c_4)/H \rceil}$ when $\Delta_e \le 12$ or $\frac{nW}{M(\Delta_e + 4) \lceil (3.63\rho^2 + c_3\rho + c_4)/H \rceil}$ when $\Delta_e > 12$, where $c_3 = \frac{8\pi}{\sqrt{3}} + \pi + 2$ and $c_4 = \frac{8\pi}{\sqrt{3}} + 2\pi + 6$.

Proof: By the similar technique in the proof of Theorem 10.4 and based on Theorem 10.5, this theorem holds.   □

From Theorem 10.5, the capacity improvement ratio of the pipeline scheduling algorithm for multi-radio WSNs compared with the single-radio-based pipeline scheduling algorithm for single-radio WSNs is $\frac{4}{3}$ when $\Delta_e \leq 12$, or $\frac{\Delta_e+4}{\Delta_e}$ when $\Delta_e > 12$.

In summary, we compare the achievable network capacity of the proposed algorithms with the most recently published algorithms for data collection, and the result is shown in Table 10.2.

10.6 Simulations and Results Analysis

We conducted simulations to verify the performances of the proposed algorithms through implementing them with the C language. For all the simulations, we assume every WSN has one sink, and all the sensor nodes of each WSN are randomly distributed in a square area and the communication radius of each node is normalized to one. Suppose the network MAC layer works with TDMA, i.e., the network time can be slotted. Every node produces one data packet in a snapshot and the size of a packet is normalized to one. Every available channel has the same bandwidth normalized to one. For any two different channels, we suppose they are orthogonal, i.e., the communications initialized over any two channels have no wireless interference. Furthermore, we assume a packet can be transmitted over a channel within a time slot.

The compared algorithms are BFS [171], SLR [182] and CDG [1]. BFS is a SDC algorithm based on a *breadth first search* tree and the scheduling is carried out path by path [171]. BFS is specifically proposed for single-radio single-channel WSNs. We extend it to the dual-radio multi-channel scenario in our simulations for fairness. SLR is a *straight-line routing* method for multi-unicast communication in multi-channel wireless networks with channel switching constraints [182]. For data collection, SLR works by setting every sensor having a unicast communication with the sink simultaneously. We also remove the channel switching constraints in SLR for fairness. Furthermore, we also implement the pipelined versions of BFS and SLR (i.e., add the pipeline technique to the data transmission in BFS and SLR), referred to as BFS-P and SLR-P respectively, when evaluating the performance of the proposed pipeline scheduling algorithm. The basic idea of CDG is discussed in Section 10.5. The proposed *multi-path scheduling* algorithm for SDC is referred to as MPS and the proposed *pipeline scheduling* algorithm for CDC is referred to as PS in the following discussions.

In the remainder of this section, we investigate the achievable capacities of MPS and PS through three groups of simulations respectively. In the simulations, H is the number of the available channels, ρ is the interference radius, n is the number of the sensors in a WSN, AR refers to the square area where a WSN is deployed, and N is the number of the snapshots in a CDC task.

Table 10.2: Comparison of Multi-path Scheduling Algorithm, Pipeline Scheduling Algorithm, and Best Existing Works.

Algorithm name	SDC/CDC	IM	RWN/AWN	Υ
Zhu's algorithm [202]	SDC	PrIM	RWN	$\Theta(W)$
Chen's algorithm [171]	SDC	PrIM	RWN, AWN	$\Theta(W)$
Luo's algorithm (CDG) [1]	SDC	PrIM/PhIM	RWN	$\Theta(W)$
Chen's Algorithm [178]	SDC/CDC	PhIM	RWN	$\Omega(W)$
Multi-path scheduling	SDC	PrIM	RWN	$\Omega\left(\dfrac{W}{2\lceil(1.81\rho^2+c_1\rho+c_2)/H\rceil}\right) = \Omega(W)$
Pipeline scheduling	CDC	PrIM	RWN	$\Omega\left(\dfrac{nW}{12M\lceil(3.63\rho^2+c_3\rho+c_4)/H\rceil}\right) = \Omega(\frac{nW}{12M})$; or $\Omega\left(\dfrac{nW}{M\Delta_e\lceil(3.63\rho^2+c_3\rho+c_4)/H\rceil}\right) = \Omega(\frac{nW}{M\Delta_e})$;

IM = interference model. PrIM = protocol interference model. PyIM = physical interference model. RWN = random wireless network. AWN = arbitrary wireless network.

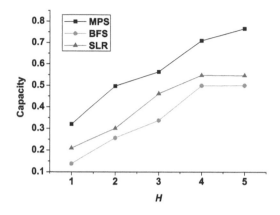

(a) Capacity vs. *H* (ρ=2, *n*=4000, AR=30 × 30)

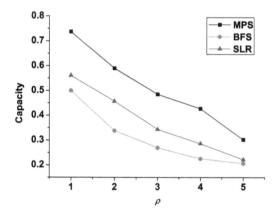

(b) Capacity vs. ρ (*H*=3, *n*=4000, AR=30 × 30)

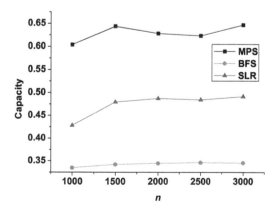

(c) Capacity vs. *n* (ρ=2, *H*=3, AR=20 × 20)

Figure 10.6: SDC capacity.

10.6.1   Performance of MPS

The SDC capacities of MPS, BFS, and SLR in different network scenarios are shown in Figure 10.6. In Figure 10.6(a), the capacity of every algorithm increases when the number of the available channels increases. This is because more available channels enable more concurrent transmissions, which accelerates the data collection process resulting in a higher capacity. After the number of the available channels arrives at four, the capacities of BFS and SLR almost maintain the same level. This is because four channels are enough to prevent channel interference. However, radio confliction becomes the main barrier of a higher capacity at this time. MPS achieves a higher capacity compared with BFS and SLR. This is because MPS simultaneously schedules all the paths without radio conflict (except at the sink). Since radio confliction on a single path can be avoided easily, MPS can simultaneously schedule all the links without radio confliction on multiple paths, which implies MPS can make use of channels in a maximum degree. Whereas, BFS just schedules links without radio confliction on one path every time and SLR schedules all the transmission links simultaneously, which leads to serious radio confliction. On average, MPS achieves 77.49% and 41.95% more capacity than BFS and SLR, respectively.

The effect of the interference radius on the capacity is shown in Figure 10.6(b). With the increase of the interference radius, more transmission interference occurs, which leads to the decrease of the capacities of all the algorithms. Nevertheless, MPS still achieves the largest capacity since it simultaneously schedules multiple paths without radio confliction, which suggests a nice tradeoff between BFS and SLR. On average, MPS achieves 67.45% and 37.37% more capacity than BFS and SLR, respectively.

The effect of the number of the sensors on the capacity is shown in Figure 10.6(c). We can see that the number of the sensors in a network has a little impact on the capacities of MPS and SLR and almost no impact on the capacity of BFS. There are two reasons for this result. First, BFS is a single-path scheduling algorithm. Whatever the number of the sensors is, it schedules only one path every time. Second, the number of the channels is fixed to two in all of these three algorithms. This implies that whatever the number of the sensors is, they can simultaneously schedule at most two interfering links without radio confliction. On average, MPS achieves 83.51% and 32.87% more capacity than BFS and SLR, respectively.

10.6.2   Performance of PS

The CDC capacities of PS, CDG, BFS-P, BFS, SLR-P, and SLR in different network scenarios are shown in Figure 10.7. Figure 10.7(a) (respectively, Figure 10.7(b) and Figure 10.7(c)) and Figure 10.7(d) (respectively, Figure 10.7(e) and Figure 10.7(f)) are the same except we do not show the achievable capacity of PS in Figure 10.7(d) (respectively, Figure 10.7(e) and Figure 10.7(f)). This is mainly for conveniently and clearly checking the achievable capacities of CDG, BFS-P, BFS, SLR-P, and SLR.

Figure 10.7(a) and (d) reflect the effect of the number of the available channels on the achievable CDC capacity. As explained before, the capacities of all the algorithms

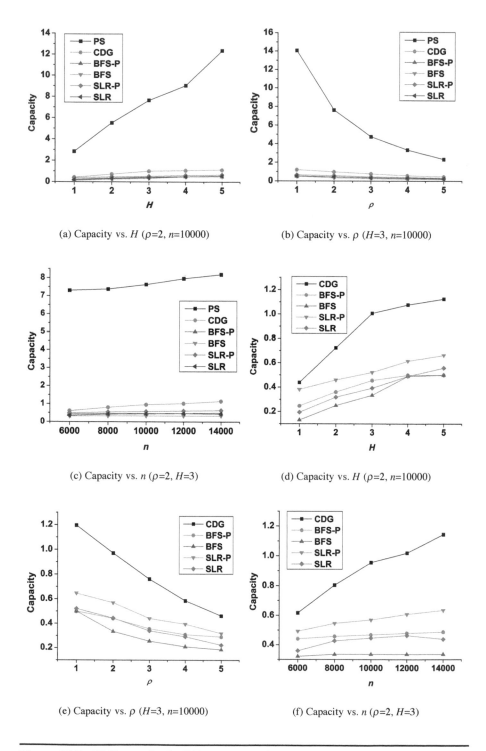

(a) Capacity vs. H (ρ=2, n=10000)

(b) Capacity vs. ρ (H=3, n=10000)

(c) Capacity vs. n (ρ=2, H=3)

(d) Capacity vs. H (ρ=2, n=10000)

(e) Capacity vs. ρ (H=3, n=10000)

(f) Capacity vs. n (ρ=2, H=3)

Figure 10.7: CDC capacity in different scenarios (AR = 50×50, N = 1000, M = 100).

increase as more and more channels are available. This is because more channels can prevent channel interferences among concurrent transmission links. PS has a much higher capacity compared with the other five algorithms. This is because by forming a CDG-based pipeline, the time overlap of gathering multiple continuous snapshots conserves a lot of time, which accelerates the data collection process directly and significantly. Furthermore, PS collects data in the CDG way, which can reduce the overall data transmission times. This also explains why CDG has a higher capacity compared with BFS and SLR. From Figure 10.7(d), we can also see that BFS-P and SLR-R have higher network capacities than BFS and SLR, respectively. This is because the use of the pipeline technique can accelerate the data collection process. On average, PS achieves a capacity 8.22 times of that of CDG, 17.1 times of that of BFS-P, 21.94 times of that of BFS, 13.17 times of that of SLR-P, and 18.39 times of that of SLR, respectively.

The effect of the interference radius on the capacity is shown in Figure 10.7(b) and (e). With the increase of the interference radius, a transmission link will interfere with other transmission links, which leads to the decrease of capacity whatever algorithm it is. PS has the highest capacity among all the algorithms since it works with pipeline and CDG. On average, PS achieves a capacity 7.31 times of that of CDG, 15.9 times of that of BFS-P, 20.43 times of that of BFS, 12.53 times of that of SLR-P, and 16.08 times of that of SLR, respectively.

The effect of the number of the sensors on the capacity is shown in Figure 10.7(c) and (f). The increase of the number of the sensors little impact on BFS and SLR, and the reasons are similar to those explained in the previous subsection, whereas, the capacities of PS and CDG have some improvement with more sensors in a WSN. This is because PS and CDG are more effective for large scale WSNs. On average, PS achieves a capacity 8.77 times of that of CDG, 15.48 times of that of BFS-P, 23.15 times of that of BFS, 12.49 times of that of SLR-P, and 18.06 times of that of SLR, respectively.

10.6.3   Impacts of N and M

In this subsection, we investigate the impacts of the number of the snapshots and the value of M to the capacities of PS and CDG. As shown in Figure 10.8(a), with the increase of N in a CDC task, PS achieves about 87.54% more capacity. This is straightforward from the analysis in Section 10.5. Since PS employs the pipeline technology, the transmissions of continuous snapshots are overlapped, which can significantly reduce the time used to collect all the snapshots data. With more snapshots in a CDC task, the capacity of PS approaches its theoretical asymptotic capacity. For CDG, the number of the snapshots has little impact on its capacity.

Since the performance of CDG is depends on the value of M, the capacities of both PS and CDG decrease about 80% with the increase of the value of M as shown in Figure 10.8(b). This is because a bigger M implies more packets have to be transmitted for every sensor and longer transmission time for each snapshot result. Nevertheless, considering that the value of M is usually much less than n, PS can still achieve a high capacity.

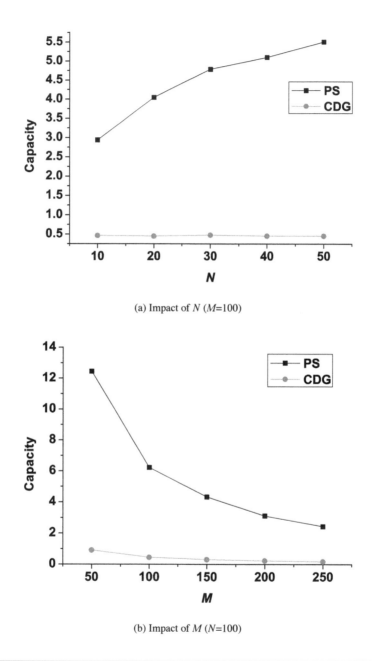

(a) Impact of N (M=100)

(b) Impact of M (N=100)

Figure 10.8: The impacts of N and M to PS and CDG ($\rho = 2$, $H = 3$, $n = 5000$, AR = 50 × 50).

10.7   Conclusion

Motivated by the fact that there exist no works dedicated studying the capacity of CDC for WSNs under the protocol interference model, we investigate this problem in dual-radio multi-channel WSNs. We first propose a *multi-path scheduling* algorithm for the SDC problem and prove that its achievable network capacity is at least $\frac{W}{2\lceil (1.81\rho^2+c_1\rho+c_2)/H \rceil}$, where W is the channel bandwidth, H is the number of available orthogonal channels, ρ is the ratio of the interference radius over the transmission radius of a node and $c_1 = \frac{2\pi}{\sqrt{3}} + \frac{\pi}{2} + 1$, and $c_2 = \frac{\pi}{\sqrt{3}} + \frac{\pi}{2} + 2$. For the CDC problem, we find that pipeline with the existing SDC methods cannot actually improve the network capacity. We explain the reason of this, and then propose a novel CDC method for dual-radio multi-channel WSNs. This method speeds up the data collection process significantly. Theoretical analysis of this method shows that the achievable asymptotic network capacity is $\frac{nW}{12M\lceil (3.63\rho^2+c_3\rho+c_4)/H \rceil}$ when $\Delta_e \leq 12$ or $\frac{nW}{M\Delta_e \lceil (3.63\rho^2+c_3\rho+c_4)/H \rceil}$ when $\Delta_e > 12$, where n is the number of sensors, M is a constant value and usually $M \ll n$, Δ_e is the maximum number of the leaf nodes having a same parent in the routing tree (i.e., data collection tree), $c_3 = \frac{8\pi}{\sqrt{3}} + \pi + 2$, and $c_4 = \frac{8\pi}{\sqrt{3}} + 2\pi + 6$. Furthermore, for completeness, we also analyze the performance of the proposed pipeline scheduling algorithm in single-radio multi-channel WSNs, which shows that for a long-run CDC, the lower bound of the achievable asymptotic network capacity is $\frac{nW}{16M\lceil (3.63\rho^2+c_3\rho+c_4)/H \rceil}$ when $\Delta_e \leq 12$ or $\frac{nW}{M(\Delta_e+4)\lceil (3.63\rho^2+c_3\rho+c_4)/H \rceil}$ when $\Delta_e > 12$. Simulation results indicate that the proposed algorithms improve the network capacity significantly compared with existing works.

Chapter 11

CDS-based Distributed Data Collection in Wireless Sensor Networks

CONTENTS

11.1 Introduction

One of the most important functions provided by wireless sensor networks (WSNs) is directly gathering data from the physical world. Generally, data gathering can be categorized as *data collection* [1][171][10], which gathers all the data from a network without any data aggregation or merging, and *data aggregation* [17]-[18], which obtains some aggregation values, e.g., MAX, MIN, SUM, etc. To evaluate network performance, *network capacity*, which reflects the data transmission/collection/broadcast rate, is usually used, e.g., *multicast capacity* [173], *unicast capacity* [175][176], *broadcast capacity* [177], and *data collection capacity* [1][171][10]. For *data collection capacity*, it is defined as the average data receiving rate at the sink, i.e., data collection capacity reflects how fast data been collected by the sink. We use data collection capacity and network capacity interchangeably throughout this chapter.

Following the seminal work [172] by Gupta and Kumar, many works emerged to study the network capacity issue under various network scenarios, e.g., multicast, unicast, broadcast, and data collection/aggregation. However, to our knowledge, most of the existing works studied the network capacity issue under an ideal assumption that *the network time is slotted, and the entire network is strictly synchronized* explicitly or implicitly, i.e., they are mainly for *centralized synchronous* wireless networks. Under the above ideal assumption, many centralized algorithms with acceptable network capacity bounds are designed and analyzed for various communication modes (e.g., multicast, unicast, broadcast, and data collection/aggregation). In the sense of providing theoretical frameworks/bounds for the design of communication protocols, these works are still sound. However, in practice, wireless networks, especially WSNs, are more likely to be distributed systems. Furthermore, for WSNs, it is difficult and not realistic to achieve ideal strict time synchronization due to the unstable deployment environments, clock drift, and other technical limits. Therefore, to comprehensively and profoundly understand the performance of practical WSNs, it is important to investigate the achievable network capacity of *distributed asynchronous* WSNs. Particularly, we study the achievable data collection capacity for distributed asynchronous WSNs in this chapter.

Different from the study in centralized synchronous WSNs, there are many new challenges arising when investigating the data collection capacity issue for distributed asynchronous WSNs. We summarize the main challenges as follows.

■ **C1:** unlike centralized synchronous WSNs, in which we can acquire the overall information of a network and make an optimized decision for data transmissions, we can only schedule data transmissions according to local information in distributed asynchronous WSNs. Due to this reason, it is very difficult to find an optimal schedule. Therefore, how to design an effective distributed algorithm for data collection is a challenge.

■ **C2:** since we cannot maintain a uniform time clock for all the sensor nodes in distributed asynchronous WSNs, every node carries out data transmissions based on its own time clock and local information. Intuitively, this kind of communication mode leads to many data collisions and retransmissions, incurring capacity degradation, unfairness among data flows, etc. Thus, how to avoid the disadvantages introduced by an asynchronous time scheme is a primary concern when designing distributed data collection algorithms.

■ **C3:** following challenges **C1** and **C2**, the third challenge is how to theoretically analyze the achievable network capacity bounds for a data collection algorithm in distributed asynchronous WSNs. Since the data collection algorithm works in a distributed manner, it is difficult, sometimes even impossible, to know the exact time when a data transmission occurs, as well as the time duration of a data transmission. Hence, both elegant analysis techniques and a carefully designed data transmission mechanism are important to obtain the achievable data collection capacity.

To address these challenges, we propose a scalable and order-optimal distributed algorithm, named *distributed data collection* (DDC), with fairness consideration and capacity analysis under the *generalized physical interference model*. To the best of our knowledge, this is the first attempt to provide detailed protocol design and rigorous capacity analysis for data collection in distributed asynchronous WSNs. DDC works in a CSMA-like manner, except for the RTS/CTS communication mode and the necessity to reply an ACK packet after receiving a data packet. In DDC, when a sensor node has some data packets for transmission, it sets up a backoff timer, and senses the wireless channel with a predefined *carrier-sensing range* (CR). If the channel is free when the backoff timer expires, this node conducts a data transmission. Under this transmission manner, DDC gathers all of the data in a network to the sink (i.e., base station). Moreover, we extend our data collection method to the case of data gathering with aggregation, and propose a *distributed data aggregation* (DDA) algorithm. We summarize the main contributions of this chapter as follows.

1. The carrier-sensing range is an important parameter in DDS, which has a significant impact on the performance of data collection. To avoid data transmission collisions/interference, especially the collisions/interference caused by the *hidden-node problems*, we derive an $\mathcal{R}_0$-*proper carrier-sensing range* ($\mathcal{R}_0$-PCR) under the *generalized physical interference model* for the nodes in a data collection WSN, where $\mathcal{R}_0$ is the *satisfied threshold of data receiving rate*. By taking $\mathcal{R}_0$-PCR as its CR, any node can initiate a data transmission with guaranteed data receiving rate as long as there is no ongoing transmissions within its CR.

2. Based on the obtained $\mathcal{R}_0$-PCR, we propose a scalable and order-optimal *distributed data collection* (DDC) algorithm with fairness consideration for asynchronous WSNs. DDC works in a CSMA-like manner, and effectively gathers all the data to the sink. Theoretical analysis of DDC surprisingly shows that its asymptotic achievable network capacity is $\mathbb{C} = \Omega(\frac{1}{2(\beta_\kappa + \beta_{\kappa+1})} \cdot W)$, where β_x ($x \in \{\kappa, \kappa+1\}$) is a constant value depends on $\mathcal{R}_0$, and W is the bandwidth of a wireless communication channel. Since the upper bound capacity of data collection is $O(W)$ [171][10], which implies the achievable data collection capacity of DDC is order-optimal. Furthermore, since $\mathbb{C}$ is independent of network size, DDC is scalable.

3. For completeness, a *distributed data aggregation* (DDA) algorithm for asynchronous WSNs is designed. We show that the number of time slots induced by DDA is upper bounded by $\log n + (\beta_\kappa + \beta_{\kappa+1} - 1)L + c_3$, where n is the number of the sensor nodes in a WSN, L is the hight of the data aggregation tree, and c_3 is a constant value depending on $\mathcal{R}_0$-PCR.

4. To be more general, we further study the delay and capacity of DDC and DDA under the Poisson node distribution model. By analysis, we demonstrate that DDC is again scalable and order-optimal, and DDA has a delay performance upper bounded by $a \log n + (\beta_\kappa + \beta_{\kappa+1} - 1)L - c_4$, where $a = \arg\min_{\varepsilon > 0} (\frac{2}{\varepsilon} + \frac{\pi \lambda R^2 (e^\varepsilon - 1)}{\varepsilon \log n})$ and $c_4 = \beta_\kappa + \beta_{\kappa+1}$ are constant values.

5. We also conduct extensive simulations to validate the performance of DDC/DDA in distributed asynchronous WSNs. The simulation results indicate that DDC/DDA can achieve comparable data collection capacity as the latest centralized and synchronized data collection algorithm.

The rest of this chapter is organized as follows. In Section 11.2, we summarize the related work and remark the differences between this chapter and the existing works. In Section 11.3, the considered network model is discussed. In Section 11.4, the proper carrier-sensing range satisfying a predefined data receiving rate for communication is derived. According to the obtained proper carrier-sensing range, a distributed asynchronous data collection algorithm is proposed in Section 11.5, followed by the theoretical analysis, which demonstrates that the proposed algorithm can achieve order-optimal data collection capacity as centralized and synchronized algorithms. Furthermore, how to applying the derived proper carrier-sensing range to data aggregation is discussed in Section 11.6. To be more general, we study the delay and capacity of DDC and DDA under the Poisson distribution model in Section 11.7. In Section 11.8, we validate the performance and scalability of DDC and DDA by simulations. Finally, this chapter is concluded and some possible future research directions are pointed out in Section 11.9.

11.2   Related Works

Following the seminal work [172], extensive works emerged to study the network capacity issue for different kinds of wireless networks. In this section, we summarize the existing works according to the communication mode.

11.2.1   Data Collection Capacity

Data collection capacity is studied in [1]–[204] for centralized synchronous wireless networks. In [1], the authors proposed a load-balanced data gathering algorithm. Considering data compression, the authors showed that the network capacity can be improved. In [202], the authors considered the minimum-latency data gathering problem and proposed a family of path scheduling algorithms. The authors in [171], [10] and [216] extended the work in [202], and derived tighter upper and lower data collection bounds. Additionally, the authors of [10] and [216] investigated the achievable capacity bound of continuous data collection in WSNs.

In [203], the authors studied data collection capacity of centralized synchronous WSNs based on a grid partition method. They also obtained the achievable data collection capacity under the protocol interference model. In [178], the authors derived the data collection capacity of WSNs under the physical interference model. Similarly, the authors in [11][217] studied the continuous data collection capacity under the physical interference model. By partitioning the network into square cells and interference zones, the nodes without interference can conduct data transmissions concurrently. The worst-case data collection capacity is studied in [204]. In [73], the authors studied the achievable data collection capacity of probabilistic WSNs. In that work, the impact of lossy links on the degradation of data collection capacity is analyzed and derived. In another work [66], the distributed data collection issue in cognitive radio networks (CRNs) is studied. The authors designed an asynchronous distributed data collection algorithm, which minimizes the data collection delay and meanwhile considers the data transmission fairness. In [218], the authors investigated the achievable data aggregation capacity for centralized synchronous WSNs in the extended network case.

11.2.2   Multicast Capacity

In [173]–[219], the multicast capacity for centralized synchronous wireless networks is studied. The multicast capacity for wireless ad hoc networks under the protocol interference model is investigated in [173]. In [173], the authors showed that the network multicast capacity is $\Theta(\sqrt{\frac{n}{\log n}} \cdot \frac{W}{k})$ when $k = O(\frac{n}{\log n})$ and is $\Theta(W)$ when $k = \Omega(\frac{n}{\log n})$, where W is the bandwidth of a wireless channel, n is the number of the nodes in a network, and k is the number of the nodes involved in one multicast session. In [220], the authors investigated the optimal multicast capacity and delay tradeoffs in mobile ad hoc networks from a global perspective. The general multicast capacity scaling law is studied and summarized under the generalized physical model in a recent work [219].

11.2.3 Uni/Broadcast Capacity

11.2.3.1 Uni/Broadcast Capacity for Random Wireless Networks

The achievable capacity of multiple unicasts in centralized synchronous wireless networks is investigated in [181]-[221]. In [181], the impact of the number of channels, the number of interfaces, and the interface switching delay on the capacity of centralized synchronous wireless networks is investigated. In [176], the authors studied the balanced unicast and multicast capacity of a wireless network consisting of n randomly placed nodes, and obtained the characterization of the scaling of the n^2-dimensional balanced unicast and $n2^n$-dimensional balanced multicast capacity regions under the Gaussian fading channel model.

In [182] and [212], the authors studied connectivity and capacity of multi-channel wireless networks. They considered a multi-channel wireless network with some constraints on channel switching, proposed some routing and channel assignment strategies for multiple unicast communications and derived the per-flow capacity. In [183], the authors first proposed a multi-channel network architecture, called MC-MDA, where each node is equipped with multiple directional antennas, and then obtained the capacity of multiple unicast communications. Similar to [183], the authors in [222] studied the local sufficient rate constraints that can be constructed at each node to ensure a feasible flow allocation for multi-radio multi-channel wireless networks. In [223], the throughput capacity of 3D regular ad hoc networks and 3D heterogeneous ad hoc networks is derived for the first time under the generalized physical interference model. The capacity scaling of multi-hop cellular networks is studied in [221], and the authors further extended their method to study the capacity of heterogeneous multi-hop cellular networks. In [177], the broadcast capacity for ad hoc networks is derived with the fixed data rate channel and the Gaussian channel, respectively.

11.2.3.2 Uni/Broadcast Capacity for Arbitrary Wireless Networks

In [189], the authors considered the scheduling problem where all the communication requests are single-hop and all the nodes transmit at a fixed power level. They proposed an algorithm to maximize the number of concurrent transmitting links in one time-slot. Unlike [189], the authors in [187] and [224] considered a power-control problem. A family of approximation algorithms were presented to maximize the capacity of arbitrary wireless networks. Considering the problem of characterizing the unicast capacity scaling in arbitrary wireless networks, the authors proposed a general cooperative communication scheme in [175].

11.2.3.3 Unicast Capacity for Mobile Wireless Networks

A general framework to characterize the capacity of wireless ad hoc networks with arbitrary mobility patterns is studied in [207]. By relaxing the "homogeneous mixing" assumption in most existing works, the capacity of a heterogeneous network is analyzed. Another work [179] studies the relationship between the capacity and the delay of mobile wireless ad hoc networks, where the authors studied how much delay must be tolerated under a certain mobile pattern to achieve an improvement of the network capacity. Similar to the work in [179] which considers that Lévy mobility and human mobility share several common features, the authors in [225] studied the delay-capacity tradeoffs for mobile wireless networks with Lévy walks and Lévy flights.

11.2.4 Remarks

Compared with the previous works, the following aspects distinguish this chapter from them. To the best of our knowledge, this chapter is the first attempt to address the distributed data collection problem with capacity analysis for asynchronous WSNs, which is more complicated but more practical. As summarized in Section 11.2.1, the existing works study the data collection capacity issue based on centralized and synchronized scheduling/algorithms. More importantly, we propose a scalable and order-optimal asynchronous distributed data collection algorithm. This demonstrates that asynchronous distributed data collection schemes can also achieve order-optimal data collection capacity as synchronized and centralized algorithms do. We study how to apply our derived proper carrier-sensing range to distributed data aggregation and propose a distributed data aggregation algorithm. We also analyze the performance of the proposed algorithm theoretically and validate its performance by simulations.

11.3 Network Model

In this chapter, we consider a connected WSN consisting of one sink node serving as the *base station* denoted by s_0, and n sensor nodes denoted by $s_1, s_2, \cdots, s_n$ respectively, deployed in an area with size $A = c_1 n$, where c_1 is a constant. Furthermore, we assume all the nodes are *independent and identically distributed (i.i.d.)*. Each node is equipped with one radio and works with a fixed power P. All the data transmissions are conducted over a common wireless channel with bandwidth W *bits/second*. The size of a data packet is B bits, and thus the transmission duration of a data packet is $\tau = B/W$ seconds. The maximum transmission radius of a sensor node is set to r (r is associated with the lowest data transmission rate determined by the following defined *generalized physical interference model*). Hence the network can be modeled as a graph $G = (V, E)$, where $V = \{s_i | i = 0, 1, 2, \cdots, n\}$ and E includes all the possible links formed by any pair of nodes in V. A node s_i ($i \in [1, n]$) is said to be *active* at time t *iff* s_i is transmitting a data packet to some other node at time t. Thus, we use $\mathcal{S}^t = \{s_k | s_k \text{ is active at time } t\}$ to denote the set of all the active nodes at time t.

To capture the wireless interference in wireless networks, the *protocol interference model* and *physical interference model* are frequently used. Furthermore, these two models abstract a data transmission as a binary function, with values *successful* or *failed*. Instead of modeling a data transmission process as a binary function, the *generalized physical interference model* (GPI) is more accurate to characterize a practical data transmission. Suppose node s_i is transmitting a data packet to node s_j at time t, i.e. $s_i \in \mathcal{S}^t$, and $\mathcal{R}_{i,j}^t$ is the data receiving rate of s_j from s_i at time t. Then, under the GPI model, $\mathcal{R}_{i,j}^t$ is determined by

$$\mathcal{R}_{i,j}^t = W \cdot \log(1 + SINR_{i,j}^t) \tag{11.1}$$

where $SINR_{i,j}^t$ is the *signal-to-interference-plus-noise ratio* (SINR) value at s_j associated with s_i and is defined as

$$SINR_{i,j}^t = \frac{P \cdot D(s_i, s_j)^{-\alpha}}{N_0 + \sum\limits_{s_k \in \mathcal{S}^t, s_k \neq s_i} P \cdot D(s_k, s_j)^{-\alpha}} \tag{11.2}$$

where N_0 is the background noise, α is the path loss exponent and usually $\alpha \geq 3$, and $D(\cdot, \cdot)$ is the Euclidian distance between two nodes.

Suppose the time consumption to gather all the n data packets produced at s_i ($1 \leq i \leq n$) is $\mathcal{T}$, then the achievable data collection capacity $\mathbb{C}$ can be defined as $nB/\mathcal{T}$, i.e., the data collection capacity reflects how fast that data can be gathered by the sink.

11.4   Carrier-sensing Range

Since we study data collection in distributed asynchronous WSNs, every node s_i ($i \in [1, n]$) in a WSN senses the activities of other nodes within its *carrier-sensing range* (CR) when it has some data packets for transmission. Only when there are no ongoing data transmissions within its CR, s_i can initiate a data transmission. Thus, how to determine the CR for each node, to make all the concurrent transmitters out of the CR of each other to simultaneously conduct data transmissions with a data rate no less than a threshold, is crucial for the performance of a distributed data collection scheme. Intuitively, a small CR implies a high degree of spatial reuse, which further implies small SINR values and followed by low data receiving rates at the receivers. On the other hand, a large CR implies a low degree of spatial reuse, which further implies large SINR values and high data receiving rates. Therefore, in this section, we study how to set a *proper carrier-sensing range* (PCR) for each node to guarantee a satisfied data receiving rate and meanwhile the highest spatial reuse degree. For clarity, we make some definitions as follows.

Definition 11.1 $\mathcal{R}_0$-**feasible state.** The set of all the active nodes $\mathcal{S}^t$ (defined in Section 11.3) is an $\mathcal{R}_0$-feasible state if all the nodes in $\mathcal{S}^t$ can simultaneously transmit data and the data receiving rate at each of their corresponding receivers is no less than $\mathcal{R}_0$. In an $\mathcal{R}_0$-feasible state $\mathcal{S}^t$, $\forall s_i \in \mathcal{S}^t$, assume s_i is transmitting a data packet to s_j, then $\mathcal{R}_{i,j}^t \geq \mathcal{R}_0$.

Based on Definition 11.1, if the lowest tolerable data transmission rate of a WSN is $\mathcal{R}_0$, then the data collection process can be represented as a series of $\mathcal{R}_0$-feasible states $\mathcal{S}^t$ ($t = \tau, 2\tau, 3\tau, \cdots, m\tau$), where $m = \lceil \mathcal{T}/\tau \rceil$.

Definition 11.2   **R-set ($\mathcal{S}_R$).** Assume R is the carrier-sensing range represented by $G = (V, E)$. An R-set, denoted by $\mathcal{S}_R$, is any maximal subset of V that satisfies $\forall s_i, s_j \in \mathcal{S}_R$ ($s_i \neq s_j$) and $D(s_i, s_j) \geq R$.

Definition 11.3   **$\mathcal{R}_0$-Proper Carrier-sensing Range ($\mathcal{R}_0$-PCR).** The carrier-sensing range R of a WSN is an $\mathcal{R}_0$-proper carrier-sensing range if for any R-set $\mathcal{S}_R$, it is always an $\mathcal{R}_0$-feasible state.

From Definition 11.3, if R is an $\mathcal{R}_0$-PCR, then s_i can initiate a data transmission with a guaranteed data receiving rate no less than $\mathcal{R}_0$ as long as there are no other active nodes within R of s_i. Then, given a threshold of data receiving rate $\mathcal{R}_0$, the $\mathcal{R}_0$-PCR can be determined by the following Theorem 11.1. In the following analysis, like that in [226], we assume the background noise is very small compared with the transmission power ($N_0 \ll P$) and thus can be ignored.

Theorem 11.1
$\mathcal{R}_0\text{-PCR} \geq \left(\sqrt[\alpha]{c_2(2^{\mathcal{R}_0/W} - 1)} + 1 \right) \cdot r$, *where c_2 is a constant.*

Proof: Let $R = \mathcal{R}_0$-PCR and $I = R - r$. To make any R-set $\mathcal{S}_R$ always an $\mathcal{R}_0$-feasible state, for $\forall s_i \in \mathcal{S}_R$, assuming its destination node is s_j, then, we have

$$\mathcal{R}_{i,j} \geq \mathcal{R}_0 \tag{11.3}$$

$$\Leftrightarrow W \cdot \log(1 + SINR_{i,j}) \geq \mathcal{R}_0 \tag{11.4}$$

$$\Leftrightarrow 1 + SINR_{i,j} \geq 2^{\mathcal{R}_0/W} \tag{11.5}$$

$$\Leftrightarrow SINR_{i,j} \geq 2^{\mathcal{R}_0/W} - 1 \tag{11.6}$$

$$\Leftrightarrow \frac{P \cdot D(s_i, s_j)^{-\alpha}}{N_0 + P \cdot \sum_{s_k \in \mathcal{S}_R, s_k \neq s_i} D(s_k, s_j)^{-\alpha}} \geq 2^{\mathcal{R}_0/W} - 1 \tag{11.7}$$

$$\Leftrightarrow \frac{D(s_i, s_j)^{-\alpha}}{\sum_{s_k \in \mathcal{S}_R, s_k \neq s_i} D(s_k, s_j)^{-\alpha}} \geq 2^{\mathcal{R}_0/W} - 1 \tag{11.8}$$

Now, we derive the lower bound of $\dfrac{D(s_i, s_j)^{-\alpha}}{\sum_{s_k \in \mathcal{S}_R, s_k \neq s_i} D(s_k, s_j)^{-\alpha}}$. Evidently, $D(s_i, s_j)^{-\alpha} \geq$

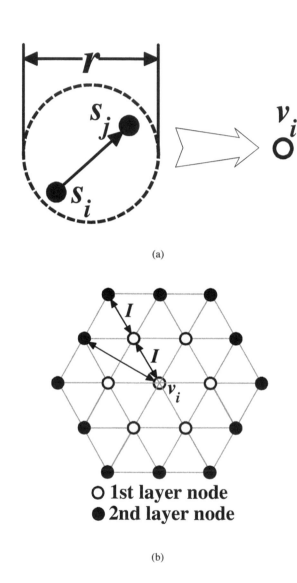

(a)

(b)

Figure 11.1: (a) Link abstraction and (b) hexagon packing.

$r^{-\alpha}$ since r is the maximum transmission range of a node (defined in Section 11.3). Furthermore, if we abstract a data transmission link as a node as shown in Figure 11.1(a), then, for the nodes in $\mathcal{S}_R$, the densest packing of nodes is the hexagon packing [226] with edge length I as shown in Figure 11.1(b). Subsequently, the nodes in $\mathcal{S}_R$ can be layered with respect to v_i (abstracted by the transmission link from s_i to s_j), with the l-th layer having at most $6l$ nodes. Furthermore, the distance between v_i and any node at the l-th layer is no less than $\frac{\sqrt{3}}{2}lI$. Then, we have

$$\sum_{s_k \in \mathcal{S}_R, s_k \neq s_i} D(s_k, s_j)^{-\alpha} \tag{11.9}$$

$$\leq 6 \cdot I^{-\alpha} + \sum_{l \geq 2} 6l \cdot (\frac{\sqrt{3}}{2}lI)^{-\alpha} \tag{11.10}$$

$$= 6 \cdot I^{-\alpha} + 6 \cdot (\frac{\sqrt{3}}{2}I)^{-\alpha} \cdot \sum_{l \geq 2} l^{-\alpha+1} \tag{11.11}$$

In Equation (11.11), $\sum_{l \geq 2} l^{-\alpha+1} = \zeta(\alpha - 1) - 1$, where $\zeta(\cdot)$ is the *Riemann zeta function*. Considering that $\alpha \geq 3$, then $\zeta(\alpha - 1) \leq \zeta(2) = \frac{\pi^2}{6}$. It follows that $\sum_{l \geq 2} l^{-\alpha+1} \leq \frac{\pi^2}{6} - 1$. Thus, we have

$$\sum_{s_k \in \mathcal{S}_R, s_k \neq s_i} D(s_k, s_j)^{-\alpha} \tag{11.12}$$

$$\leq 6 \cdot I^{-\alpha} + 6 \cdot (\frac{\sqrt{3}}{2}I)^{-\alpha} \cdot (\frac{\pi^2}{6} - 1) \tag{11.13}$$

$$= (6 + (\pi^2 - 6)(\frac{\sqrt{3}}{2})^{-\alpha}) \cdot I^{-\alpha} \tag{11.14}$$

$$= c_2 \cdot I^{-\alpha}, \tag{11.15}$$

where $c_2 = (6 + (\pi^2 - 6)(\frac{\sqrt{3}}{2})^{-\alpha})$. It follows that

$$\frac{D(s_i, s_j)^{-\alpha}}{\sum\limits_{s_k \in \mathcal{S}_R, s_k \neq s_i} D(s_k, s_j)^{-\alpha}} \geq \frac{r^{-\alpha}}{c_2 \cdot I^{-\alpha}}. \tag{11.16}$$

Therefore, to make Equation (11.8) valid, it is sufficient to have

$$\frac{r^{-\alpha}}{c_2 \cdot I^{-\alpha}} \geq 2^{\mathcal{R}_0/W} - 1 \tag{11.17}$$

$$\Leftrightarrow I^{-\alpha} \leq \frac{r^{-\alpha}}{c_2(2^{\mathcal{R}_0/W} - 1)} \tag{11.18}$$

$$\Leftrightarrow I \geq (\frac{1}{c_2(2^{\mathcal{R}_0/W} - 1)})^{-1/\alpha} \cdot r \tag{11.19}$$

$$\Leftrightarrow I \geq \sqrt[\alpha]{c_2(2^{\mathcal{R}_0/W} - 1)} \cdot r. \tag{11.20}$$

Therefore,

$$\mathcal{R}_0\text{-PCR} = R \tag{11.21}$$

$$= I + r \geq \sqrt[\alpha]{c_2(2^{\mathcal{R}_0/W} - 1)} \cdot r + r \tag{11.22}$$

$$= (\sqrt[\alpha]{c_2(2^{\mathcal{R}_0/W} - 1)} + 1) \cdot r. \tag{11.23}$$

□

From Theorem 11.1, we know that given a threshold of data receiving rate $\mathcal{R}_0$, we can determine an $\mathcal{R}_0$-PCR, which is at least a constant times r. Since a small CR implies a high degree of spatial reuse, we set $\mathcal{R}_0$-PCR $= (\sqrt[\alpha]{c_2(2^{\mathcal{R}_0/W} - 1)} + 1) \cdot r$. Furthermore, Figure 11.2 depicts the relation between $\mathcal{R}_0$ and $\mathcal{R}_0$-PCR, where the X-axis represents the threshold of data receiving rate $\mathcal{R}_0$, and the Y-axis represents the corresponding $\mathcal{R}_0$-PCR. From Figure 11.2, we can tell with the increase of $\mathcal{R}_0$, the associated $\mathcal{R}_0$-PCR increases accordingly for every α value. This is because a high data receiving rate requires that CR should be sufficiently large to avoid interferences, which also implies a low degree of spatial reuse. Additionally, a large α also implies a small $\mathcal{R}_0$-PCR. This is because the interference impact decreases quickly with the increase of α, which can also be derived from Equation (11.2).

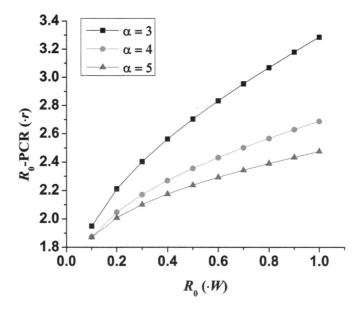

Figure 11.2: $\mathcal{R}_0$ versus $\mathcal{R}_0$-**PCR.**

11.5 Distributed Data Collection and Capacity

According to the obtained $\mathcal{R}_0$-PCR in Section 11.4, if we set the CR of a WSN as $\mathcal{R}_0$-PCR, then all the nodes in an R-set ($R = \mathcal{R}_0$-PCR) can simultaneously transmit data at a guaranteed data receiving rate without interference by letting each node work on the re-start (RS) mode [226]. Thus, in this section, we propose a CSMA-like data collection algorithm for distributed asynchronous WSNs, which has an order-optimal capacity.

11.5.1 Distributed Data Collection

Before presenting the distributed data collection algorithm, for a WSN represented by $G = (V, E)$, we construct a *connected dominating set* (CDS)-based data collection tree, denoted by T, according to the following steps [17]: (*i*) Construct a *breadth-first-search* (BFS) tree in G beginning at the sink s_0, and obtain a *maximal independent set* (MIS) $\mathcal{D}$ according to the search sequence. Note that $\mathcal{D}$ is a *dominating set* (DS) of G. For the nodes in $\mathcal{D}$, we call them *dominators*. Taking the network shown in Figure 11.3(a) as an example, the black nodes are dominators. (*ii*) To form a CDS, we choose the nodes, named *connectors*, as few as possible in V to connect all the nodes in $\mathcal{D}$. The set of all the connectors is denoted by $\mathcal{C}$. As shown in Figure 11.3(a), the blue nodes are the connectors chosen to connect all the dominators. Then, each dominator, except for s_0, has a connector as its parent node in T. On the other hand, each connector has a dominator as its parent node in T. (*iii*) For any other node in $V \setminus (\mathcal{D} \cup \mathcal{C})$, called a *dominatee*, randomly choose a dominator within its communication range as its parent node. Then, the CDS-based data collection tree T rooted at s_0 is formed. For the network shown in Figure 11.3(a), the constructed CDS-based data collection tree is shown in Figure 11.3(b).

Assume L is the height of T, i.e. the maximum number of hops from s_0 to any node, and $L(s_i)$ is the number of hops from node s_i to s_0 in T. Evidently, according to the construction process of T, $\forall s_i \in \mathcal{D}$, $L(s_i)$ is an even number, and $\forall s_j \in \mathcal{C}$, $L(s_j)$ is an odd number. Furthermore, we define $\mathcal{L}_\iota = \{s_i | L(s_i) = \iota\}$ $(0 \leq \iota \leq L)$. Then, the following lemma [17] shows some properties of T.

Lemma 11.1
[17] (*i*) s_0 *is adjacent to at most* 12 *connectors in* $\mathcal{C}$; (*ii*) $\forall s_i \in \mathcal{D}$, $s_i \neq s_0$, s_i *is adjacent to at most* 11 *connectors in* $\mathcal{L}_{L(s_i)+1}$.

Based on T, we propose a *distributed data collection* (DDC) algorithm for asynchronous WSNs as shown in Algorithm 16. In Algorithm 16, $counter(s_i)$ is a counter that denotes the number of data packets transmitted by s_i, τ_w is the *backoff contention window*, and t_i^j $(1 \leq j \leq counter(s_i))$ is the backoff time set for the transmission of the j-th data packet at node s_i. As in [226] and for the same reasons, we assume (*i*) $\tau_w \ll \tau$ such that τ_w is negligible compared with the data transmission time, and (*ii*)

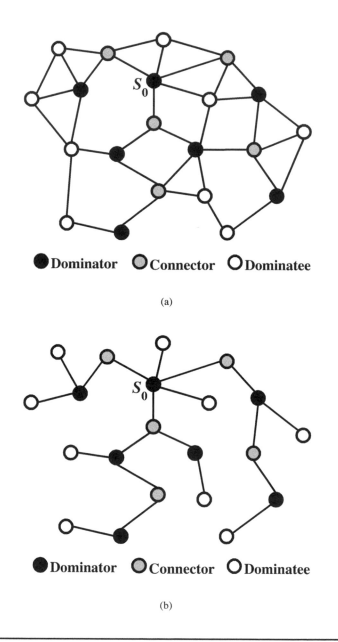

Figure 11.3: A CDS-based data collection tree.

no two transmitters within the CR of each other have their backoff timers expired at the same time instant.[1]

According to Algorithm 16, DDC runs in a CSMA-like manner, except for the RTS/CTS working mode and the necessity to reply an ACK packet after receiving a data packet. By properly setting the CR and working in the RS mode, a transmission with satisfied data receiving rate can be guaranteed as shown in Section 11.4.

In Algorithm 16 (here, taking the algorithm running process at node s_i as an example), Lines 1-5 are basic settings. Line 6 randomly sets the backoff time for each data transmission. In Lines 7-8, the backoff time for each transmission is reset to $(\tau_w - t_i^{j-1}) + t_i^j$, and this is mainly for fairness (any node will not wait too long when it has some data to transmit) as shown in Theorem 11.2 and Corollary 11.2 (see Section 11.5.2). Under this setting, a node cannot transmit multiple data packets in a short time period. Actually, each node can transmit up to one data packet during each backoff contention window. In Lines 9-14, s_i begins the countdown process and keeps sensing the channel with $\mathcal{R}_0$-PCR. If the wireless channel is busy sensed by s_i, the countdown process at s_i will be frozen. In this way, when a data transmission is ongoing, all the other nodes having data packets within the CR of the transmitter will stop their countdown process, i.e., they can share the waiting time. In Lines 15-16, s_i transmits the j-th data packet when the backoff timer expires. Since no two transmitters within the CR of each other have their backoff timers expired at the same time instant, the transmission of the j-th data packet can carried out successfully.

11.5.2 Capacity Analysis

In this subsection, we analyze the achievable data collection capacity of the DDC algorithm. Since the upper bound capacity of data collection is $O(W)$ [171][10], we investigate the lower bound capacity of DDC in this subsection. First, we study the upper bound time consumption to collect all data packets at dominatees to the CDS, i.e., the upper bound time consumption to collect data packets at $V \setminus (\mathcal{D} \cup \mathcal{C})$ to $\mathcal{D} \cup \mathcal{C}$.

Let $R = \mathcal{R}_0\text{-PCR} = (\sqrt[\alpha]{c_2(2^{\mathcal{R}_0/W} - 1)} + 1) \cdot r$, where $\mathcal{R}_0$-PCR is the CR used in DDC. Then, we have the following lemma which indicates the average/upper bound number of the sensor nodes, denoted by $\mathbb{A}/\mathbb{U}$, within the CR of a node.

Lemma 11.2

Let the random variable X denote the number of sensor nodes within the carrier-sensing area of a node. Then,

(i) $\mathbb{A} = \mathbf{E}[X] = \frac{\pi R^2}{c_1}$.

(ii) $\Pr[X > \log n + \frac{\pi R^2(e^2-1)}{2c_1}] \le \Pr[X \ge \log n + \frac{\pi R^2(e^2-1)}{2c_1}] \le \frac{1}{n^2}$. Thus, it is almost impossible that the carrier-sensing area of a node contains more than $\log n + \frac{\pi R^2(e^2-1)}{2c_1}$ nodes, i.e., it is almost sure that $\mathbb{U} = \log n + \frac{\pi R^2(e^2-1)}{2c_1}$.

---

[1]Collisions due to simultaneous countdown-to-zero can be tackled by an exponential backoff mechanism in which the transmission probability of each node is adjusted in a dynamic way based on the network busyness [226].

Algorithm 16: The DDC Algorithm

---

input : CDS-based data collection tree T, $\mathcal{R}_0$-PCR
output: a distributed asynchronous data collection plan

1   $counter(s_i) \leftarrow 0$;

2   $s_i(i \in [1,n])$ sets its CR as $\mathcal{R}_0$-PCR according to the required threshold of data receiving rate $\mathcal{R}_0$;

3   **while** s_i *has some data packets for transmission* **do**

4      $counter(s_i) \leftarrow counter(s_i) + 1$;

5      $j \leftarrow counter(s_i)$;

6   s_i randomly sets a backoff time t_i^j for the transmission of the j-th packet in window $(0, \tau_w]$;

7   **if** $j > 1$ **then**

8         $t_i^j \leftarrow (\tau_w - t_i^{j-1}) + t_i^j$;

9   **while** t_i^j *is not countdown to 0* **do**

10   s_i senses the channel with $\mathcal{R}_0$-PCR;

11   **if** s_i *senses that the channel is busy* **then**

12   s_i stops the countdown process (the backoff timer is frozen) until the channel becomes free again;

13   **if** s_i *senses that the channel is free* **then**

14           $t_i^j --$;

15   **if** $t_i^j == 0$, i.e. *the backoff timer expires* **then**

16   s_i transmits the j-th data packet to its parent node

Proof: Since all the wireless nodes are *i.i.d.* in an area with size $A = c_1 n$, then for any node, it is located at the carrier-sensing area of a particular node with probability $p = \frac{\pi R^2}{c_1 n}$. Then, X satisfies the *binomial distribution* with parameters (n, p). Thus, the average number of the nodes within the carrier-sensing area of a node is $\mathbb{A} = np = \frac{\pi R^2}{c_1}$.

Now, we prove the second statement. Let $a = \log n + \frac{\pi R^2 (e^2 - 1)}{2c_1}$. Then, applying the Chernoff bound and for any $\varepsilon > 0$, we have

$$\Pr[X > a] \leq \Pr[X \geq a] \tag{11.24}$$

$$\leq \min_{\varepsilon > 0} \frac{\mathbf{E}[e^{\varepsilon X}]}{e^{\varepsilon a}} \tag{11.25}$$

$$= \min_{\varepsilon > 0} \frac{[1 + (e^\varepsilon - 1)p]^n}{e^{\varepsilon a}} \tag{11.26}$$

$$\leq \min_{\varepsilon > 0} \frac{e^{(e^\varepsilon - 1)pn}}{e^{\varepsilon a}} \tag{11.27}$$

$$= \min_{\varepsilon > 0} \exp[(e^\varepsilon - 1)pn - \varepsilon a] \tag{11.28}$$

$$= \min_{\varepsilon > 0} \exp[(e^\varepsilon - 1)\mathbb{A} - \varepsilon a]. \tag{11.29}$$

Particularly, let $\varepsilon = 2$, then

$$\Pr[X > a] \tag{11.30}$$

$$\leq \exp[(e^2 - 1)\mathbb{A} - 2a] \tag{11.31}$$

$$= \exp[(e^2 - 1) \cdot \frac{\pi R^2}{c_1} - 2(\log n + \frac{\pi R^2 (e^2 - 1)}{2c_1})] \tag{11.32}$$

$$= \exp[-2\log n] \leq \exp[-2\ln n] \tag{11.33}$$

$$= \frac{1}{n^2}. \tag{11.34}$$

$\sum_{n>0} \frac{1}{n^2}$ is the *Riemann zeta function* with parameter 2, and $\sum_{n>0} \frac{1}{n^2} = \frac{\pi^2}{6} < \infty$. It follows that $\Pr[X \leq a] \approx 1$ according to the Borel-Cantelli lemma, i.e. it is almost sure that the carrier-sensing area of a node contains no more than $\log n + \frac{\pi R^2 (e^2 - 1)}{2c_1}$ nodes. Thus, it is reasonable to use $\log n + \frac{\pi R^2 (e^2 - 1)}{2c_1}$ as the upper bound of the number of the nodes within the carrier-sensing area of a node, i.e. $\mathbb{U} = \log n + \frac{\pi R^2 (e^2 - 1)}{2c_1}$. $\square$

Based on Lemma 11.2, we can derive the upper bound time consumption to collect all the data packets at $V \setminus (\mathcal{D} \cup \mathcal{C})$ to $\mathcal{D} \cup \mathcal{C}$ in DDC.

Theorem 11.2

Any node s_i with data packets for transmission can transmit at least one data packet to its parent node within time $2\mathbb{U}\tau = (2\log n + \frac{\pi R^2 (e^2 - 1)}{c_1})\tau$.

Proof. According to the DDC algorithm, any node s_i with data packets for transmission will carrier-sense the node activities within its CR. When the backoff timer of s_i expires and meanwhile the channel sensed by s_i is free, s_i can transmit a data packet successfully. Thus, the problem now is how long it takes until s_i actually initiates a data transmission in the worst case, i.e., the waiting time of s_i in the worst case. For convenience, assume s_j is any other node within the CR of s_i having data packets for transmission, $t_i, t_j \in (0, \tau_w](t_i \neq t_j)$ are the backoff times for the current data transmissions of s_i and s_j respectively, and $T(U)$, $T(s_i)$ and $T(s_j)$ are the universal time (standard time) and the system time maintained at s_i and s_j respectively. Furthermore, if s_j has more than one data packet for transmission, the backoff time for s_j to transmit a subsequent data packet is denoted by t_{j+1}. Evidently, the transmission sequence of s_i and s_j follows one of the following three cases. (Note that no two transmitters within the CR of each other have their backoff timers expired at the same time instant.)

Case 1: s_i and s_j share a synchronized backoff contention window. In this case, as shown in Figure 11.4(a), s_i will transmit a data packet before/after s_j transmits a data packet. This is because

$$t_{j+1} = t_j + (\tau_w - t_j) + t'_{j+1} \tag{11.35}$$

$$= \tau_w + t'_{j+1} \tag{11.36}$$

$$> t_i, \tag{11.37}$$

where $t'_{j+1} \in (0, \tau]$ is the backoff time chosen by s_j for the subsequent data transmission according to the DDC algorithm.

Case 2: s_i and s_j share an asynchronous backoff contention window and $t_i < t_j$. In this case, as shown in Figure 11.4(b), s_i will transmit a data packet before s_j according to DDC.

Case 3: s_i and s_j share an asynchronous backoff contention window and $t_i > t_j$. In this case, as shown in Figure 11.4(c), when s_i tries to transmit a data packet, it sets a backoff time t_i for the packet and carrier-senses the channel. It turns out that the channel is busy since s_j is transmitting some data. Therefore, we conclude that $0 < t_i - t_j < 2\tau_w$ because the time slots of s_i and s_j have some overlap (otherwise, s_i cannot know that the channel is occupied by s_j when it tries to transmit the data packet). Since $0 < t_i - t_j < 2\tau_w$, it is possible that

$$t_{j+1} = t_j + (\tau_w - t_j) + t'_{j+1} \tag{11.38}$$

$$= \tau_w + t'_{j+1} \tag{11.39}$$

$$< t_i. \tag{11.40}$$

This implies that s_j may transmit two data packets before s_i transmits one data packet. On the other hand, according to the DDC algorithm, we have

$$t_{j+2} = t_j + (\tau_w - t_j) + t'_{j+1} + (\tau_w - t'_{j+1}) + t'_{j+2} \tag{11.41}$$

$$= 2\tau + t'_{j+2} \tag{11.42}$$

$$> t_i, \tag{11.43}$$

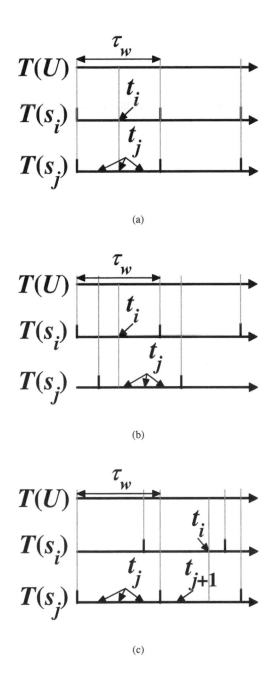

(a)

(b)

(c)

Figure 11.4: Transmission sequence of s_i **and** s_j.

where t_{j+2} is the time that s_j transmits its third data packet and t'_{j+2} is the backoff time set by s_j for its third data packet transmission. Consequently, s_i will transmit one data packet before s_j transmits the third data packet.

In summary, s_j can transmit at most two data packets before s_i transmits one data packet in the worst case. Considering that there are at most $\mathbb{U}$ sensor nodes within the carrier-sensing area of s_i according to Lemma 11.2, s_i can transmit at least one data packet to its parent node within time $2\mathbb{U}\tau$ in the worst case in DDC.   □

Corollary 11.1
In DDC, the time consumption of collecting all the data packets at $V \setminus (\mathcal{D} \cup \mathcal{C})$ to $\mathcal{D} \cup \mathcal{C}$ is at most $2\mathbb{U}\tau$.

Proof: Based on the construction process of the data collection tree T, every node in $V \setminus (\mathcal{D} \cup \mathcal{C})$ has a parent node in $\mathcal{D} \cup \mathcal{C}$. Thus, all the data packets at $V \setminus (\mathcal{D} \cup \mathcal{C})$ can be transmitted to the nodes in $\mathcal{D} \cup \mathcal{C}$ within time $2\mathbb{U}\tau$ according to Theorem 11.2.   □

After time $2\mathbb{U}\tau$, all the data packets at $V \setminus (\mathcal{D} \cup \mathcal{C})$ will be collected to $\mathcal{D} \cup \mathcal{C}$ according to Corollary 11.1. Subsequently, we investigate the time consumption to collect all the data packets at $(\mathcal{D} \cup \mathcal{C}) \setminus \{s_0\}$ to the sink s_0.

Lemma 11.3
[17] *Assume that $\mathcal{X}$ is a disk of radius r_d and $\mathcal{M}$ is a set of points with mutual distance of at least 1. Then $|\mathcal{X} \cap \mathcal{M}| \leq \frac{2\pi r_d^2}{\sqrt{3}} + \pi r_d + 1$.*

Let $\kappa = \sqrt[\alpha]{c_2(2^{\mathcal{R}_0/W} - 1)} + 1$. It follows that $\mathcal{R}_0\text{-PCR} = \kappa \cdot r$. Then, we can obtain the following lemma by applying Lemma 11.3.

Lemma 11.4
Assume that $\mathcal{X}$ is a disk of radius $\mathcal{R}_0\text{-PCR}$, then $|\mathcal{X} \cap (\mathcal{D} \cup \mathcal{C})| \leq \beta_\kappa + \beta_{\kappa+1}$, where $\beta_x = \frac{2\pi x^2}{\sqrt{3}} + \pi x + 1$, i.e. the number of dominators and connectors within the CR of a node is at most $\beta_\kappa + \beta_{\kappa+1}$ in DDC.

Proof: Since $\mathcal{X}$ is a disk of radius $\mathcal{R}_0\text{-PCR}$, it is possible for some connectors in $\mathcal{X}$ only connecting some dominators out of disk $\mathcal{X}$ as shown in Figure 11.5. On the other hand, all the dominators adjacent to the connectors in $\mathcal{X} \cap \mathcal{C}$ must locate in a concentric disk of $\mathcal{X}$ with radius $\mathcal{R}_0\text{-PCR} + r = (\kappa + 1)r$, denoted by $\mathcal{X}'$ as shown in Figure 11.5.

Now, if r is normalized to 1, then $\mathcal{X}$ (respectively, $\mathcal{X}'$) is a disk of radius κ (respectively, $\kappa + 1$), and $\mathcal{D}$ is a set of nodes with mutual distance of at least 1. Then, by Lemma 11.3, we have $|\mathcal{X} \cap \mathcal{D}| \leq \beta_\kappa = \frac{2\pi \kappa^2}{\sqrt{3}} + \pi \kappa + 1$ (respectively, $|\mathcal{X}' \cap \mathcal{D}| \leq \beta_{\kappa+1} = \frac{2\pi(\kappa+1)^2}{\sqrt{3}} + \pi(\kappa + 1) + 1$), i.e., the number of the dominators within $\mathcal{X}$ (respectively, $\mathcal{X}'$) is at most β_κ (respectively, $\beta_{\kappa+1}$). Additionally, according to the aforementioned discussion and the CDS-based data collection tree construction

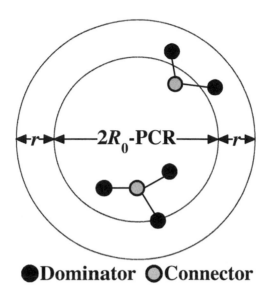

●Dominator ○Connector

Figure 11.5: The number of dominators and connectors within the CR of a node.

---

process, each connector in $\mathcal{X} \cap \mathcal{C}$ must have a dominator parent located at disk $\mathcal{X}'$, which implies $|\mathcal{X} \cap \mathcal{C}| \leq |\mathcal{X}' \cap \mathcal{D}| \leq \beta_{\kappa+1}$. It follows that $|\mathcal{X} \cap (\mathcal{D} \cup \mathcal{C})| \leq \beta_\kappa + \beta_{\kappa+1}$ is proven. □

From Lemma 11.4, we can obtain the following corollary.

Corollary 11.2

After time $2\mathbb{U}\tau$, every node in $(\mathcal{D} \cup \mathcal{C}) \setminus \{s_0\}$ with data packets for transmission can transmit at least one data packet to its parent node within time $2(\beta_\kappa + \beta_{\kappa+1})\tau$ in DDC.

Proof: According to Lemma 11.4, there are at most $\beta_\kappa + \beta_{\kappa+1}$ dominators and connectors within the CR of a node. Furthermore, all the nodes in $V \setminus (\mathcal{D} \cup \mathcal{C})$ have no data packets for transmission after time $2\mathbb{U}\tau$ according to Corollary 11.1. Then, by the same technique used to prove Theorem 11.2, the conclusion of this corollary can be obtained. □

Based on Lemma 11.4 and Corollary 11.2, we can obtain the time consumption to collect all the data packets at $(\mathcal{D} \cup \mathcal{C}) \setminus \{s_0\}$ to the sink s_0 as shown in Theorem 11.3.

Theorem 11.3

After time $2\mathbb{U}\tau$, it takes at most $2(n - \Delta_0) \cdot (\beta_\kappa + \beta_{\kappa+1}) \cdot \tau$ time to collect all the data packets at $(\mathcal{D} \cup \mathcal{C}) \setminus \{s_0\}$ to the sink s_0 in DDC, where Δ_0 is the degree of s_0 in the data collection tree T.

Proof: As shown in Corollary 11.1, after time $2\mathbb{U}\tau$, all the nodes in $V \setminus (\mathcal{D} \cup \mathcal{C})$ have no data packets for transmission, and meanwhile, s_0 has received at least Δ_0 data packets according to Theorem 11.2, since it has Δ_0 child nodes in T. Subsequently, s_0 receives at least one data packet in every $2(\beta_\kappa + \beta_{\kappa+1})\tau$ time according to Corollary 11.2. Thus, it takes at most $(n - \Delta_0) \cdot 2(\beta_\kappa + \beta_{\kappa+1})\tau$ time to collect all the data packets at $(\mathcal{D} \cup \mathcal{C}) \setminus \{s_0\}$ to the sink s_0 after time $2\mathbb{U}\tau$. □

Theorem 11.4

The lower bound of data collection capacity achieved by DDC is $\Omega(\frac{1}{2(\beta_\kappa + \beta_{\kappa+1})} \cdot W)$, which is scalable and order-optimal.

Proof: According to Theorem 11.2 and Theorem 11.3, to collect all the n data packets to the sink, the time consumption

$$\mathcal{T} \leq 2\mathbb{U}\tau + 2(n - \Delta_0) \cdot (\beta_\kappa + \beta_{\kappa+1}) \cdot \tau \tag{11.44}$$

$$= [(2\log n + \frac{\pi R^2 (e^2 - 1)}{c_1}) + 2(n - \Delta_0) \cdot (\beta_\kappa + \beta_{\kappa+1})] \cdot \tau \tag{11.45}$$

$$\leq [2\log n + \frac{\pi R^2 (e^2 - 1)}{c_1} + 2(\beta_\kappa + \beta_{\kappa+1})n] \cdot \tau \tag{11.46}$$

$$= O(2(\beta_\kappa + \beta_{\kappa+1})n \cdot \tau). \tag{11.47}$$

Thus, the achievable data collection capacity of DDC is

$$\mathbb{C} = \frac{nB}{\mathcal{T}} \tag{11.48}$$

$$\geq \frac{nB}{O(2(\beta_\kappa + \beta_{\kappa+1})n \cdot \tau)} \tag{11.49}$$

$$= \Omega(\frac{1}{2(\beta_\kappa + \beta_{\kappa+1})} \cdot W). \tag{11.50}$$

As mentioned before, the upper bound capacity of data collection is $O(W)$ [171, 10], and $\beta_{\kappa+1}$ is a constant value depending on $\mathcal{R}_0$, which implies the achievable data collection capacity of the DDC algorithm is order-optimal. Furthermore, since $\mathbb{C}$ is independent of network size n, DDC is scalable. □

11.6 $\mathcal{R}_0$-PCR-based Distributed Data Aggregation

As introduced in Section 11.1, data gathering can be categorized as data collection and data aggregation. Therefore, for completeness, we in this section discuss how to apply the derived proper carrier-sensing range $\mathcal{R}_0$-PCR to distributed data aggregation in WSNs.

In data aggregation, multiple data packets can be aggregated into one data packet by applying an aggregation function, e.g. MAX, MIN, SUM, etc. Formally, the data aggregation problem can be defined as follows. Let $X, Y \subseteq V$ and $X \cap Y = \emptyset$. The data of the nodes in X is said to be *aggregated* to the nodes in Y in a time slot, if all the nodes in X can transmit their data packets to the nodes in Y concurrently and interference-freely during a time slot. Here, X is called a *transmitter set*. Then, the data aggregation problem can be defined as seeking a *data aggregation schedule* which consists of a sequence of transmitter sets $X_1, X_2, \cdots, X_M$, such that

1. $\forall 1 \leq i \neq j \leq M, X_i \cap X_j = \emptyset$;

2. $\bigcup_1^M X_i = V \setminus \{s_0\}$, where M is the latency of this data aggregation schedule;

3. Data can be aggregated from X_i to $V \setminus \bigcup_{j=1}^i X_j$ during time slot i for $i = 1, 2, \cdots, M$.

Ever since the data aggregation problem was raised, extensive research has been conducted on this issue ([17],[109]-[227], and references therein), especially for the minimum-latency aggregation schedule (MLAS) problem, which tries to obtain a data aggregation schedule with the objective to minimize the latency (minimize M). In [109], [113] and [17], several centralized data aggregation algorithms are proposed under the unit disk graph (UDG) model and the protocol interference model. Chen et al. [109] proved that the MLAS problem is NP-hard. Furthermore, they designed a $(\Delta - 1)$-approximation algorithm for this problem, where Δ is the maximum degree of the topological graph of a network. Subsequently, Huang et al. [113] proposed another data aggregation algorithm which has a better performance. By analysis, they showed that the delay of their algorithm is upper bounded by $23\Re + \Delta - 18$ ($\Re \sim L$ and L is defined in Section 11.5), where $\Re$ is the network radius. Recently, Wan et al. [17] proposed three data aggregation algorithms of latency upper bounded by $15\Re + \Delta - 4$, $2\Re + O(\log \Re) + \Delta$, and $(1 + O(\log \Re / \sqrt[3]{\Re}))\Re$, respectively. Xu et al. [164] studied periodic query scheduling for data aggregation with the minimum delay consideration. They designed the centralized aggregation scheduling algorithms under various wireless interference models, and analyzed the induced delay of each algorithm. As explained in Section 11.1, centralized algorithms have many shortcomings in distributed wireless networks. To overcome these shortcomings, some state-of-the-art distributed algorithms are proposed under the UDG model and the protocol interference model [155][228][227]. In [155], Yu et al. proposed a distributed CDS-based data aggregation schedule algorithm with latency upper bounded by $24\mathfrak{D} + 6\Delta + 16$, where $\mathfrak{D}$ is the network diameter. Xu et al. [228] also proposed a distributed data aggregation algorithm with a better latency bound of $16\Re' + 6\Delta - 14$, where $\Re'$ is the inferior network radius which satisfies $\Re' \leq \Re \leq \mathfrak{D} \leq 2\Re'$. The most recently published distributed data aggregation algorithm is [227], in which Li et al. proposed an aggregation scheme of latency upper bounded by $16\Re' + \Delta - 14$.

Unlike the previous works, we design an $\mathcal{R}_0$-PCR-based *distributed data aggregation* (DDA) algorithm. The main differences between this DDA and the previous works can be summarized as follows. First, DDA is a distributed and asynchronous algorithm while many previous algorithms (e.g. [17], [109]-[164]) are centralized. Since WSNs tend to be distributed systems, distributed and asynchronous algorithms are more practical and suitable. Second, DDA runs under the generalized physical interference model while most of the previous works are under the UDG model or the protocol interference model. Compared with the generalized physical interference model, the protocol interference model is simplified and can make the analysis process much easier. On the other hand, the generalized physical interference model considers the aggregated interference in a WSN, which is more practical as well as more complicated.

The description is shown in Algorithm 17. DDA is similar to DDC. The main difference is that each node s_i $(1 \leq i \leq n)$ only transmits one data packet to its parent node, while in DDC, it may have to transmit multiple data packets to its parent node, i.e., the traffic load of a data collection task is much heavier than that of a data aggregation task.

Algorithm 17: The DDA Algorithm

> **input** : CDS-based data collection tree T, $\mathcal{R}_0$-PCR
> **output**: a distributed asynchronous data aggregation plan

1   s_i $(1 \leq i \leq n)$ sets its CR as $\mathcal{R}_0$-PCR according to the required threshold of data receiving rate $\mathcal{R}_0$;

2   **while** s_0 *has not received the aggregation data* **do**

3   **if** s_i *is a leaf node in T or s_i has received the aggregation data from all of its children in T* **then**

4   **if** s_i *is a non-leaf node* **then**

5   s_i obtains the aggregation value of its data and the data of its children by applying the aggregation function;

6   s_i randomly sets a backoff time t_i for its data transmission in window $(0, \tau_w]$;

7   **while** t_i *is not countdown to* 0 **do**

8   s_i senses the channel with $\mathcal{R}_0$-PCR;

9   **if** s_i *senses that the channel is busy* **then**

10   s_i stops the countdown process (the backoff timer is frozen) until the channel becomes free again;

11   **if** s_i *senses that the channel is free* **then**

12        $t_i - -$;

13   **if** $t_i == 0$, i.e. *the backoff timer expires* **then**

14   s_i transmits the aggregation data to its parent node in T.

In Algorithm 17, the routing structure is a CDS-based data collection tree T as in DDC, and we also assume no two transmitters within the CR of each other have their backoff timers expired at the exactly same time instant.

Now, we analyze the delay performance of DDA. Similar to the delay of DDC, we can obtain the upper bound of the time consumption of DDA as shown in Theorem 11.5.

Theorem 11.5
The induced delay of DDA is upper bounded by $\log n + (\beta_\kappa + \beta_{\kappa+1} - 1)L + c_3$ *time slots, where L is the hight of the data collection (aggregation) tree T, and $c_3 = \frac{\pi R^2 (e^2 - 1)}{2c_1} - \beta_\kappa - \beta_{\kappa+1}$ is a constant value depending on $\mathcal{R}_0$-PCR.*

Proof: From Lemma 11.2, the upper bound of the number of nodes within a disk of radius $\mathcal{R}_0$-PCR is $\mathbb{U} = \log n + \frac{\pi R^2 (e^2 - 1)}{2c_1}$. Therefore, any node waits at most $\mathbb{U} - 2$ time slots before transmitting its data to its parent node (minus two means the transmitter and its parent node are not counted). Therefore, it takes at most $(\mathbb{U} - 1) \cdot \tau$ time to aggregate all the data at $V \setminus (\mathcal{C} \cup \mathcal{D})$ to $\mathcal{C} \cup \mathcal{D}$ according to the schedule strategy in DDA. After $(\mathbb{U} - 1) \cdot \tau$ time, there is no data for transmission at nodes in $V \setminus (\mathcal{C} \cup \mathcal{D})$. Based on Lemma 11.4, the number of dominators and connectors within a disk of radius $\mathcal{R}_0$-PCR is upper bounded by $\beta_\kappa + \beta_{\kappa+1}$. Consequently, according to DDA, a node in $\mathcal{C} \cup \mathcal{D}$ has an opportunity to transmit one data packet within time $(\beta_\kappa + \beta_{\kappa+1} - 1) \cdot \tau$. Considering the hight of the data collection (aggregation) tree T is L (which implies the number of hops from the sink to any node in $\mathcal{C} \cup \mathcal{D}$ is at most $L - 1$), the number of time slots consumed by DDA is upper bounded by

$$(\mathbb{U} - 1) + (\beta_\kappa + \beta_{\kappa+1} - 1)(L - 1) \tag{11.51}$$

$$= \mathbb{U} + (\beta_\kappa + \beta_{\kappa+1} - 1)L - \beta_\kappa - \beta_{\kappa+1} \tag{11.52}$$

$$= \log n + (\beta_\kappa + \beta_{\kappa+1} - 1)L + \frac{\pi R^2 (e^2 - 1)}{2c_1} - \beta_\kappa - \beta_{\kappa+1} \tag{11.53}$$

$$= \log n + (\beta_\kappa + \beta_{\kappa+1} - 1)L + c_3, \tag{11.54}$$

where $c_3 = \frac{\pi R^2 (e^2 - 1)}{2c_1} - \beta_\kappa - \beta_{\kappa+1}$. □

11.7 Data Collection and Aggregation under Poisson Distribution Model

In Section 11.3, we assume that all the sensor nodes are independent and identically distributed. Based on that network distribution model, we obtain the achievable capacity of the proposed data collection method DDC, which is order-optimal, and the delay upper bound of the designed data aggregation method DDA. To be more general, in this section, we consider another frequently employed non-i.i.d. model, named the *Poisson distribution* model, and analyze the performances of DDC and DDA.

Under the Poisson distribution model, we assume that one sink node s_0 and n sensor nodes $s_1, s_2, \ldots, s_n$ are distributed according to a two-dimensional Poisson point process with density λ in some area with size $A = c_1 n$. To make data collection and aggregation meaningful, we also assume that the network is connected. Then, by the same method in Section 11.5, a CDS-based data collection tree T can be constructed. Therefore, we can still exploit DDC and DDA to finish data collection and aggregation tasks under the Poisson distribution model. Now, we analyze the delay performance of DDC and DDA.

Let $R = \mathcal{R}_0\text{-PCR} = (\sqrt[\alpha]{c_2(2^{\mathcal{R}_0/W} - 1)} + 1) \cdot r = \kappa \cdot r$. We first analyze the average/upper bound of the number of sensor nodes within the CR of a node as shown in the following lemma.

Lemma 11.5

Let the random variable X denote the number of sensor nodes within the CR of a node. Then, we have

(i) $\mathbf{E}[X] = \pi \lambda R^2$;

(ii) it is almost sure that the number of sensor nodes within the CR of a node is upper bounded by $a \log n$, where $a = \arg\min_{\varepsilon > 0} (\frac{2}{\varepsilon} + \frac{\pi \lambda R^2 (e^\varepsilon - 1)}{\varepsilon \log n})$.

Proof: (*i*) Since the sensor nodes are distributed according to a two-dimensional Poisson point process with density λ, we have

$$\mathbf{E}[X] = \sum_{k=1}^{+\infty} \Pr(X = k) \cdot k \tag{11.55}$$

$$= \sum_{k=1}^{+\infty} \frac{(\pi \lambda R^2)^k}{k!} \exp(-\pi \lambda R^2) \cdot k \tag{11.56}$$

$$= \pi \lambda R^2. \tag{11.57}$$

(*ii*) Similar to the proof in Lemma 11.2, applying the Chernoff bound and for any $\varepsilon > 0$, we have

$$\Pr[X > a \log n] \leq \Pr[X \geq a \log n] \tag{11.58}$$

$$\leq \min_{\varepsilon > 0} \frac{\mathbf{E}[e^{\varepsilon X}]}{e^{\varepsilon a \log n}} \tag{11.59}$$

$$= \min_{\varepsilon > 0} \frac{\exp(\pi \lambda R^2 (e^\varepsilon - 1))}{e^{\varepsilon a \log n}} \tag{11.60}$$

$$= \min_{\varepsilon > 0} \exp(\pi \lambda R^2 (e^\varepsilon - 1) - \varepsilon a \log n) \tag{11.61}$$

$$= \exp(-2 \log n) \leq \frac{1}{n^2}. \tag{11.62}$$

Since $\sum_{n>0} \frac{1}{n^2}$ is upper bounded by $\frac{\pi^2}{6}$, it follows that the number of sensor nodes within the CR of a node is upper bounded by $a \log n$ almost surely, where $a = \arg\min_{\varepsilon > 0} (\frac{2}{\varepsilon} + \frac{\pi \lambda R^2 (e^\varepsilon - 1)}{\varepsilon \log n})$. $\qquad\square$

Based on Lemma 12.6, it is reasonable to take $a \log n$ as the upper bound of the number of sensor nodes within the CR of a node. Then, we have the following theorem, which indicates the upper bound of the induced delay of our data collection algorithm DDC under the Poisson distribution model.

Theorem 11.6

Under the Poisson distribution model, the induced delay of DDC to collect all the data (n data packets) to the sink is upper bounded by $2(a \log n + (n - \Delta_0)(\beta_\kappa + \beta_{\kappa+1})) \cdot \tau$, *where* $a = \arg\min_{\varepsilon>0} \left(\frac{2}{\varepsilon} + \frac{\pi \lambda R^2 (e^\varepsilon - 1)}{\varepsilon \log n}\right)$.

Proof: Based on Lemma 12.6 and by similar methods to Theorem 11.2 and Corollary 11.1, it can be proven that the time consumption to collect all the data packets at $V \setminus (\mathcal{D} \cup \mathcal{C})$ to $\mathcal{D} \cup \mathcal{C}$ is upper bounded by $2a \log n\tau$. Subsequently, similar to Theorem 11.3, the time consumption to collect all the n data packets to the sink is

$$\mathcal{T} \leq 2a \log n\tau + 2(n - \Delta_0)(\beta_\kappa + \beta_{\kappa+1})\tau \tag{11.63}$$

$$= 2(a \log n + (n - \Delta_0)(\beta_\kappa + \beta_{\kappa+1})) \cdot \tau, \tag{11.64}$$

where $a = \arg\min_{\varepsilon>0} \left(\frac{2}{\varepsilon} + \frac{\pi \lambda R^2 (e^\varepsilon - 1)}{\varepsilon \log n}\right)$. □

Based on Theorem 11.6, the achievable data collection capacity of DDC can be obtained as shown in the following theorem.

Theorem 11.7

Under the Poisson distribution model, the achievable data collection capacity of DDC is lower bounded by $\Omega\left(\frac{1}{2(\beta_\kappa + \beta_{\kappa+1})} \cdot W\right)$, *which is scalable and order-optimal.*

Proof: By a similar method to Theorem 11.4, this theorem can be proven. □

Now, we analyze the induced delay of DDA under the Poisson distribution model, which is shown in Theorem 11.8

Theorem 11.8

Under the Poisson distribution model, the induced delay of DDA is upper bounded by $a \log n + (\beta_\kappa + \beta_{\kappa+1} - 1)L - c_4$, *where* $a = \arg\min_{\varepsilon>0} \left(\frac{2}{\varepsilon} + \frac{\pi \lambda R^2 (e^\varepsilon - 1)}{\varepsilon \log n}\right)$ *and* $c_4 = \beta_\kappa + \beta_{\kappa+1}$ *are constant values.*

Proof: By a similar method to Theorem 11.5, this theorem can be proven. □

11.8 Simulation Results

In this section, we present simulation results to validate the performances of DDC and DDA. In all the simulations, we consider the WSNs consisting of one sink node and n sensor nodes which are randomly deployed in a square area with size $A = c_1 n$. Thus the node density is $\frac{1}{c_1}$. Since our primary concern is the achievable capacity and scalability (respectively, induced delay) of DDC (respectively, DDA), we make some simplification and normalization on the simulation settings. The maximum transmission radius of a node is normalized to one and any node can work on the re-start (RS) mode with the IPCS technique [226]. During the data collection period, every node produces a data packet whose size is also normalized to one. Furthermore, all the nodes work with the same transmission power $P = 1$ and over a common wireless channel with bandwidth normalized to one, which implies the transmission time of a data packet τ is 1 in the ideal case. Then, we set the backoff contention window $\tau_w = \frac{1}{10}$ for DDC and DDA in all the simulations. For a data transmission, the background noise is negligible compared with the interference brought by concurrent transmissions. Hence, we do not consider the background noise. For other important system parameters, e.g., the network size A, the node density $\frac{1}{c_1}$, the number of nodes n, the path loss exponent α, etc., we specify them later in each group of simulations.

The compared algorithm for DDC is the *multi-path scheduling* (MPS) algorithm proposed in [10], which is the most recently published centralized and synchronized data collection method under the simplified *protocol interference model* for WSNs. In MPS, the interference radius $R_I = \eta \cdot r (\eta \geq 1)$, where η is a constant and r is the communication radius of a node. Thus, in the following simulations, we set $R_I = \mathcal{R}_0$-PCR, which guarantees that MPS can also initiate data transmissions with a satisfied data receiving rate $\mathcal{R}_0$. The compared algorithm for DDA is the *enhanced pipelined aggregation scheduling* (E-PAS) algorithm [17], which is the best and latest centralized data aggregation algorithm. Since E-PAS is also designed under the protocol interference model, we set the interference radius of E-PAS to $\mathcal{R}_0$-PCR according to different $\mathcal{R}_0$ values. In the following, each group of simulations is repeated 100 times and the results are the average values.

11.8.1 DDC Capacity versus $\mathcal{R}_0$ and α

In this subsection, we consider the WSNs deployed in a square area with size $A = 20 \times 20$ and the node density is 3. The impacts of $\mathcal{R}_0$ and α on the capacities of DDC and MPS are shown in Figure 11.6. From Figure 11.6(a)-(c), we can see that with the increase of $\mathcal{R}_0$, the achievable capacities of both DDC and MPS increase. Although a large $\mathcal{R}_0$ implies a large $\mathcal{R}_0$-PCR (shown in Figure 11.2), which further implies that fewer nodes can conduct transmissions concurrently, on the other hand a large $\mathcal{R}_0$-PCR also implies short transmission time of a data packet. Furthermore, with the increase of $\mathcal{R}_0$, the decrease of the transmission time of a data packet is faster than the increase of $\mathcal{R}_0$-PCR, i.e., $\mathcal{R}_0$ dominates the achievable data collection capacity. It follows that a large $\mathcal{R}_0$ leads to a high capacity for both DDC and MPS.

From Figure 11.6(d)-(f), we can see that with the increase of α, the achievable capacities of DDC and MPS also increase. This is because, for any transmission, the interference impact from other concurrent transmissions decreases quickly with the increase of α. Thus, a large α implies a small $\mathcal{R}_0$-PCR and results in more nodes

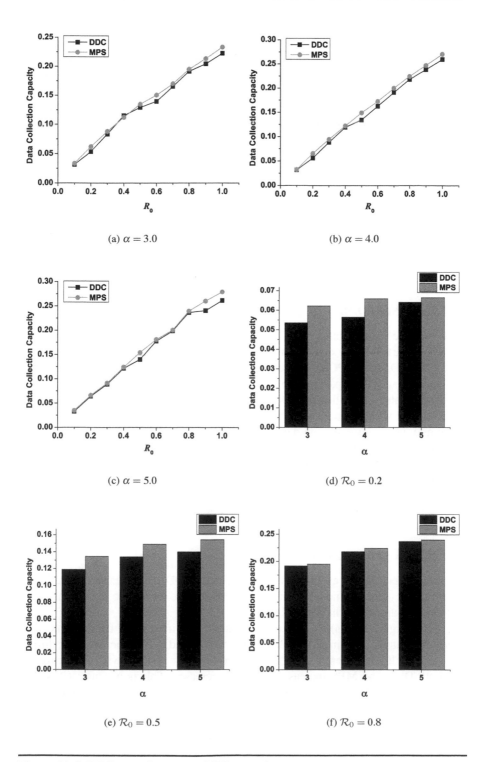

Figure 11.6: DDC capacity versus MPS capacity.

being able to initiate transmissions concurrently. Therefore, the achievable data collection capacities of DDC and MPS increase when α increases.

From Figure 11.6, we can also see that DDC achieves similar data collection capacity to the centralized and synchronous MPS, although DDC is a distributed and asynchronous data collection algorithm. The reason is that we set a proper CR for DDC. By setting the CR of each node as $\mathcal{R}_0$-PCR, as many as possible nodes can initiate data transmissions concurrently with a guaranteed data receiving rate at the receivers. This can also be seen from Theorem 11.1. From the proof of Theorem 11.1, by packing all the possible concurrent data transmissions in the densest manner, we obtain a small proper CR maximizing the number of concurrent transmissions. Consequently, as many as possible transmissions can be scheduled simultaneously without interference at any time, inducing high achievable data collection capacity of DDC. Particularly, the average capacity differences between DDC and MPS are 5.25%, 4.99%, and 4.27% when $\alpha = 3$, $\alpha = 4$, and $\alpha = 5$, respectively, which indicates that DDC achieves comparable capacity to centralized and synchronized MPS.

11.8.2   Scalability of DDC

We examine the scalability of DDC with respect to the number of sensor nodes in a WSN. In the following simulations, we set the path loss exponent α to 4, $\mathcal{R}_0$ to 1 (i.e., the CR for DDC is 1-PCR), the default network size to 10×10, and the default node density is four. The impacts of the node density and the network size on the scalability and achievable capacities of DDC and MPS are shown in Figure 11.7. where we can see that with the increase of the number of sensor nodes (by fixing the network size and increasing the node density in Figure 11.7(a) and fixing the node density and increasing the network size in Figure 11.7(b)), the acheivable capacity of DDC remains as stable as that of centralized and synchronized MPS, which implies DDC is scalable with respect to n, the number of sensor nodes in a WSN. This is because the capacity of DDC only depends on $\mathcal{R}_0$-PCR, which is a distance-dependent parameter. Thus, DDC is scalable for WSNs with different network sizes and node densities.

11.8.3   Performance of DDA

In this subsection, we examine the performance of DDA with respect to α, $\mathcal{R}_0$, and the number of sensor nodes n. In all the simulations, we set the node density to 4. The results are shown in Figure 11.8.

From Figure 11.8(a)-(c), we can see that with the increase of the guaranteed data receiving rate $\mathcal{R}_0$, the induced delay by both DDA and E-PAS increases for different α values. This is different from the data collection situation, where the capacities of both DDC and MPS increase when $\mathcal{R}_0$ increases. This is because: (*i*) with the increase of $\mathcal{R}_0$, the corresponding $\mathcal{R}_0$-PCR increases as well (which can be seen from Figure 11.2). It follows that fewer data transmissions can be conducted simultaneously in DDA and E-PAS. On the other hand, even a larger $\mathcal{R}_0$ implies more data can be transmitted during one data transmission, i.e., fewer transmission times.

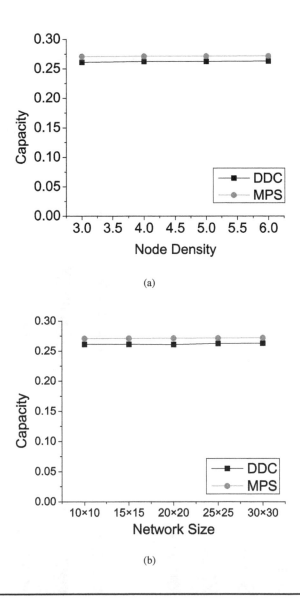

(a)

(b)

Figure 11.7: DDC/MPS capacity versus node density/network size.

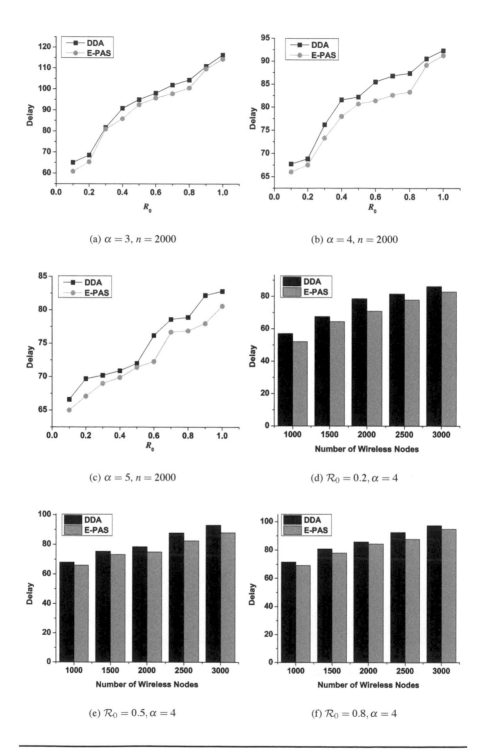

Figure 11.8: Data aggregation delay of DDA and E-PAS.

The induced delay of DDA and E-PAS still increases with the increase of $\mathcal{R}_0$ since $\mathcal{R}_0$-PCR now plays the dominating role in data aggregation; (*ii*) data collection has much more traffic (which is of order $O(n^2)$) than data aggregation (which is of order $O(n)$). Therefore, the data transmission rate (decided by $\mathcal{R}_0$) has more impacts on the delay (as well as capacity) of data collection, while the data transmission concurrency (decided by $\mathcal{R}_0$-PCR) has more impacts on the delay of data aggregation, i.e., the guaranteed data receiving rate ($\mathcal{R}_0$) will dominate the delay increasing of data collection while the carrier-sensing (interference) range ($\mathcal{R}_0$-PCR) will dominate the delay increasing of data aggregation. From Figure 11.8(a)-(c), we can also see that DDA has similar delay performance to E-PAS although DDA schedules data transmission in a distributed and asynchronous manner. On average, the delay differences between DDA and E-PAS in Figure 11.8(a)-(c) are around 3.1%, 3.2%, and 2.6% respectively, which are quite small.

The data aggregation delay of DDA and E-PAS in WSNs with different sizes is shown in Figure 11.8(d)-(f). From Figure 11.8(d)-(f), we can see that the induced delay of DDA and E-PAS increases when the network becomes larger. The reason is straightforward since more sensor nodes imply heavier traffic load. From Figure 11.8(d)-(f), we can also see that the delay difference between DDA and E-PAS is very small. Particularly, in Figure 11.8(d)-(f), the average delay differences between DDA and E-PAS are about 6.1%, 4.4%, and 3.3% respectively, which implies DDA has comparable delay performance as the best centralized data aggregation algorithm.

11.9 Conclusion

Since WSNs in practice tend to be distributed asynchronous systems and most of the existing works study the network capacity issues for centralized synchronized WSNs, we investigate the achievable data collection capacity for distributed asynchronous WSNs in this chapter. To avoid data transmission collisions/interferences, we derive an $\mathcal{R}_0$-proper carrier-sensing range ($\mathcal{R}_0$-PCR) under the generalized physical interference model. By taking $\mathcal{R}_0$-PCR as its carrier-sensing range, any node can initiate a data transmission with a guaranteed data receiving rate. Subsequently, based on the obtained $\mathcal{R}_0$-PCR, we propose a scalable distributed data collection (DDC) algorithm with fairness consideration for asynchronous WSNs. Theoretical analysis of DDC surprisingly shows that its achievable data collection capacity is also order-optimal as that of centralized synchronized algorithms. Moreover, we study how to apply $\mathcal{R}_0$-PCR to distributed data aggregation in asynchronous WSNs, and propose a distributed data aggregation (DDA) algorithm. By analysis, the delay bound of DDA is present. To be more general, we investigate the delay and capacity of DDC and DDA under the Poisson node distribution model. The analysis again shows that DDC is order-optimal and scalable with respect to achievable data collection capacity. The extensive simulation results demonstrate that DDC has comparable data collection capacity compared with the most recently published centralized and synchronized data collection algorithm, and DDC is scalable in WSNs with different network sizes and node densities. DDA also has similar performance to the latest and best centralized data aggregation algorithm.

Chapter 12

CDS-based Broadcast Scheduling in Cognitive Radio Networks

CONTENTS

12.1   Introduction

Wireless spectrum is one of the most precious resources. With the rapid growth of the number of wireless devices, communications over the unlicensed spectrum bands become very crowded. On the other hand, the utilization of the spectrum assigned to licensed users varies from 15% to 85% according to the report from the Federal Communications Commission (FCC) [229], which is very inefficient. To alleviate the interference and collisions on the unlicensed spectrum, as well as to improve the efficiency of the licensed spectrum, researchers propose a new communication paradigm named Cognitive Radio Networks (CRNs), which enable unlicensed users to sense and learn the communication environment with an equipped cognitive radio, and opportunistically access the licensed spectrum without causing any unacceptable interference to the activities of licensed users [230].

Under the CRN model, a *secondary network* consisting of secondary users (SUs) (unlicensed users) coexists with a *primary network* consisting of primary users (PUs) (licensed users). They share the same time, space, and spectrum. The SU senses and learns its local wireless environment before initializing a data transmission. If there is a spectrum opportunity, this SU can conduct the data transmission without hurting any activity of the primary network.

Ever since the CRN communication model was introduced, many efforts have been spent on various issues in CRNs. In this chapter, we study the minimum-latency broadcast scheduling (MLBS) problem for CRNs. Broadcast is one of the most fundamental operations in wireless networks, as well as in CRNs and aims to deliver a message from a source to all the other nodes. In multi-hop CRNs, the *broadcast latency* is defined as the time consumption by which all the nodes in the network have received the broadcast message from the source directly or via a multi-hop manner. Since time is a major concern for many CRN operations, we study the MLBS problem and we try to seek a broadcast scheduling strategy with the minimum broadcast latency.

The MLBS problem is NP-hard in traditional wireless networks even under the simple unit disk graph (UDG) model [231]. Therefore, many approximate algorithms are proposed for MLBS in traditional wireless networks [231]-[232]. However, it is not trivial to adopt these algorithms to CRNs due to the following reasons. CRN is a new communication paradigm, which consists of two networks with different priorities. The activity of the primary network introduces many constraints on the spectrum accessibility of SUs. Unlike traditional wireless networks, where nodes can access the spectrum freely, SUs have to sense and learn the local wireless environment. Only when there are spectrum opportunities for SUs, i.e., the activity of SUs does not cause any unacceptable interference to the primary network, they can access the spectrum. Furthermore, the activities of PUs also induce more interference to SUs besides the interference produced in the secondary network. Therefore, when design broadcast scheduling algorithms for CRNs, elegant techniques are required to consider spectrum opportunities and wireless interference together.

Recently, a few works [233][234][235][236] considered the broadcast scheduling issue for CRNs. In [233], Song and Xie proposed a distributed broadcast protocol without employing a common control channel. The proposed protocol is also unaware of global network topology or time synchronization information. Nevertheless, explicit broadcast latency bound is not discussed in that work. In [234], Kondareddy and Agrawal studied the selective broadcasting problem in CRNs, in which a message is transmitted over a pre-selected set of channels. They proposed a simple heuristic solution and validated their method by simulations. In [235], a broadcasting algorithm is designed for CRNs to maximize the message reachability and reduce the data redundancy and propagation latency. However, analysis for the algorithm is not provided and only simulations are conducted to examine the algorithm. For the MLBS problem, Arachchige *et al.* [236] recently proposed two heuristics with time complexities of $O(RMN^3 \log N + LN^3 \log N)$ and $O(LMN^3 \log N)$ respectively, where R is the network radius, M is the number of available channels in a CRN, N is the number of SUs, and L is the number of time slots in one-time schedule. However, the time performances of these two heuristics are far from the optimum solution ($O(R)$). In this work, we significantly improve these two results.

Considering that existing works for the broadcast problem in CRNs are either heuristic solutions without performance guarantee or with performance far from the optimal solution, we first propose a mixed broadcast scheduling (MBS) algorithm under the *unit disk graph* (UDG) model, which tries to finish the broadcast scheduling by wisely exploiting unicast and broadcast collaboratively. Subsequently, we theoretically analyze the proposed algorithm and obtain its induced broadcast latency and redundancy, which remedies the gap of existing works. To be more general, we further extend MBS to the CRNs under the *protocol interference model* (PrIM), and also analyze its latency and redundancy performance. Finally, we conduct extensive simulations to validate the performance of MBS. In summary, the main contributions of this chapter are as follows.

1. For the MLBS problem in CRNs, we propose a mixed broadcast scheduling (MBS) algorithm under the UDG model, denoted by MBS-UDG. MBS-UDG consists of two phases. In the first phase, the broadcast packet is delivered to all the backbone SUs of the broadcasting tree. Subsequently, all the dominator SUs of the broadcasting tree are partitioned into disjoint and interference-free subsets by hexagonal tessellation and coloring. Then, in Phase II, these subsets are scheduled concurrently and repeatedly to finish the broadcast task by employing mixed unicast and broadcast communication modes. By theoretical analysis, we show that the broadcast latency of MBS-UDS is $(367(\hbar - 1) + 12\Delta_T)/p$ time slots if $\Delta_T \leq 1/p$, and $367(\hbar - 1)/p + 12/p^2 + 12e^{\pi\lambda r^2} \log_{1-p} \frac{1}{p\Delta_T}$ time slots if $\Delta_T > 1/p$, where $\hbar$ is the height of the broadcasting tree (similar as the network radius), Δ_T is the maximum number

of dominatee children (leaf children) connected by a dominator in the broadcasting tree, λ is the transmission intensity of PUs, r (resp., R) is the transmission range of SUs (respectively, PUs), and $p = \exp(-\pi\lambda(r^2 + R^2))$ is the spectrum opportunity for a secondary communication. Evidently, this result is a significant improvement over [236]. We also analyze the broadcast redundancy of MBS-UDG, which is defined as the maximum transmission time of the broadcast packet by any SU in the broadcast scheduling. The broadcast redundancy of MBS-UDG is upper bounded by $12 + \Delta_T$ when $\Delta_T \le 1/p$, and $12 + 1/p + \log_{1-p} \frac{1}{p\Delta_T}$ when $\Delta_T > 1/p$.

2. We extend MBS-UDG to CRNs under the PrIM, denoted by MBS-PrIM. By similar techniques employed in MBS-UDG, we also show the broadcast latency and redundancy of MBS-PrIM.

3. Finally, extensive simulations are conducted to validate the performance of MBS. Simulation results indicate that MBS significantly improves both the broadcast latency and the broadcast redundancy over existing algorithms. Specifically, MBS induces over 290% and 840% less latency and redundancy than existing algorithms, respectively.

The rest of this chapter is organized as follows. In Section 12.2, we summarize existing works and discuss the differences. In Section 12.3, the network model, interference model and problem definition are presented. The broadcasting tree is constructed in Section 12.4. In Section 12.5, the mixed broadcast scheduling (MBS) algorithm under the UDG model is designed and analyzed. To be more general, we extend the proposed broadcast scheduling algorithm to CRNs under the protocol interference model in Section 12.6. In Section 12.7, extensive simulations are conducted to examine the performance of MBS. Finally, we conclude this chapter in Section 12.8.

12.2   Related Work

Broadcast scheduling, especially MLBS, is a fundamental problem in wireless networks. In this section, we will summarize some existing works on MLBS for traditional wireless networks. Subsequently, the works on the broadcasting issue for CRNs will also be discussed. Finally, we make some remarks on the differences between this work and existing works.

12.2.1   Broadcast Scheduling in Traditional Wireless Networks

In [231][237], Gandhi *et al.* first studied the MLBS problem under the UDG model for ad hoc networks. They proved the NP-hardness of the MLBS problem and subsequently proposed an approximation broadcast scheduling algorithm. However, the approximation guarantee of their algorithm is greater than 400. Later on, Huang *et al.* in [238] improved the algorithm in [231][237]. They designed two algorithms with approximation guarantees of 51 and 24 respectively, and one more efficient broadcast scheduling algorithm of latency $R + O(\sqrt{R}\log^{1.5} R)$, where R is the network radius. Since R is a trivial lower latency bound of any broadcast scheduling algorithm and $O(\sqrt{R}\log^{1.5} R) < O(R)$ when $R \to \infty$, the third algorithm in [238] is nearly optimal with respect to order. Huang *et al.* further improved the ratio bounds of their algorithms in [239], in which three approximation algorithms are designed with latency upper bounded by $24R - 23$, $16R - 15$, and $R + O(\log R)$, respectively. In [240], Gandhi *et al.* gave another one-to-all broadcast scheduling algorithm, which yields a scheduling strategy of approximation radio 12. They also designed two all-to-all broadcast scheduling algorithms with approximation ratios of 20 and 34, respectively.

In [241], Chen *et al.* investigated the MLBS problem for traditional wireless networks under the protocol interference model. They designed a broadcast scheduling algorithm of approximation ratio 26 under the UDG model and $2\pi\alpha^2$ under the protocol interference model, where α is the ratio between the interference range and the transmission range of a node. Similarly, Mahjourian *et al.* [242] also studied the MLBS problem for wireless ad hoc networks under the protocol interference model. They designed a broadcast scheduling algorithm of latency upper bounded by $O((\max(\alpha,\beta))^2)R$, where β is the ratio between the carrier sensing range and the transmission range of a node. In [232], Huang *et al.* studied the MLBS problem for traditional wireless networks under the more precise physical interference model. They first designed a constant approximation broadcast scheduling algorithm under the two-disk model and then extend it to the physical interference model. Besides the aforementioned determined broadcast scheduling algorithms, Huang *et al.* proposed a randomized broadcast scheduling algorithm for traditional wireless networks in [243].

12.2.2 Broadcast Scheduling in CRNs

As a new promising communication paradigm, CRNs attracted lots of attention recently. However, for the broadcast scheduling problem, as one of the most fundamental issues in CRNs, only a few existing works consider it [234]-[236]. In [234], Kondareddy and Agrawal studied the selective broadcasting problem in multi-hop CRNs, where the broadcast message is transmitted over a pre-selected set of channels. By introducing the concepts of neighbor graphs and minimal neighbor graphs, a heuristic selective broadcasting algorithm is designed. Furthermore, the authors also took simulations to verify the proposed algorithm. In [235], a broadcasting algorithm is proposed with objectives to maximize the reachability of the broadcast message, and reduce the redundancy and propagation latency. However, no quantitative analysis is provided except for simulations. Recently, Arachchige *et al.* [236] studied the MLBS problem for CRNs under a simple network interference model. They first formulated the MLBS problem as an integer linear programming (ILP) problem and then proposed two heuristics. However, the time complexities of their heuristics are $O(RMN^3 \log N + LN^3 \log N)$ and $O(LMN^3 \log N)$ respectively, which are far from the optimum solution.

12.2.3 Remarks

As we discussed in Section 12.1, the broadcasting algorithms for traditional wireless networks cannot be applied to CRNs trivially. This is because the only major concern for broadcast scheduling in traditional wireless networks is to deal with wireless interference. Nevertheless, besides wireless interference, the spectrum opportunity is another important concern when design broadcast scheduling algorithms for CRNs. Existing works for the broadcast scheduling issue in CRNs only provide some simple heuristic algorithms without theoretical performance guarantees. To remedy this gap, we first design a mixed broadcast scheduling (MBS) algorithm under the UDG model with theoretical analysis both on the broadcast latency and on the data redundancy. Subsequently, we extend MBS to the more general protocol interference model, and also analyze the performance of MBS under that model. Finally, extensive simulations are conducted to verify MBS.

12.3 System Model and Problem Definition

In this section, we give the network model and interference model. We also formally define the studied minimum-latency broadcast scheduling problem in CRNs.

12.3.1 Network Model

We consider a secondary network coexisting with a primary network. They share the same time, space, and spectrum. Particularly, we consider the general single-spectrum model, which is reasonable for protocol design and analysis in CRNs as indicated by many existing works [244][245].

Primary Network: The primary network consists of N Poisson distributed primary users (PUs) denoted by set $V_p = \{S_1, S_2, \cdots, S_N\}$. The transmission radius and interference radius are defined as R and R_I, respectively. The network time is assumed to be slotted with each time slot of length τ, and a PU can transmit one data packet during a time slot. At the very beginning of each time slot, each PU decides to transmit some data or keeps silent in that time slot according to the working protocol of the primary network (a PU to be a receiver can be treated as silent since it will not cause any interference to other PUs or SUs). Furthermore, during each time slot, the primary transmitters are distributed according to a two-dimensional Poisson point process X_T with density λ. Therefore, according to the *displacement theorem* [246], the distribution of primary receivers during each time slot is correlated with X_T, and forms another two-dimensional Poisson point process X_R also with density λ.

Secondary Network: The considered secondary network coexists with the primary network, which consists of one randomly distributed *broadcasting source* secondary user (SU) denoted by s_0, and n randomly distributed SUs denoted by $s_1, s_2, \cdots, s_n$. The transmission radius and interference radius of SUs are represented by r and r_I, respectively. For a pair of SUs s_i and s_j, there is a link between them *iff* the Euclidean distance $D(s_i, s_j)$ between them satisfies $D(s_i, s_j) \leq r$.

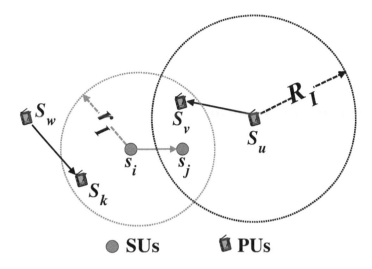

Figure 12.1: Conditions for successful communications from SU s_i to SU s_j.

Therefore, the secondary network can be modeled by a graph $G = (V_s, E_s)$, where $V_s = \{s_0, s_1, s_2, \cdots, s_n\}$ is the node set and E_s is the set of all the possible links formed by SUs in V_s. To make the broadcast scheduling problem meaningful, we assume G is connected.

Note that in a CRN, the link between SUs s_i and s_j does not imply s_i can transmit a data packet to s_j during a time slot because of lacking spectrum opportunities. Therefore, as shown in Figure 12.1, to successfully transmit a data packet to s_j from s_i, three more conditions need to be satisfied:

1. There are no PUs within the interference range of s_i and receiving some data;

2. There are no PUs such that s_j is within the interference range of them and they are transmitting some data;

3. The communication between s_i and s_j is interference-free. Consequently, when we design broadcast scheduling algorithms, we have to consider interference avoidance and spectrum opportunities together.

12.3.2 Interference Model

In this chapter, we consider two frequently exploited interference models: the UDG model and PrIM.

UDG Model: The UDG model is a widely used interference model in existing works [231]-[232], which is simple and very convenient in performance analysis. In the UDG model, the interference range is equal to the transmission range, which implies $R_I = R$ and $r_I = r$ in our network model.

PrIM: The PrIM is a general version of the UDG model, in which the interference range of a node is assumed to be greater than or equal to the transmission range of that node. Hence, in our network model, we assume $R_I = \beta \cdot R$ and $r_I = \beta \cdot r$, where $\beta \geq 1$ is a constant value.

12.3.3 Problem Definition

Based on the defined network model and interference model, we can formally define the MLBS problem for CRNs as follows. Given a secondary network represented by graph $G = \{V_s = \{s_0, s_1, s_2, \cdots, s_n\}, E_s\}$, where s_0 is the broadcasting source and s_0 holds a data packet which needs to send to all the other SUs in the network, a broadcast scheduling of latency ℓ is a sequence of subsets $\langle V_0, V_1, V_2, \cdots, V_\ell \rangle$ satisfying that

1. $V_0 = \{s_0\}$ and $\forall 1 \leq i \leq \ell, V_i \subseteq V_s$;

2. During time slot $1 \leq i \leq \ell$, all the SUs in V_i can receive the broadcast data packet from some SUs in $\bigcup_{j=0}^{i-1} V_j$ successfully;

3. $\bigcup_{i=0}^{\ell} = V_s$, which implies all the SUs have a copy of the broadcast data packet at the end of the scheduling.

The MLBS problem is to seek a broadcast scheduling strategy of minimum ℓ. Its *broadcast latency* is defined as ℓ, and its *broadcast redundancy* is defined as $\max\{d_i | s_i \ (0 \leq i \leq n) \ transmits \ d_i \ times \ of \ the \ broadcast \ data \ packet \ during \ the \ entire \ MLBS \ process\}$. Evidently, when designing a MLBS algorithm, in addition to minimizing the latency ℓ, we also want to have low broadcast redundancy, which implies lower energy consumption and spectrum requirements. The MLBS problem is NP-hard even in conventional wireless networks under the UDG model. Therefore, we design an approximation algorithm with latency and redundancy analysis under the UDG model and the PrIM for CRNs.

12.4   Broadcasting Tree and Coloring

In this section, we construct a *connected dominating set* (CDS) based broadcasting tree, denoted by T, which serves as the scheduling infrastructure. Subsequently, we study the tessellation and coloring of a plane, which is useful in the designed interference-free scheduling algorithm.

12.4.1   CDS-based Broadcasting Tree

For the secondary network represented by graph $G = (V_s, E_s)$, a *dominating set* (DS) of G is a subset $\mathcal{D}$ of V_s such that $\forall s_i \in V_s$, either $s_i \in \mathcal{D}$ or s_i is adjacent to some SU in $\mathcal{D}$. If the induced subgraph $G[\mathcal{D}]$ of G on $\mathcal{D}$ is connected, $\mathcal{D}$ is a *connected*

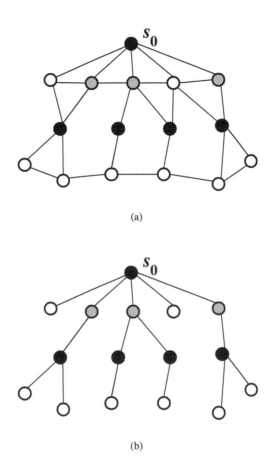

(a)

(b)

Figure 12.2: Construction of the CDS-based broadcasting tree.

dominating set (CDS) of *G*. A *maximal independent set* (MIS) $\mathcal{M}$ of *G* is a subset of V_s such that (*i*) $\forall s_i \in V_s$, either $s_i \in \mathcal{M}$ or $\exists s_j \in \mathcal{M}$, $(s_i, s_j) \in E_s$; and (*ii*) $\forall s_i, s_j \in \mathcal{M}$, $(s_i, s_j) \notin E_s$. Clearly, an MIS is also a DS.

Employing a CDS as the infrastructure of a wireless network is widely adopted in existing literature. In this chapter, we construct a CDS-based broadcasting tree *T* according to the following three steps. First, make a breadth first searching (BFS) on *G* starting from s_0, and identify an MIS, denoted by $\mathcal{D}$ of *G*. As shown in Figure 12.2(a), the set of black nodes is an MIS of that network. Evidently, $\mathcal{D}$ is also a DS. Usually, the black nodes in $\mathcal{D}$ are called *dominators*. Second, find a minimal set, denoted by $\mathcal{C}$, of nodes to connect the dominators in $\mathcal{D}$ to make $\mathcal{D} \cup \mathcal{C}$ a CDS of *G*. As shown in Figure 12.2 (a), $\mathcal{C}$ consists of gray nodes, which are also named *connectors*. The black nodes and blue nodes together form a CDS in Figure 12.2(a). Finally, every remaining white node in $V_s \setminus (\mathcal{D} \cup \mathcal{C})$, also called a *dominatee*, assigns its neighboring dominator with the shortest distance to s_0 as its parent node. For every connector (respectively, dominator except for s_0) in $\mathcal{C}$ (respectively, $\mathcal{D}$), assign its neighboring dominator (respectively, connector) with the shortest distance to s_0 as its parent. Then, a CDS-based broadcasting tree *T* is constructed, e.g. Figure 12.2(b) shows the broadcasting tree constructed for the secondary network in Figure 12.2(a).

For $s_i \in V_s$ in *T*, define $p(s_i)$ $(s_i \neq s_0)$ as its parent node and $c(s_i) = \{s_j | p(s_j) = s_i\}$ as its children set. Let $\Delta(s_i) = |c(s_i) \cap (V_s \setminus (\mathcal{D} \cup \mathcal{C}))|$ be the number of dominatee children s_i has in *T*. Clearly, if s_i is a connector or dominatee, $\Delta(s_i) = 0$. We also define $\Delta_T = \max\{\Delta(s_i) | s_i \in V_s\}$, which indicates the maximum number of dominatee children that a dominator may have. The *height* of s_0 in *T* is defined to be $h(s_0) = 0$ and for the other node $s_i \neq s_0$, its *height* in *T* is defined as $h(s_i) = h(p(s_i)) + 1$. Furthermore, the *height* of *T* is defined as $\hbar = \max\{h(s_i) | s_i \in V_s\}$.

Now we introduce some important properties of *T*, which will be used in our algorithm, in Lemma 12.1. For Lemma 12.1, Lemma 12.1.1 has been proven in [17], Lemma 12.1.2 can be easily proven based on the *Wegner theorem* [247], and Lemma 12.1.3 is a result in [17].

Lemma 12.1

1) $\forall s_i \in \mathcal{C}$, s_i *is adjacent to at most 5 dominators in* $\mathcal{D}$, *of which one is its parent node.* $\forall s_i \in \mathcal{D}$, $|c(s_i) \cap \mathcal{C}| \leq 11$ *if* $s_i \neq s_0$ *and* $|c(s_0) \cap \mathcal{C}| \leq 12$, *i.e.,* s_i *has at most 11 connector children if* s_i *is a non-root (not* s_0*) dominator and* s_0 *has at most 12 connector children; 2) There are at most 21 (respectively, 42) dominators within a disk of radius 2r (respectively, 3r); 3) Generally, suppose the number of dominators within a disk of radius* θr *is* ϕ_θ. *Then* $\phi_\theta \leq \frac{2\pi}{\sqrt{3}}\theta^2 + \pi\theta + 1$.

12.4.2 Tessellation and Coloring

A *tessellation* of a plane means covering a plane with a pattern of flat shapes so that there are no overlaps or gaps. A *regular tessellation* is a pattern made by repeating a *regular polygon*. In existing literature [232], *hexagonal tessellation* is a frequently used regular tessellation method to partition a network into hexagons. Similarly, for a CRN, we can use half-open half-closed hexagons of radius $r/2$ as shown in Figure 12.3(a) to partition the network into hexagons as shown in Figure 12.3(c). For a hexagonal tessellation, many coloring methods are proposed. For instance, we give a 12-coloring unit of hexagons in Figure 12.3(b), and its corresponding coloring of the tessellation is shown in Figure 12.3(c). There are also many other 12-coloring methods for a hexagonal tessellation [232][248]. From Figure 12.3(b) and (c), we can see that the hexagons of the same color are separated by at least the distance of $4 \cdot \frac{r}{2} = 2r$. Generally, we have a conclusion as shown in Lemma 12.2 [232][248].

Lemma 12.2

In a $3k^2$-coloring of a hexagonal tessellation, hexagons with the same color are separated by distance of at least $(3k-2) \cdot \gamma$, where γ is the radius of the hexagons.

Now, we tessellate the considering CRN with half-open half closed hexagons of radius $\gamma = \frac{r}{2}$. Consequently, each hexagon contains at most one dominator of $\mathcal{D}$ in the broadcasting tree T. We further make a $3k^2$-coloring of the hexagonal tessellation, and assign each dominator the same color as the hexagon it belongs to. Then, $\mathcal{D}$ can be partitioned into $3k^2$ disjoint subsets $\mathcal{D}_1, \mathcal{D}_2, \cdots, \mathcal{D}_{3k^2}$ with $\mathcal{D}_i = \{s_i | s_i \in \mathcal{D}, $ and s_i is assigned color $i\}$ for $1 \leq i \leq 3k^2$. Based on Lemma 12.2 and considering that all the hexagons are half-open half closed, we can conclude that $\forall s_u, s_v \in \mathcal{D}_i$, $D(s_u, s_v) > (3k-2) \cdot \gamma = \frac{(3k-2)r}{2}$.

12.5 Broadcast Scheduling under UDG Model

In this section, we first consider the MLBS problem under the UDG model and propose a broadcast scheduling algorithm. We then analyze its latency and redundancy.

12.5.1 MLBS under UDG Model

Before giving the broadcast scheduling algorithm, we make some assumption on the size of the broadcast data packet first. In Section 12.3, we assume the network is slotted with unit τ, and a PU can transmit one data packet during a time slot. For the broadcast data packet in the secondary network, we assume its size is smaller than the data packet transmitted in the primary network.[1] Therefore, we can partition a time slot into two parts τ_s, which is the *sensing window* for SUs to sense local wireless environments, and τ_d, which is the *data transmission window* and longer enough to transmit the broadcast data packet as shown in Figure 12.4. Then, in our designed broadcast scheduling algorithm, for a SU s_i with data for transmission, it can randomly choose a *sensing point* τ_p in the sensing window, e.g., τ_p in Figure 12.4, to

---

[1]If the broadcast data packet has a large size, we can break it into several smaller packets to accommodate our algorithm.

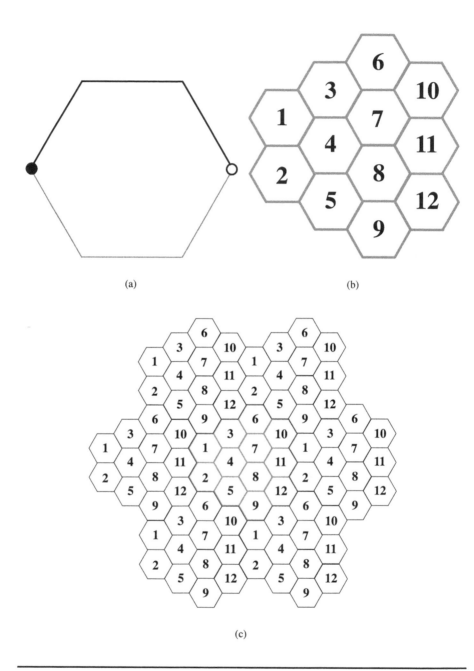

(a)

(b)

(c)

Figure 12.3: (a) A half-open (the bottom half and the right-most vertex) half-closed (the top half and the left-most vertex) hexagon; (b) a basic 12-coloring unit of hexagons where each color is represented by a positive integer number; and (c) a hexagonal tessellation and its 12-coloring unit.

Figure 12.4: Partition of a time slot for SUs.

sense the local communication environment. If there is a spectrum opportunity, s_i can initialize a data transmission immediately after the sensing point according to the scheduling protocol. Furthermore, if multiple SUs try to transmit data during the same time slot, we assume no two SUs will chose the exactly same sensing point in the sensing window,[2] which implies no two SUs will start to transmit data at the exactly same time point.

Algorithm 18: MBS-UDG

---

    input : CDS-based broadcasting tree T
    output: a broadcast scheduling plan

1   $\mathcal{B}_0 \leftarrow (\mathcal{C} \cup \mathcal{D}) \setminus \{s_0\}, \mathcal{B}_1 \leftarrow \{s_0\}$;
2   **while** $\mathcal{B}_0 \neq \emptyset$ **do**
3   **for** *every* $s_i \in \mathcal{B}_1$, *if* $c(s_i) \cap \mathcal{B}_0 \neq \emptyset$ **do**
4   **for** *every* $s_j \in c(s_i) \cap \mathcal{B}_0$ **do**
5   s_i transmits the broadcast data packet to s_j by unicast when there is a spectrum opportunity for this transmission;
6          $\mathcal{B}_1 \leftarrow \mathcal{B}_1 \cup \{s_j\}, \mathcal{B}_0 \leftarrow \mathcal{B}_0 \setminus \{s_j\}$;

7   Tessellate the CRN with half-open half closed hexagons of radius $\frac{r}{2}$ and give a $3k^2$-coloring to the tessellation, where $k = 2$;
8   Partition the dominators $\mathcal{D}$ into $3k^2$ disjoint subsets $\mathcal{D}_1, \mathcal{D}_2 \cdots, \mathcal{D}_{3k^2}$ according to the method in Section 12.4, and let $\mathcal{B}_0 \leftarrow V_s \setminus \mathcal{B}_1$;
9   **while** $\mathcal{B}_0 \neq \emptyset$ **do**
10   Schedule $\mathcal{D}_1, \mathcal{D}_2 \cdots, \mathcal{D}_{3k^2}$ repeatedly each for one time slot by calling procedure BROAD-UNI-CAST($\mathcal{D}_i$)

Now, we are ready to present the broadcast scheduling algorithm. Based on the constructed broadcasting tree in Section 12.4, we propose a mixed broadcast scheduling (MBS) algorithm under the UDG model, named MBS-UDG, as shown in Al-

---

[2]The situation that two SUs choose the same sensing points can be tackled by assigning them different sensing points in a centralized manner instead of randomly choosing, or by an exponential backoff mechanism.

gorithm 18. From Algorithm 18, we can see that both the unicast communication mode and the broadcast communication mode are exploited to finish the broadcast scheduling. Basically, Algorithm 18 consists of two phases: Phase I (lines 1-6) is used to broadcast the data packet to all the dominators and connectors of T, i.e., $\{s_0\} \rightarrow \mathcal{D} \cup \mathcal{C}$; and Phase II (lines 7-10) is used to broadcast the data packet to all the dominatees of T, i.e. $\mathcal{D} \rightarrow V_s \setminus (\mathcal{D} \cup \mathcal{C})$. In Phase I, the broadcast scheduling is finished by unicast. In Phase II, we first tessellate the network plane with hexagons. Subsequently, we partition the dominator set $\mathcal{D}$ into disjoint subsets by coloring the tessellation. Finally, since every dominatee has a dominator parent in T, these dominator subsets are scheduled repeatedly until all the dominatees receive the data packet. When scheduling each dominator subset, the procedure BROAD-UNI-CAST($\mathcal{D}_i$) (Algorithm 19) is called. From Algorithm 19, we can see that BROAD-UNI-CAST($\mathcal{D}_i$) finishes the broadcast scheduling by employing mixed unicast and broadcast. For a SU $s_i \in \mathcal{D}_i$, if the number of its children that have not received the data packet is no less than a threshold $\exp(\pi\lambda(r^2 + R^2))$, it will broadcast the data packet. Otherwise, it will transmit data to the children that wait to receive the broadcast packet by unicast.

Algorithm 19: BROAD-UNI-CAST($\mathcal{D}_i$)

---

input : CDS-based broadcasting tree T
output: a broadcast scheduling plan

1   **if** $\forall s_u \in \mathcal{D}_i,\ c(s_u) \cap \mathcal{B}_0 = \emptyset$ **then**
2   **return**
3   $\forall s_u \in \mathcal{D}_i$, **if** $|c(s_u) \cap \mathcal{B}_0| > \exp(\pi\lambda(r_I^2 + R_I^2))$ **then**
4   s_u broadcasts the data packet to its children in $c(s_u) \cap \mathcal{B}_0$ when its transmission does not cause any unacceptable interference to the primary network;
5   **else if** $0 < |c(s_u) \cap \mathcal{B}_0| \leq \exp(\pi\lambda(r_I^2 + R_I^2))$ **then**
6   For every $s_v \in (c(s_u) \cap \mathcal{B}_0)$, s_u transmits the data packet to it by unicast when there is a spectrum opportunity for this transmission;
7   $\forall s_v \in (c(s_u) \cap \mathcal{B}_0)$, **if** s_v *has successfully received the broadcast data packet from* s_u **then**
8     $\mathcal{B}_0 \leftarrow \mathcal{B}_0 \setminus \{s_v\}$

In MBS-UDG, we employ unicast sometimes instead of broadcast. This is mainly because the spectrum opportunities of the neighbors of a SU vary over time. Consequently, a broadcast operation may not have more effective receivers than a unicast operation. In some cases, a broadcast may not have any effective receiver due to lack of spectrum opportunities, which is common in CRNs. We further take an example to analyze the reasons. For a SU s_i carrying the broadcast packet, suppose it has three children s_1, s_2, and s_3 waiting for receiving the data from s_i. Assume s_1, s_2, and s_3 have spectrum opportunities to receive the data at the 5-th time slot, at the 10-th time slot, and at the 15-th time slot, respectively. Then, if s_i keeps broadcasting the data packet, 15 time slots are needed to let all its children receive the data, and s_i broadcasts the data packet 15 times. On the other hand, if s_i exploits unicast to finish this broadcast task, s_i only has to transmit the packet 3 times without increasing the latency. Evidently, in these situations, unicast has advantages, e.g., shorter data transmission times, less energy consumption, less interference to other ongoing communications, and etc., over broadcast. Therefore, by carefully designing, we properly employ unicast or broadcast based on different situations in MBS-UDG. This can also be seen from the following lemma, which indicates how many dominatee children can successfully receive the broadcast packet from s_u during one transmission (broadcast or unicast) of s_u.

Lemma 12.3

In Algorithm 19, the expected number of dominatee children that can successfully receive the broadcast packet from each data transmission of $s_u \in \mathcal{D}_i$ is at least 1.

Proof. According to Algorithm 19, this lemma is true if the data transmission of s_u is unicast. If this data transmission is broadcast, then we have $|c(s_u) \cap \mathcal{B}_0| > \exp(\pi\lambda(r_I^2 + R_I^2))$ dominatee children of s_u waiting for receiving the broadcast packet. Then, during this broadcast transmission, the expected number of children that can successfully receive the broadcast packet is $|c(s_u) \cap \mathcal{B}_0| \cdot p > \exp(\pi\lambda(r_I^2 + R_I^2)) \cdot p = 1$, where p is the spectrum opportunity for a secondary communication as shown in Lemma 12.4. □

12.5.2 Analysis of MBS-UDG

In this subsection, we analyze the latency and redundancy performance of MBS-UDG.

12.5.2.1 Broadcast Latency of MBS-UDG

Since MBS-UDG consists of two phases, we will analyze the latency of these two phases respectively. First, based on the defined network model, we can obtain the spectrum opportunity for a pair of SUs as shown in Lemma 12.4.

Lemma 12.4

Suppose $(s_u, s_v) \in E_s$ in $G = (V_s, E_s)$. The spectrum opportunity for s_u to transmit a data packet to s_v is $p = \exp(-\pi\lambda(r_I^2 + R_I^2))$.

Proof. A spectrum opportunity happens from s_u to s_v only when two events happen, i.e., **E1:** *there is no primary receiver within the interference range of s_u* and **E2:** *there is no primary transmitter within the disk centered at s_v of radius R_I.* Further-

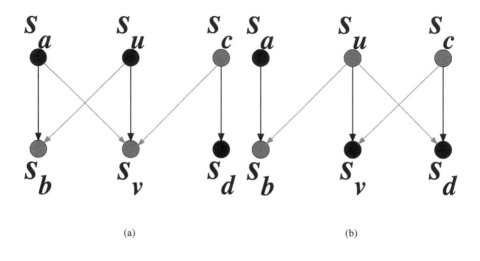

(a)   (b)

Figure 12.5: (a) Interference analysis in Phase I.

more, from Section 12.3, we know that the distribution of both the primary transmitters and the primary receivers satisfies a two-dimensional Poisson point process with density λ. Therefore, we have

$$p(\mathbf{E1}) = \frac{e^{-\lambda \pi r_I^2}(-\lambda \pi r_I^2)^0}{0!} \tag{12.1}$$

$$= e^{-\lambda \pi r_I^2}. \tag{12.2}$$

Similarly, $p(\mathbf{E2}) = e^{-\lambda \pi R_I^2}$. Therefore, the probability that a spectrum opportunity exists from s_u to s_v during a time slot is

$$p = p(\mathbf{E1}) \cdot p(\mathbf{E2}) \tag{12.3}$$

$$= \exp(-\pi\lambda(r_I^2 + R_I^2)). \tag{12.4}$$

$\square$

Since $r_I = r$ and $R_I = R$ under the UDG model, let $p = \exp(-\pi\lambda(r^2 + R^2))$ in this section. In Phase I, the task is to broadcast the data packet to all the other dominators and connectors of T from the data source (root) s_0. According to the construction process of T, we know that for $\forall s_v \in \mathcal{C}$, its parent node in T is a dominator, and $\forall s_v \in \mathcal{D}$, if $s_v \neq s_0$, its parent node in T is a connector. For convenience, without specification, we use (s_u, s_v) to represent the communication that s_u transmits the broadcast data packet to s_v in this chapter. Then, we have the following lemma, which indicates the time consumption to transmit the broadcast packet from a parent dominator to a child connector.

Lemma 12.5

Suppose $s_v \in \mathcal{C}$ is waiting to receive the broadcast packet, and $s_u = p(s_v) \in \mathcal{D}$ is currently holding the broadcast packet. Then, the expected time for s_v to receive the broadcast packet from s_u in MBS-UDG is at most $262\tau/p$.

Proof: To prove this lemma, we first ignore the primary network and suppose the secondary network is a stand-alone network. Then, in Phase I of MBS-UDG, communication (s_u, s_v) may interfere with two other kind of ongoing communications as shown in Figure 12.5(a), i.e., **Type 1:** communications like (s_a, s_b), where $s_a \in \mathcal{D}$ and $s_b \in \mathcal{C}$. For this kind of communications, s_a may produce interference to s_v, and similarly s_u may produce interference to s_b. We have $D(s_a, s_u) \le 2r$. Based on Lemma 12.1, there are at most 20 dominators other than s_u within a disk of radius $2r$ and each dominator has at most 11 connector children that are waiting for data transmission. Consequently, there are at most 220 communications of **Type 1**. **Type 2:** communications like (s_c, s_d) where $s_c \in \mathcal{C}$ and $s_d \in \mathcal{D}$. In this case, only s_c may produce interference to s_v and $D(s_u, s_d) \le 3r$ under the UDG model. Based on Lemma 12.1, there are at most 41 dominators other than s_u within a disk of radius $3r$. Therefore, there are at most 41 communications of **Type 2**. In summary, if the secondary network is a stand-alone network, to successfully transmit a data packet to s_v from s_u, the time consumption is at most $220 + 41 + 1 = 262$ time slots by MBS-UDG, where plus 1 is the time consumption for communication (s_u, s_v). Now, take the primary network into consideration and based on Lemma 12.4, the expected time for s_v to receive the broadcast packet from s_u in MBS-UDG is upper bounded by $262\tau/p$. □

By similar technique, we can obtain the time consumption to transmit the broadcast packet from a connector parent to a dominator child as shown in Lemma 12.6.

Lemma 12.6

Suppose $s_v \in \mathcal{D}$ is waiting for receiving the broadcast packet, and $s_u = p(s_v) \in \mathcal{C}$ is currently holding the broadcast packet. Then, the expected time for s_v to receive the broadcast packet from s_u in MBS-UDG is at most $472\tau/p$.

Proof: As in Lemma 12.5, we first assume the secondary network is a stand-alone network. Then, communication (s_u, s_v) also may interfere with two type of communications as shown in Figure 12.5(b), i.e. **Type 1:** communications like (s_a, s_b) where $s_a \in \mathcal{D}$ and $s_b \in \mathcal{C}$; and **Type 2:** communications like (s_c, s_d) where $s_c \in \mathcal{C}$ and $s_d \in \mathcal{D}$. In communications of **Type 1**, only s_u may produce interference to s_b. Therefore, we have $D(s_v, s_a) \le 3r$ under the UDG model. According to Lemma 12.1, there are at most 41 dominators like s_a around (s_u, s_v), and each dominator has at most 11 connector children waiting for receiving the broadcast packet. It follows that the number of communications of **Type 1** is at most 451. In communications of **Type 2**, s_u and s_c may interfere the data transmission of each other. Therefore, $D(s_v, s_d) \le 2r$ under the UDG model. Then, there are at most 20 communications of **Type 2** around communication (s_u, s_v) based on Lemma 12.1. In summary, at most $451 + 20 + 1 = 472$ time slots are needed for s_u to transmit a data packet to s_v in MBS-UDG if the secondary network is a stand-alone network. If we take the primary network into consideration and based on Lemma 12.4, the expected time for communication (s_u, s_v) is upper bounded by $472\tau/p$ in MBS-UDG. □

Based on Lemma 12.5 and Lemma 12.6, we can obtain the time consumption of Phase I as shown in Theorem 12.1.

Theorem 12.1
In MBS-UDG, the expected time consumption of Phase I is at most $367(\hbar - 1)\tau/p$.

Proof: Suppose s_v is the last SU to receive the broadcast packet in Phase I of MBS-UDG. Then, s_v must be a dominator. This is because each connector must have a dominator parent and at least one dominator child. Otherwise, that connector will not be selected to the minimal connector set based on the construction process of T. Now, suppose the data transmission path from s_0 to s_v is $\mathcal{P} : s_0 \to s_a \to \cdots \to s_u \to s_v$. Then, the transmissions on $\mathcal{P}$ are *dominator→connector communications* and *connector→dominator communications* alternately. Therefore, based on Lemma 12.5 and Lemma 12.6, the expected time consumption for s_v to receive the broadcast packet is at most

$$\frac{h(s_v)}{2} \cdot (262\tau/p + 472\tau/p) = 367h(s_v)\tau/p, \tag{12.5}$$

where $h(s_v)$ is the height of s_v in T. Furthermore, $h(s_v) \leq \hbar - 1$, where $\hbar$ is the height of T. It follows this theorem holds. □

Now, we analyze the time consumption of Phase II in MBS-UDG. In Phase II, only dominators and dominatees are involved in the broadcasting process. We partition dominators into $3k^2 = 3 \times 2^2 = 12$ disjoint subsets $\mathcal{D}_i$ $(1 \leq i \leq 12)$, and schedule these subsets repeatedly by calling procedure BROAD-UNI-CAST($\mathcal{D}_i$) until all the dominatees received the broadcast packet.

To obtain the time consumption of Phase II, we first analyze the time consumption for $\mathcal{D}_i$ to transmit the broadcast packet to all the SUs in $\bigcup_{s_u \in \mathcal{D}_i} c(s_u) \setminus \mathcal{C}$. From Lemma 12.2 and the discussion in Section 12.4.2, we know that for $\forall s_u, s_v \in \mathcal{D}_i$ $(1 \leq i \leq 12)$, $D(s_u, s_v) > \frac{(3 \times 2 - 2)r}{2} = 2r$, which implies all the SUs in $\mathcal{D}_i$ can transmit data simultaneously under the UDG model as long as they have spectrum opportunities. Then, we have the following Lemma 12.7 which indicates the time consumption of procedure BROAD-UNI-CAST($\mathcal{D}_i$).

Lemma 12.7
The expected number of time slots consumed by dominators in $\mathcal{D}_i$ to transmit the broadcast packet to all the dominatees in $\bigcup_{s_u \in \mathcal{D}_i} c(s_u) \setminus \mathcal{C}$ by BROAD-UNI-CAST($\mathcal{D}_i$) is at most Δ_T/p if $\Delta_T \leq 1/p$, and $1/p^2 + \exp(\pi\lambda r^2)\log_{1-p}\frac{1}{p\Delta_T}$ if $\Delta_T > 1/p$.

Before giving the proof of Lemma 12.7, we prove another two useful lemmas as follows.

Lemma 12.8

If a dominator $s_u \in \mathcal{D}_i$ executes the broadcast operation (lines 3-4) of procedure BROAD-UNI-CAST($\mathcal{D}_i$) κ times, then the expected number of dominatee children that receive the broadcast packet from s_u is $a_\kappa = \Delta(s_u)(1 - (1-p)^\kappa)$, where $\Delta(s_u)$ is the number of dominatee children that s_u has in T.

Proof: For $s_u \in \mathcal{D}_i$, let a_x be the expected number of its dominatee children that can receive the broadcast packet at the end of the x-th broadcast operation of s_u. Then, we have $a_1 = \Delta(s_u) \cdot p$, and recursively,

$$a_2 = a_1 + (\Delta(s_u) - a_1)p \tag{12.6}$$
$$= \Delta(s_u)p + (1-p)a_1, \tag{12.7}$$
$$a_3 = a_2 + (\Delta(s_u) - a_2)p \tag{12.8}$$
$$= \Delta(s_u)p + (1-p)a_2, \tag{12.9}$$
$$\vdots, \tag{12.10}$$
$$a_\kappa = a_{\kappa-1} + (\Delta(s_u) - a_{\kappa-1})p \tag{12.11}$$
$$= \Delta(s_u)p + (1-p)a_{\kappa-1}. \tag{12.12}$$

To obtain the general term of a_κ, we have

$$a_\kappa = \Delta(s_u)p + (1-p)a_{\kappa-1} \tag{12.13}$$
$$= \Delta(s_u)p + \Delta(s_u)p(1-p) + (1-p)^2 a_{\kappa-2} \tag{12.14}$$
$$= \cdots \tag{12.15}$$
$$= \Delta(s_u)p \cdot (1 + (1-p) + (1-p)^2 + \cdots + (1-p)^{\kappa-2} + (1-p)^{\kappa-1}) \tag{12.16}$$
$$= \Delta(s_u)(1 - (1-p)^\kappa). \tag{12.17}$$

□

Lemma 12.9

Suppose a dominator $s_u \in \mathcal{D}_i$ executes the broadcast operation (lines 3-4) of procedure BROAD-UNI-CAST($\mathcal{D}_i$) at most κ times in MBS-UDG. Then, $\kappa = 0$ if $\Delta(s_u) \leq 1/p$, and $\kappa = \log_{1-p} \frac{1}{p\Delta(s_u)}$ otherwise.

Proof: If $\Delta(s_u) \leq 1/p$, it is straightforward that $\kappa = 0$ according to Algorithm 19 (line 3). Otherwise, suppose $\Delta(s_u) > 1/p$. Then, the number of s_u's dominatee children that receive the broadcast packet during the broadcast operations of s_u is $a_\kappa = \Delta(s_u)(1 - (1-p)^\kappa)$ according to Lemma 12.8. Based on Algorithm 19, the broadcast operation (lines 3-4) will not be executed when $|c(s_u) \cap \mathcal{B}_0| \leq 1/p$ (the number of s_u's dominatee children that wait for receiving the broadcast packet is less than or equal to $1/p$). Therefore, we have

$$a_\kappa + 1/p = \Delta(s_u) \Rightarrow \kappa = \log_{1-p} \frac{1}{p\Delta(s_u)}. \tag{12.18}$$

□

Now, we are ready to prove Lemma 12.7 as follows.

Proof of Lemma 12.7: Suppose s_u is the last SU in $\mathcal{D}_i$ to transmit the broadcast packet to its dominatee children. Then, if $\Delta(s_u) \leq 1/p$, s_u only employs unicast to transmit the broadcast packet. Therefore, the expected number of time slots for $s_u \to c(s_u) \setminus C$ is at most $\Delta(s_u)/p \leq \Delta_T/p$. On the other hand, if $\Delta(s_u) > 1/p$, s_u will conduct at most $1/p$ unicast operations and $\kappa = \log_{1-p} \frac{1}{p\Delta(s_u)}$ broadcast operations according to Algorithm 19 and Lemma 12.9. Furthermore, the condition in Algorithm 19 for a SU to initialize a broadcast operation is that this broadcast operation should not cause any unacceptable interference to the primary network. Then, by similar technique in Lemma 12.4, we can conclude that the opportunity for a broadcast operation is $\exp(-\pi\lambda r^2)$, i.e., the probability that there is no PU receiving some data within the interference range of the broadcasting SU. Consequently, the time slots needed for s_u in this case are

$$1/p^2 + \kappa/\exp(-\pi\lambda r^2) \tag{12.19}$$

$$= 1/p^2 + \exp(\pi\lambda r^2)\log_{1-p}\frac{1}{p\Delta(s_u)} \tag{12.20}$$

$$\leq 1/p^2 + \exp(\pi\lambda r^2)\log_{1-p}\frac{1}{p\Delta_T}. \tag{12.21}$$

□

Based on Lemma 12.7, we can obtain the time consumption of Phase II of MBS-UDG as shown in Theorem 12.2.

Theorem 12.2

The expected number of time slots consumed by Phase II of MBS-UDG is at most

$$\begin{cases} 12\exp(\pi\lambda(r^2+R^2))\Delta_T, & \text{if } \Delta_T \leq \exp(\pi\lambda(r^2+R^2)); \\ 12\exp(2\pi\lambda(r^2+R^2)) \\ +12\exp(\pi\lambda r^2)\log_{1-p}\frac{\exp(\pi\lambda(r^2+R^2))}{\Delta_T}, & \text{if } \Delta_T > \exp(\pi\lambda(r^2+R^2)). \end{cases} \tag{12.22}$$

Proof: Since $\mathcal{D}$ can be partitioned into 12 disjoint and interference-free subsets, this theorem can be obtained based on the conclusion of Lemma 12.7. □

MBS-UDG consists of Phase I and Phase II, and these two phases are executed sequentially. Consequently, the time consumption of MBS-UDG can be obtained directly from Theorem 12.1 and Theorem 12.2 as follows.

Theorem 12.3

The expected time consumption of MBS-UDG is upper bound by

$$\begin{cases} (367(\hbar-1)+12\Delta_T)\tau/p, & \text{if } \Delta_T \leq 1/p; \\ (367(\hbar-1)/p+12/p^2+12e^{\pi\lambda r^2}\log_{1-p}\frac{1}{p\Delta_T})\tau, & \text{if } \Delta_T > 1/p. \end{cases} \tag{12.23}$$

12.5.2.2   Broadcast Redundancy of MBS-UDG

Now, we analyze the broadcast redundancy of MBS-UDG, which is defined as the maximum transmission time of the broadcast packet by any SU during the entire broadcast scheduling (see Section 12.3). According to Algorithm 18, we know that for any dominatee, its broadcast redundancy is 0 since it does not transmit the broadcast packet. For any $s_u \in \mathcal{C}$, it transmits the broadcast packet to its children, which are dominators (see the construction process of T), by unicast according to Algorithm 18. Therefore, the broadcast redundancy for $\forall s_u \in \mathcal{C}$ is at most 4 since it has at most 4 dominator children based on Lemma 12.1. For any $s_u \in \mathcal{D}$, it has at most 12 connector children which have to receive the broadcast packet by unicast according to Algorithm 18. For its dominatee children, it will transmit at most $\Delta(s_u) \leq \Delta_T$ times by unicast if $\Delta(s_u) \leq 1/p$ according to Algorithm 19. Otherwise, it will broadcast for $\log_{1-p} \frac{1}{p\Delta(s_u)} \leq \log_{1-p} \frac{1}{p\Delta_T}$ times and unicast for $1/p$ times. Therefore, for $s_u \in \mathcal{D}$, its broadcast redundancy is at most $12 + \Delta_T$ when $\Delta(s_u) \leq 1/p$, and $12 + 1/p + \log_{1-p} \frac{1}{p\Delta_T}$ when $\Delta(s_u) > 1/p$. In summary, we use the following theorem to show the broadcast redundancy of MBS-UDG.

Theorem 12.4
The broadcast redundancy of MBS-UDG is at most $12 + \Delta_T$ when $\Delta_T \leq 1/p$, and $12 + 1/p + \log_{1-p} \frac{1}{p\Delta_T}$ when $\Delta_T > 1/p$.

12.6   Broadcast Scheduling under PrIM

Under the UDG model, the interference range of a SU/PU is assumed to be same as its transmission range. In reality, the interference range of a node is usually larger than its transmission range. Therefore, a more general interference model for CRNs is the *PrIM*, where the interference range of a SU/PU is β times of its transmission range as defined in Section 12.3, i.e., $r_I = \beta r$ and $R_I = \beta R$, where $\beta \geq 1$ is a constant. Based on Lemma 12.4, let $p_\beta = \exp(-\pi\lambda\beta^2(r^2 + R^2))$ denote the unicast opportunity under the PrIM in this section.

To make proposed MBS algorithm work under the PrIM, we need to change the k in MBS-UDG to $k = \lceil \frac{2\beta+4}{3} \rceil$. Then, according to Lemma 12.2, for $\forall s_u, s_v \in \mathcal{D}_i$, $D(s_u, s_v) > \frac{(3k-2)r}{2} \geq (\beta+1)r$, which implies all the SUs in $\mathcal{D}_i$ can conduct data transmission simultaneously as long as they have spectrum opportunities. The new MBS algorithm under the PrIM is denoted by MBS-PrIM. In the following, we analyze the delay and redundancy performance of MBS.

As MBS-UDG, MBS-PrIM also consists of two phases: Phase I (lines 1-6) and Phase II (lines 7-10). First, we have the following lemmas, which indicate the maximum time consumption for a *dominator→connector* communication and a *connector→dominator* communication in Phase I of MBS-PrIM, respectively.

Lemma 12.10

For a communication $(s_u \in \mathcal{D}, s_v \in \mathcal{C})$ in Phase I of MBS-PrIM, the expected number of time slots to successfully transmit the broadcast packet from s_u to s_v is at most $(11\phi_{\beta+1} + \phi_{\beta+2} - 11)/p_\beta$.

Proof: This lemma can be proven by similar techniques as in Lemma 12.5. First, assume the secondary network is a stand-alone network. Then, the communication (s_u, s_v) may interfere with two types of communications like $(s_a \in \mathcal{D}, s_b \in \mathcal{C})$ and $(s_c \in \mathcal{C}, s_d \in \mathcal{D})$ respectively as shown in Figure 12.5(a). For the first type of interfering communications, we have $D(s_u, s_v) \leq r_I + r = (\beta + 1)r$. Then, based on Lemma 12.1, there are at most $\phi_{\beta+1} - 1$ dominators like s_a which may interfere with (s_u, s_v), and each dominator may have at most 11 connector children. Therefore, the number of communications like (s_a, s_b) is at most $11(\phi_{\beta+1} - 1)$. Similarly, for the second type of interfering communications, we have $D(s_u, s_d) \leq r_I + 2r = (\beta + 2)r$, which implies the number of this kind of communications is at most $\phi_{\beta+2} - 1$. In summary, at most $11(\phi_{\beta+1} - 1) + \phi_{\beta+2} - 1 + 1 = 11\phi_{\beta+1} + \phi_{\beta+2} - 11$ time slots are consumed for communication (s_u, s_v) if the secondary network is a stand-alone network.

Considering the primary network, the spectrum opportunity for a unicast transmission of SUs under the PrIM is p_β. Consequently, the maximum expected time consumption for communication (s_u, s_v) is at most $(11\phi_{\beta+1} + \phi_{\beta+2} - 11)/p_\beta$ time slots. □

Lemma 12.11

For a communication $(s_u \in \mathcal{C}, s_v \in \mathcal{D})$ in Phase I of MBS-PrIM, the expected number of time slots to successfully transmit the broadcast packet from s_u to s_v is at most $(\phi_{\beta+1} + 11\phi_{\beta+2} - 11)/p_\beta$.

Proof: By similar techniques in Lemma 12.6 and Lemma 12.10, this lemma can be proven. □

Based on Lemma 12.10 and Lemma 12.11, we can obtain the time consumption of Phase I in MBS-PrIM as follows.

Theorem 12.5

Phase I of MBS-PrIM consumes at most $(6\phi_{\beta+1} + 6\phi_{\beta+2} - 11)(\hbar - 1)/p_\beta$ time slots.

Proof: By similar techniques of Theorem 12.1, this theorem can be proven. □

Now, we analyze the time consumption for Phase II of MBS-PrIM, which is shown in Theorem 12.6.

Theorem 12.6

Let $k = \lceil \frac{2\beta + 4}{3} \rceil$. The expected number of time slots consumed by Phase II of MBS-PrIM is at most $3k^2 \Delta_T / p_\beta$ if $\Delta_T \leq 1/p_\beta$, and $3k^2/p_\beta^2 + 3k^2 \exp(\pi \lambda \beta^2 r^2) \log_{1-p_\beta} \frac{1}{p_\beta \Delta_T}$ if $\Delta_T > 1/p_\beta$.

Proof: By similar techniques of Lemma 12.8 and Lemma 12.9, we can prove that the number of broadcast operations that a dominator can conduct in MBS-PrIM is $\log_{1-p_\beta}\frac{1}{p_\beta\Delta_T}$ when $\Delta_T > 1/p_\beta$, and 0 otherwise. Furthermore, the broadcast opportunity for a SU under the PrIM is $\exp(-\pi\lambda\beta^2 r^2)$. Therefore, to schedule these $3k^2$ disjoint dominator sets in Phase II, the time consumption is at most $3k^2\Delta_T/p_\beta$ time slots if $\Delta_T \leq 1/p_\beta$, and $3k^2/p_\beta^2 + 3k^2\exp(\pi\lambda\beta^2 r^2)\log_{1-p_\beta}\frac{1}{p_\beta\Delta_T}$ time slots if $\Delta_T > 1/p_\beta$. □

Based on Theorem 12.5 and Theorem 12.6, we can obtain the broadcast latency of MBS-PrIM as shown in Theorem 12.7.

Theorem 12.7
Let $k = \lceil\frac{2\beta+4}{3}\rceil$. The expected number of time slots consumed by Phase II of MBS-PrIM is at most $((6\phi_{\beta+1} + 6\phi_{\beta+2} - 11)(\hbar - 1) + 3k^2\Delta_T)/p_\beta$ if $\Delta_T \leq 1/p_\beta$, and $(6\phi_{\beta+1} + 6\phi_{\beta+2} - 11)(\hbar - 1)/p_\beta + 3k^2/p_\beta^2 + 3k^2\exp(\pi\lambda\beta^2 r^2)\log_{1-p_\beta}\frac{1}{p_\beta\Delta_T}$ if $\Delta_T > 1/p_\beta$.

12.6.1 Redundancy of MBS-PrIM

Now, we analyze the broadcast redundancy of MBS-PrIM. By similar techniques of analyzing the redundancy of MBS-PrIM, we can obtain the broadcast redundancy of MBS-PrIM as follows.

Theorem 12.8
The broadcast redundancy of MBS-PrIM is at most $12 + \Delta_T$ when $\Delta_T \leq 1/p_\beta$, and $12 + 1/p_\beta + \log_{1-p_\beta}\frac{1}{p_\beta\Delta_T}$ when $\Delta_T > 1/p_\beta$.

12.7 Simulation and Analysis

In this section, we examine the latency and redundancy performance of MBS by simulations. In all the simulations, we assume a secondary network coexists with a primary network distributed in a square area of size $X \times Y$, and they share a common channel (detailed network model can be found in Section 12.3). The primary network is Poisson distributed and the activity of PUs satisfies a two-dimensional Poisson point process with density λ (see Section 12.3). In the secondary network, a broadcast source SU which carries a broadcast packet and n other SUs are randomly distributed. The *network density*, denoted by ρ, of the secondary network is defined as the ratio between n and the network size. The network time is slotted with each time slot of length 1 millisecond (ms). For other network settings, e.g., r, R, β etc., defined in Section 12.3, we will specify them in each group of simulations.

Since the UDG model is a special case of the PrIM, we consider the PrIM in all the simulations. Furthermore, we use MBS to denote our algorithm. The compared algorithms are the two heuristic algorithms proposed in [236], denoted by H1 and H2 respectively, which are the only algorithms for the MLBS problem in CRNs. In both H1 and H2, only the broadcast communication mode is employed. The basic idea of H1 is that each SU broadcasts the data packet to its child which has the path to the farthest SU (from the source SU) first. On the other hand, H2 tries to maximize the number of children that can receive the broadcast packet during each transmission at each step. In the following, we will compare MBS with H1 and H2 for both broadcast latency and redundancy. Each group of simulations is repeated 10 times, and the results are the average values of these 10 times.

12.7.1   Broadcast Latency of MBS

In this subsection, we check the latency performance of MBS, H1, and H2 under different scenarios. If we fix the density of SUs as $\rho = 4.0$, the impacts of the network size on the broadcast latency of MBS, H1, and H2 are shown in Figure 12.6(a). From Figure 12.6(a), we can see that the induced latency of all the three algorithms increases when the network size increases. The reason is straightforward since more SUs are involved in the broadcast task and the height of the broadcasting tree increases. MBS has better performance than H1 and H2, especially in large CRNs. This is because a successful broadcast opportunity (the opportunity that all the neighbors of a SU can receive the broadcast packet from that SU by one broadcast communication) is usually much rarer than a successful unicast opportunity. By taking this advantage, MBS first transmits the broadcast packet to the selected backbone SUs by unicast. Subsequently, MBS schedules these backbone SUs concurrently by employing mixed unicast and broadcast communication modes to achieve maximum transmission efficiency (see the analysis in Section 12.5.1), which significantly accelerates the broadcast scheduling. Furthermore, H1 and H2 have similar performance, which is consistent with the results in [236]. On average, MBS induces 298.04% and 298.16% less time compared with H1 and H2 respectively.

When we fix the network size and change the network density ρ of SUs, the latency performance of MBS, H1, and H2 is shown in Figure 12.6(b). From Figure 12.6(b), we can see that with the increase of ρ, the latency of the three algorithms increases. This is different from broadcasting in traditional wireless networks where network density is usually not a factor to affect the broadcast latency. The reason is that large ρ implies an inner node in the broadcasting tree has more children. Furthermore, because of lacking of spectrum opportunities, a SU may have to transmit more times to let all its neighbors to receive the broadcast packet. Therefore, if we only increase the node density, the induced broadcast latency will also increase, which can also be seen from Lemma 12.8. Again, MBS has better performance than H1 and H2 since it exploits unicast and broadcast elegantly. On average, MBS induces 251.02% and 251.09 less latency than H1 and H2, respectively.

The impacts of the intensity of primary activities λ on the latency performance of MSB, H1, and H2 are shown in Figure 12.6(c). From Figure 12.6(c), we can see that

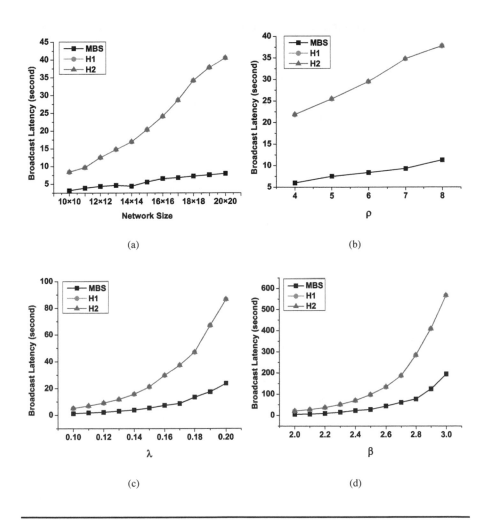

Figure 12.6: Broadcast latency of MBS. Without specifications, the default settings are $X = 15, Y = 15, \rho = 4.0, \lambda = 0.15, r = 1.0, R = 1.0,$ **and** $\beta = 2.0.$

when λ increases, the primary network will conduct more transmission and reception activities, which implies the spectrum opportunities for SUs become less. Therefore, SUs have to wait for longer time to acquire a spectrum opportunity to transmit data, followed by the increase of broadcast latency of the three algorithms. From Figure 12.6(c), we can also see that the latency increase of H1 and H2 is much faster than that of MBS. This is because successful broadcast opportunities are fewer than successful unicast opportunities. To maximize the transmission efficiency and reduce the latency, MBS exploits unicast and broadcast communication modes carefully based on different network scenarios. Consequently, MBS has a better performance and induces 282.11% and 282.15% less latency on average compared with H1 and H2, respectively.

When increase β (the coefficient in the PrIM), the changes of the latency of the three algorithms are shown in Figure 12.6(b). From Figure 12.6(b), we can see that when β increases, the induced latency of MBS, H1, and H2 increases. This is because a large β implies large interference area of both SUs and PUs. Hence, both the transmission concurrency and the spectrum opportunities for all the algorithms decrease, and followed by more time consumption. Again, MBS takes less time than H1 and H2 to finish the broadcast scheduling since it exploits mixed unicast and broadcast communication modes depending on different network environments.

12.7.2 Broadcast Redundancy of MBS

We examine the broadcast redundancy performance of MBS, H1, and H2 in this subsection and the results are shown in Figure 12.7. Figure 12.7(a) shows the impacts of network size on the redundancy of MBS, H1, and H2. From Figure 12.7(a), we can see that if we fix the network density and change the network size, the broadcast redundancy of all the three algorithms keeps stable. This is because the network density is fixed, which implies the number of children of an inner node in the broadcasting tree keep stable no matter the size of the network/tree. On the other hand, the broadcast redundancy depends on the number of children that a node has and the available spectrum opportunities. Therefore, network size has a little impact on the redundancy of all the three algorithms. Furthermore, MBS has a better performance than H1 and H2. This is because H1 and H2 employ the broadcast communication mode only. As we analyzed in Section 12.5.1, a broadcast transmission in CRNs may not have any receiver due to the lack of spectrum opportunities. Therefore, a SU in H1 and H2 may broadcast many times to guarantee its children to receive the broadcast packet, which also wastes the energy of a SU and causes more interference to both the primary network and the secondary network. In MBS, the unicast and broadcast communication modes are all employed depending on network scenarios to maximize the transmission efficiency. Consequently, MBS is more efficient on broadcast redundancy. On average, MBS induces 63.61% and 63.84% less redundancy than H1 and H2, respectively.

The impacts of the density of SUs ρ on the broadcast redundancy of MBS, H1, and H2 are shown in Figure 12.7(b). From Figure 12.7(b), we can see that when ρ increases, the redundancy of H1 and H2 remains almost stable. This is because H1 and

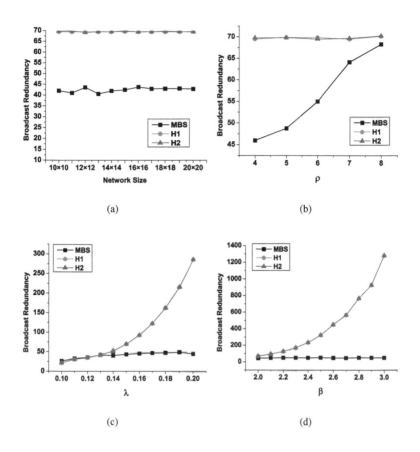

(a)

(b)

(c)

(d)

Figure 12.7: Broadcast redundancy of MBS. Without specifications, the default settings are $X = 15, Y = 15, \rho = 4.0, \lambda = 0.15, r = 1.0, R = 1.0,$ **and** $\beta = 2.0.$

H2 only exploit the broadcast communication mode and the broadcast opportunity depends on the primary activity (e.g. λ) and wireless interference (e.g. $\beta, r,$ and R). Therefore, if the parameters related to primary activity and wireless interference are fixed, the broadcast opportunity for H1 and H2 will keep stable, which implies their redundancy keeps stable. On the other hand, when ρ increases, a backbone node has more children, which implies MBS will exploit the broadcast communication mode more times based on Algorithm 19. Therefore, the induced redundancy of MBS increases and gets close to H1 and H2 for large ρ. This further implies MBS has more advantages in sparse CRNs than H1 and H2 with respect to broadcast redundancy.

Figure 12.7(c) shows the impacts of the primary activity λ on the redundancy of MBS, H1, and H2. From Figure 12.7(c), we can see that λ has a little impact on the redundancy of MBS. This is because even large λ implies less spectrum opportunities, MBS can choose to exploit the broadcast communication mode or the unicast communication mode, to guarantee at least one children SU can receive the broadcast packet from the parent SU (see Lemma 12.3). Consequently, although large λ induces more latency, it has little impact on the redundancy of MBS. On the other hand, since H1 and H2 exploit the broadcast communication mode which highly depends on λ, their redundancy increases significantly when λ increases. This also explains why H1 and H2 have similar or even a little better redundancy performance compared with MBS for small λ, e.g. $\lambda = 0.1$. On average, MBS has 148.86% and 149.31% less redundancy than H1 and H2, respectively.

The impacts of β on the redundancy of MBS, H1, and H2 are shown in Figure 12.7 (b). From Figure 12.7(b), we can see that when β increases, the redundancy of H1 and H2 increases significantly. This is because their redundancy highly depends on the broadcast spectrum opportunity, which is further affected by β. A large β implies large interference area for both SUs and PUs. Therefore, large β induces less spectrum opportunities and followed by high redundancy in H1 and H2. On the other hand, the dominating factor to decide the redundancy of MBS is the number of children a backbone SU has. Therefore, a large β has significant impacts on the latency of MBS and small impacts on the redundancy of MBS. Consequently, the redundancy of MBS keeps almost stable. On average, H1 and H2 induce 8.4637 and 8.467 times of more redundancy than that of MBS, respectively.

12.8   Conclusion

Since existing broadcasting algorithms for CRNs are either heuristic solutions without performance guarantee or with performance far from the optimal solution, we study the MLBS problem for CRNs in this chapter. First, we propose a mixed broadcast scheduling (MBS) algorithm under the UDG model, denoted by MBS-UDG. We obtain the latency performance of MBS-UDG, which significantly improves existing results. Furthermore, the redundancy performance of MBS-UDG is also obtained. Second, we generalize MBS to CRNs under the PrIM, and analyze the algorithm latency and redundancy performance under that model. Finally, simulations are conducted to examine the performance of MBS, which demonstrate that MBS significantly improves both the broadcast latency and the broadcast redundancy over existing algorithms.

References

[1] C. Luo, F. Wu, J. Sun, and C.W. Chen. Compressive data gathering for large-scale wireless sensor networks. In *Proceedings of the 15th Annual International Conference on Mobile Computing and Networking*, pages 145–156. ACM, 2009.

[2] M. Zuniga and B. Krishnamachari. Analyzing the transitional region in low power wireless links. In *Sensor and Ad Hoc Communications and Networks, First Annual IEEE Communications Society Conference*, pages 517–526. 2004.

[3] G. Zhou, T. He, S. Krishnamurthy, and J.A. Stankovic. Impact of radio irregularity on wireless sensor networks. In *Proceedings of the 2nd International Conference on Mobile Systems, Applications, and Services*, pages 125–138. ACM, 2004.

[4] A. Cerpa, J.L. Wong, M. Potkonjak, and D. Estrin. Temporal properties of low power wireless links: modeling and implications on multi-hop routing. In *Proceedings of the 6th ACM International Symposium on Mobile Ad Hoc Networking and Computing*, pages 414–425. 2005.

[5] A. Cerpa, J.L. Wong, L. Kuang, M. Potkonjak, and D. Estrin. Statistical model of lossy links in wireless sensor networks. In *Information Processing in Sensor Networks, Fourth International Symposium*, pages 81–88. 2005.

[6] Y. Liu, Q. Zhang, and L.M. Ni. Opportunity-based topology control in wireless sensor networks. *Parallel and Distributed Systems, IEEE Transactions on*, 21(3):405–416, 2010.

[7] R. Rajaraman. Topology control and routing in ad hoc networks: a survey. *ACM SIGACT News*, 33(2):60–73, 2002.

[8] W. Di, Q. Yan, and T. Ning. Connected dominating set based hybrid routing algorithm in ad hoc networks with obstacles. *IEEE International Conference on Communications*, volume 9, pages 4008–4013. 2006.

[9] W. El-Hajj, A. Al-Fuqaha, M. Guizani, and H.H. Chen. On efficient network planning and routing in large-scale manets. *Vehicular Technology, IEEE Transactions on*, 58(7):3796–3801, 2009.

[10] S. Ji, Y. Li, and X. Jia. Capacity of dual-radio multi-channel wireless sensor networks for continuous data collection. In *INFOCOM, Proceedings of IEEE*, pages 1062–1070. 2011.

[11] S. Ji, R. Beyah, and Y. Li. Continuous data collection capacity of wireless sensor networks under physical interference model. In *Mobile Ad Hoc and Sensor Systems, IEEE 8th International Conference*, pages 222–231. 2011.

[12] M. Li, L. Ding, Y. Shao, Z. Zhang, and B. Li. On reducing broadcast transmission cost and redundancy in ad hoc wireless networks using directional antennas. *Vehicular Technology, IEEE Transactions on*, 59(3):1433–1442, 2010.

[13] B.K. Polat, P. Sachdeva, M.H. Ammar, and E.W. Zegura. Message ferries as generalized dominating sets in intermittently connected mobile networks. *Pervasive and Mobile Computing*, 7(2):189–205, 2011.

[14] B. Deb, S. Bhatnagar, and B. Nath. Multi-resolution state retrieval in sensor networks. In *Sensor Network Protocols and Applications, 2003. Proceedings of the First IEEE International Workshop*, pages 19–29. 2003.

[15] K.P. Shih, D.J. Deng, R.S. Chang, and H.C. Chen. On connected target coverage for wireless heterogeneous sensor networks with multiple sensing units. *Sensors*, 9(7):5173–5200, 2009.

[16] H.M. Ammari and J. Giudici. On the connected k-coverage problem in heterogeneous sensor nets: the curse of randomness and heterogeneity. In *Distributed Computing Systems, 29th IEEE International Conference on*, pages 265–272. 2009.

[17] P.J. Wan, S.C.H. Huang, L. Wang, Z. Wan, and X. Jia. Minimum-latency aggregation scheduling in multihop wireless networks. In *Proceedings of the 10th ACM International Symposium on Mobile Ad Hoc Networking and Computing*, pages 185–194. 2009.

[18] M. Yan, J. He, S. Ji, and Y. Li. Minimum latency scheduling for multi-regional query in wireless sensor networks. In *Performance Computing and Communications Conference, IEEE 30th International*, pages 1–8. 2011.

[19] A. Ephremides, J.E. Wieselthier, and D.J. Baker. A design concept for reliable mobile radio networks with frequency hopping signaling. *Proceedings of the IEEE*, 75(1):56–73, 1987.

[20] S. Guha and S. Khuller. Approximation algorithms for connected dominating sets. *Algorithmica*, 20(4):374–387, 1998.

[21] M.R. Gary and D.S. Johnson. Computers and intractability: A guide to the theory of np-completeness, 1979.

[22] J. Schleich, G. Danoy, P. Bouvry, and L. Thi Hoai An. Blackbone2: an efficient deterministic algorithm for creating 2-connected m-dominating set-based backbones in ad hoc networks. In *Proceedings of the 7th ACM international Symposium on Mobility Management and Wireless Access*, pages 91–98. 2009.

[23] D. Kim, W. Wang, X. Li, Z. Zhang, and W. Wu. A new constant factor approximation for computing 3-connected m-dominating sets in homogeneous wireless networks. In *INFOCOM, 2010 Proceedings IEEE*, pages 1–9. 2010.

[24] F. Dai and J. Wu. On constructing k-connected m-dominating set in wireless ad hoc and sensor networks. *Journal of Parallel and Distributed Computing*, 66(7):947–958, 2006.

[25] L. Ding, X. Gao, W. Wu, W. Lee, X. Zhu, and D.Z. Du. Distributed construction of connected dominating sets with minimum routing cost in wireless networks. In *Distributed Computing Systems (ICDCS), 2010 IEEE 30th International Conference on*, pages 448–457. 2010.

[26] D. Kim, Y. Wu, Y. Li, F. Zou, and D.Z. Du. Constructing minimum connected dominating sets with bounded diameters in wireless networks. *Parallel and Distributed Systems, IEEE Transactions on*, 20(2):147–157, 2009.

[27] T.N. Nguyen and DT Huynh. Energy-efficient connected d-hop dominating sets in wireless sensor networks. In *Pervasive Computing and Communications, 2009. PerCom 2009. IEEE International Conference on*, pages 1–6. 2009.

[28] J. He, S. Ji, M. Yan, Y. Pan, and Y. Li. Genetic-algorithm-based construction of load-balanced cdss in wireless sensor networks. In *Military Communications Conference*, pages 667–672. 2011.

[29] B.N. Clark, C.J. Colbourn, and D.S. Johnson. Unit disk graphs. *Discrete Mathematics*, 86(1):165–177, 1990.

[30] S. Guha and S. Khuller. Approximation algorithms for connected dominating sets. *Algorithms ESA'96*, pages 179–193, 1996.

[31] D. Du and X. Hu. *Steiner Tree Problems in Computer Communication Networks*. World Scientific Publishing Company Incorporated, 2008.

[32] L. Ruan, H. Du, X. Jia, W. Wu, Y. Li, and K.I. Ko. A greedy approximation for minimum connected dominating sets. *Theoretical Computer Science*, 329(1):325–330, 2004.

[33] J. Wu, M. Gao, and I. Stojmenovic. On calculating power-aware connected dominating sets for efficient routing in ad hoc wireless networks. In *Parallel Processing, International Conference on*, pages 346–354. IEEE, 2001.

[34] F. Dai and J. Wu. An extended localized algorithm for connected dominating set formation in ad hoc wireless networks. *Parallel and Distributed Systems, IEEE Transactions on*, 15(10):908–920, 2004.

[35] D. Zhou, M.T. Sun, and T.H. Lai. A timer-based protocol for connected dominating set construction in ieee 802.11 multihop mobile ad hoc networks. In *Applications and the Internet, Proceedings of Symposium on*, pages 2–8. IEEE, 2005.

[36] K. Sakai, F. Shen, K.M. Kim, M.T. Sun, and H. Okada. Multi-initiator connected dominating set construction for mobile ad hoc networks. In *Communications, IEEE International Conference on*, pages 2431–2436. 2008.

[37] P.J. Wan, K.M. Alzoubi, and O. Frieder. Distributed construction of connected dominating set in wireless ad hoc networks. *Mobile Networks and Applications*, 9(2):141–149, 2004.

[38] W. Wu, H. Du, X. Jia, Y. Li, and S.C.H. Huang. Minimum connected dominating sets and maximal independent sets in unit disk graphs. *Theoretical Computer Science*, 352(1):1–7, 2006.

[39] P.J. Wan, L. Wang, and F. Yao. Two-phased approximation algorithms for minimum cds in wireless ad hoc networks. In *Distributed Computing Systems, 28th International Conference on*, pages 337–344. IEEE, 2008.

[40] M. Li, P.J. Wan, and F. Yao. Tighter approximation bounds for minimum CDS in wireless ad hoc networks. *Algorithms and Computation*, pages 699–709, 2009.

[41] X. Li, X. Gao, and W. Wu. A better theoretical bound to approximate connected dominating set in unit disk graph. *Wireless Algorithms, Systems, and Applications*, pages 162–175, 2008.

[42] X. GAO, Y. WANG, X. LI, and W. WU. Analysis on theoretical bounds for approximating dominating set problems. *Discrete Mathematics, Algorithms and Applications*, 1(01):71–84, 2009.

[43] R. Misra and C. Mandal. Rotation of CDS via connected domatic partition in ad hoc sensor networks. *Mobile Computing, IEEE Transactions on*, 8(4):488–499, 2009.

[44] F. Wang, M.T. Thai, and D.Z. Du. On the construction of 2-connected virtual backbone in wireless networks. *Wireless Communications, IEEE Transactions on*, 8(3):1230–1237, 2009.

[45] D. Mandala, X. Du, F. Dai, and C. You. Load balance and energy efficient data gathering in wireless sensor networks. *Wireless Communications and Mobile Computing*, 8(5):645–659, 2008.

[46] M. Song, Y. Zhao, J. Wang, and EK Park. A high throughput load balance algorithm for multichannel wireless sensor networks. In *Communications, IEEE International Conference on*, pages 1–5. 2009.

[47] Y. Deng and Y. Hu. A load balance clustering algorithm for heterogeneous wireless sensor networks. In *E-Product E-Service and E-Entertainment, International Conference on*, pages 1–4. IEEE, 2010.

[48] G. Cao and F. Yu. The analysis of load balance for wireless sensor network using compressive sensing. In *Computational Science and Engineering (CSE), IEEE 14th International Conference on*, pages 100–105. 2011.

[49] M. Zhao and Y. Yang. A framework for mobile data gathering with load balanced clustering and mimo uploading. In *INFOCOM, Proceedings*, pages 2759–2767. IEEE, 2011.

[50] M. Kalantari, M. Haghpanahi, and M. Shayman. A p-norm flow optimization problem in dense wireless sensor networks. In *INFOCOM 27th Conference on Computer Communications. IEEE*, pages 341–345. 2008.

[51] S.R. Kulkarni and P. Viswanath. A deterministic approach to throughput scaling in wireless networks. *Information Theory, IEEE Transactions on*, 50(6):1041–1049, 2004.

[52] S. Hadim and N. Mohamed. Middleware challenges and approaches for wireless sensor networks. *Distributed Systems Online, IEEE*, 7(3):1–1, 2006.

[53] S.Y. Ni, Y.C. Tseng, Y.S. Chen, and J.P. Sheu. The broadcast storm problem in a mobile ad hoc network. In *Proceedings of the 5th Annual International Conference on Mobile Computing and Networking*, pages 151–162. ACM, 1999.

[54] B. Das and V. Bharghavan. Routing in ad-hoc networks using minimum connected dominating sets. In *Communications, Towards the Knowledge Millennium, International Conference on*, Volume 1, pages 376–380. IEEE, 1997.

[55] D.E. Goldberg. Genetic algorithms in search, optimization, and machine learning. 1989.

[56] A. Bari, S. Wazed, A. Jaekel, and S. Bandyopadhyay. A genetic algorithm based approach for energy efficient routing in two-tiered sensor networks. *Ad Hoc Networks*, 7(4):665–676, 2009.

[57] X.M. Hu, J. Zhang, Y. Yu, H.S.H. Chung, Y.L. Li, Y.H. Shi, and X.N. Luo. Hybrid genetic algorithm using a forward encoding scheme for lifetime maximization of wireless sensor networks. *Evolutionary Computation, IEEE Transactions on*, 14(5):766–781, 2010.

[58] J.R. Koza, M.A. Keane, M.J. Streeter, W. Mydlowec, J. Yu, and G. Lanza. *Genetic Programming IV: Routine Human-Competitive Machine Intelligence.* Springer, 2005.

[59] M. Cardei, X. Cheng, X. Cheng, and D.Z. Du. Connected domination in multihop ad hoc wireless networks. In *International Conference on Computer Science and Informatics*, Volume 140. Citeseer, 2002.

[60] Y. Li, M.T. Thai, F. Wang, C.W. Yi, P.J. Wan, and D.Z. Du. On greedy construction of connected dominating sets in wireless networks. *Wireless Communications and Mobile Computing*, 5(8):927–932, 2005.

[61] K.M. Alzoubi, P.J. Wan, and O. Frieder. Message-optimal connected dominating sets in mobile ad hoc networks. In *Proceedings of the Third ACM International Symposium on Mobile Ad Hoc Networking & Computing*, Volume 9, pages 157–164, 2002.

[62] Zhu S. Thai M. Li, Y. and D.Z. Du. Localized construction of connected dominating set in wireless networks. 2004.

[63] C. Adjih, P. Jacquet, L. Viennot, et al. Computing connected dominated sets with multipoint relays. 2002.

[64] X. Cheng, M. Ding, D.H. Du, and X. Jia. Virtual backbone construction in multihop ad hoc wireless networks. *Wireless Communications and Mobile Computing*, 6(2):183–190, 2006.

[65] J. H. Holland. *Adaptation in Natural and Artificial System*, Volume 1. University of Michigan Press, 1975.

[66] Z. Cai, S. Ji, J. He, and AG Bourgeois. Optimal distributed data collection for asynchronous cognitive radio networks. In *Distributed Computing Systems, 32nd International Conference on*, pages 245–254. 2012.

[67] M. Yan, J.S. He, S. Ji, and Y. Li. Multi-regional query scheduling in wireless sensor networks with minimum latency. *Wireless Communications and Mobile Computing*, 2012.

[68] J. He, S. Ji, M. Yan, Y. Pan, and Y. Li. Load–balanced CDS construction in wireless sensor networks via genetic algorithm. *International Journal of Sensor Networks*, 11(3):166–178, 2012.

[69] M.T.C.S.E.S. JIS. Computers and intractability a guide to the theory of NP completeness. 1979.

[70] S. Gueye and P. Michelon. A linearization framework for unconstrained quadratic (0-1) problems. *Discrete Applied Mathematics*, 157(6):1255–1266, 2009.

[71] S. Ji, A.S. Uluagac, R. Beyah, and Z. Cai. Practical unicast and convergecast scheduling schemes for cognitive radio networks. *Journal of Combinatorial Optimization*, pages 1–17, 2012.

[72] A. Helmy, S. Garg, P. Pamu, and N. Nahata. Card: A contact-based architecture for resource discovery in ad hoc networks.

[73] S. Ji, R. Beyah, and Z. Cai. Snapshot/continuous data collection capacity for large-scale probabilistic wireless sensor networks. In *INFOCOM, 2012 Proceedings, IEEE*, pages 1035–1043. 2012.

[74] S. Ji and Z. Cai. Distributed data collection and its capacity in asynchronous wireless sensor networks. In *INFOCOM, 2012 Proceedings, IEEE*, pages 2113–2121. 2012.

[75] P.J. Wan, K.M. Alzoubi, and O. Frieder. Distributed construction of connected dominating set in wireless ad hoc networks. In *INFOCOM 2002. Twenty-First Annual Joint Conference of the IEEE Computer and Communications Societies*. Volume 3, pages 1597–1604. 2012.

[76] K.W. Chin, J. Judge, A. Williams, and R. Kermode. Implementation experience with manet routing protocols. *ACM SIGCOMM Computer Communication Review*, 32(5):49–59, 2002.

[77] A. Agrawal and R.E. Barlow. A survey of network reliability and domination theory. *Operations Research*, 32(3):478–492, 1984.

[78] M. Al-Obaidy, A. Ayesh, and A.F. Sheta. Optimizing the communication distance of ad hoc wireless sensor networks by genetic algorithms. *Artificial Intelligence Review*, 29(3):183–194, 2008.

[79] J. Wang, C. Niu, and R. Shen. Priority-based target coverage in directional sensor networks using a genetic algorithm. *Computers & Mathematics with Applications*, 57(11):1915–1922, 2009.

[80] J. He, S. Ji, P. Fan, Y. Pan, and Y. Li. Constructing a load-balanced virtual backbone in wireless sensor networks. In *Computing, Networking and Communications (ICNC), 2012 International Conference on*, pages 959–963. IEEE, 2012.

[81] J.S. He, S. Ji, Y. Pan, and Y. Li. Greedy construction of load-balanced virtual backbones in wireless sensor networks. *Wireless Communications and Mobile Computing*, 2012.

[82] S. Biswas and R. Morris. Opportunistic routing in multi-hop wireless networks. *ACM SIGCOMM Computer Communication Review*, 34(1):69–74, 2004.

[83] S. Chachulski, M. Jennings, S. Katti, and D. Katabi. *Trading Structure for Randomness in Wireless Opportunistic Routing*, Volume 37. ACM, 2007.

[84] J. Ma, Q. Zhang, C. Qian, and L.M. Ni. Energy-efficient opportunistic topology control in wireless sensor networks. In *Proceedings of the 1st International MobiSys Workshop on Mobile Opportunistic Networking*, pages 33–38. ACM, 2007.

[85] J. Ma, C. Qian, Q. Zhang, and L.M. Ni. Opportunistic transmission-based QoS topology control in wireless sensor networks. In *Mobile Ad Hoc and Sensor Systems, 2008. 5th IEEE International Conference on*, pages 422–427. 2008.

[86] Y. Liu, L. Ni, and C. Hu. A generalized probabilistic topology control for wireless sensor networks. *Selected Areas in Communications, IEEE Journal on*, 30(9):1780–1788, 2012.

[87] S. Lin, J. Zhang, G. Zhou, L. Gu, J.A. Stankovic, and T. He. adaptive transmission power control for wireless sensor networks. In *Proceedings of the 4th International Conference on Embedded Networked Sensor Systems*, pages 223–236. ACM, 2006.

[88] D. Son, B. Krishnamachari, and J. Heidemann. Experimental study of concurrent transmission in wireless sensor networks. In *Proceedings of the 4th International Conference on Embedded Networked Sensor Systems*, pages 237–250. ACM, 2006.

[89] B. Karp and H.T. Kung. Gpsr: Greedy perimeter stateless routing for wireless networks. In *Proceedings of the 6th Annual International Conference on Mobile Computing and Networking*, pages 243–254. ACM, 2000.

[90] S. Yang and R. Tinós. A hybrid immigrants scheme for genetic algorithms in dynamic environments. *International Journal of Automation and Computing*, 4(3):243–254, 2007.

[91] X. Yu, K. Tang, and X. Yao. An immigrants scheme based on environmental information for genetic algorithms in changing environments. In *Evolutionary Computation, IEEE Congress on*, pages 1141–1147. 2008.

[92] X. Yu, K. Tang, T. Chen, and X. Yao. Empirical analysis of evolutionary algorithms with immigrants schemes for dynamic optimization. *Memetic Computing*, 1(1):3–24, 2009.

[93] J.J. Grefenstette et al. *Genetic Algorithms for Changing Environments*. Navy Research Laboratory, Navy Center for Applied Research in Artificial Intelligence, 1992.

[94] S. Basagni, M. Mastrogiovanni, A. Panconesi, and C. Petrioli. Localized protocols for ad hoc clustering and backbone formation: A performance comparison. *Parallel and Distributed Systems, IEEE Transactions on*, 17(4):292–306, 2006.

[95] C.A. Coello Coello. Evolutionary multi-objective optimization: a historical view of the field. *Computational Intelligence Magazine*, 1(1):28–36, 2006.

[96] K. Deb, A. Pratap, S. Agarwal, and T. Meyarivan. A fast and elitist multiobjective genetic algorithm: Nsga-ii. *Evolutionary Computation, IEEE Transactions on*, 6(2):182–197, 2002.

[97] D.A.V. Veldhuizen and G.B. Lamont. Multiobjective evolutionary algorithms: Analyzing the state-of-the-art. *Evolutionary Computation*, 8(2):125–147, 2000.

[98] N. Srinivas and K. Deb. Muiltiobjective optimization using nondominated sorting in genetic algorithms. *Evolutionary Computation*, 2(3):221–248, 1994.

[99] J. Horn, N. Nafpliotis, and D.E. Goldberg. A niched Pareto genetic algorithm for multiobjective optimization. In *Evolutionary Computation, 1994. Proceedings of the First IEEE Conference on*, pages 82–87. IEEE, 1994.

[100] S. Agrawal, BK Panigrahi, and M.K. Tiwari. Multiobjective particle swarm algorithm with fuzzy clustering for electrical power dispatch. *Evolutionary Computation, IEEE Transactions on*, 12(5):529–541, 2008.

[101] M. Gong, L. Jiao, H. Du, and L. Bo. Multiobjective immune algorithm with nondominated neighbor-based selection. *Evolutionary Computation, IEEE Transactions on*, 16(2):225–255, 2008.

[102] Q. Zhang, A. Zhou, and Y. Jin. Rm-meda: A regularity model-based multiobjective estimation of distribution algorithm. *Evolutionary Computation, IEEE Transactions on*, 12(1):41–63, 2008.

[103] A.J. Nebro, F. Luna, E. Alba, B. Dorronsoro, J.J. Durillo, and A. Beham. Abyss: Adapting scatter search to multiobjective optimization. *Evolutionary Computation, IEEE Transactions on*, 12(4):439–457, 2008.

[104] T. Back, D.B. Fogel, and Z. Michalewicz. *Handbook of Evolutionary Computation*. IOP Publishing Ltd., 1997.

[105] R. Cristescu, B. Beferull-Lozano, and M. Vetterli. On network correlated data gathering. In *Twenty-third Annual Joint Conference of the IEEE Computer and Communications Societies*, Volume 4, pages 2571–2582. 2004.

[106] S. Madden, R. Szewczyk, M.J. Franklin, and D. Culler. Supporting aggregate queries over ad-hoc wireless sensor networks. In *Mobile Computing Systems and Applications, 2002. Proceedings Fourth IEEE Workshop on*, pages 49–58. IEEE, 2002.

[107] H.Ö. Tan and I. Körpeoglu. Power efficient data gathering and aggregation in wireless sensor networks. *ACM SIGMOD Record*, 32(4):66–71, 2003.

[108] H.O. Tan, I. Korpeoglu, and I. Stojmenovic. Computing localized power-efficient data aggregation trees for sensor networks. *Parallel and Distributed Systems, IEEE Transactions on*, 22(3):489–500, 2011.

[109] X. Chen, X. Hu, and J. Zhu. Minimum data aggregation time problem in wireless sensor networks. *Mobile Ad-hoc and Sensor Networks*, pages 133–142, 2005.

[110] Y. Xue, Y. Cui, and K. Nahrstedt. Maximizing lifetime for data aggregation in wireless sensor networks. *Mobile Networks and Applications*, 10(6):853–864, 2005.

[111] H.C. Lin, F.J. Li, and K.Y. Wang. Constructing maximum-lifetime data gathering trees in sensor networks with data aggregation. In *Communications (ICC), 2010 IEEE International Conference on*, pages 1–6. 2010.

[112] Y. Wu, S. Fahmy, and N.B. Shroff. On the construction of a maximum-lifetime data gathering tree in sensor networks: Np-completeness and approximation algorithm. In *27th Conference on Computer Communications. IEEE*, pages 356–360. 2008.

[113] S.C.H. Huang, P.J. Wan, C.T. Vu, Y. Li, and F. Yao. Nearly constant approximation for data aggregation scheduling in wireless sensor networks. In *26th IEEE International Conference on Computer Communications. IEEE*, pages 366–372. 2007.

[114] X. Xu, X.Y. Li, X. Mao, S. Tang, and S. Wang. A delay-efficient algorithm for data aggregation in multihop wireless sensor networks. *Parallel and Distributed Systems, IEEE Transactions on*, 22(1):163–175, 2011.

[115] Y. Li, L. Guo, and S.K. Prasad. An energy-efficient distributed algorithm for minimum-latency aggregation scheduling in wireless sensor networks. In *Distributed Computing Systems (ICDCS), 2010 IEEE 30th International Conference on*, pages 827–836. IEEE, 2010.

[116] K. Kalpakis and S. Tang. A combinatorial algorithm for the maximum lifetime data gathering with aggregation problem in sensor networks. *Computer Communications*, 32(15):1655–1665, 2009.

[117] D. Virmani and S. Jain. Construction of decentralized lifetime maximizing tree for data aggregation in wireless sensor networks. *World Academy of Science, Engineering and Technology*, 52:54–63, 2009.

[118] D. Luo, X. Zhu, X. Wu, and G. Chen. Maximizing lifetime for the shortest path aggregation tree in wireless sensor networks. In *INFOCOM, 2011 Proceedings, IEEE*, pages 1566–1574. 2011.

[119] S. Boyd and L. Vandenberghe. *Convex Optimization*. Cambridge university press, 2004.

[120] I.F. Akyildiz, W. Su, Y. Sankarasubramaniam, and E. Cayirci. A survey on sensor networks. *Communications Magazine, IEEE*, 40(8):102–114, 2002.

[121] S. Slijepcevic and M. Potkonjak. Power efficient organization of wireless sensor networks. In *Communications, 2001. ICC 2001. IEEE International Conference on*, Volume 2, pages 472–476. 2001.

[122] K. Kar, S. Banerjee, et al. Node placement for connected coverage in sensor networks. In *Modeling and Optimization in Mobile, Ad Hoc and Wireless Networks*, 2003.

[123] Y.C. Wang, C.C. Hu, and Y.C. Tseng. Efficient placement and dispatch of sensors in a wireless sensor network. *Mobile Computing, IEEE Transactions on*, 7(2):262–274, 2008.

[124] C.F. Huang and Y.C. Tseng. The coverage problem in a wireless sensor network. *Mobile Networks and Applications*, 10(4):519–528, 2005.

[125] T. Yan, T. He, and J.A. Stankovic. Differentiated surveillance for sensor networks. In *Proceedings of the 1st International Conference on Embedded Networked Sensor Systems*, pages 51–62. ACM, 2003.

[126] M. Cardei, M.T. Thai, Y. Li, and W. Wu. Energy-efficient target coverage in wireless sensor networks. In *24th Annual Joint Conference of the IEEE Computer and Communications Societies Proceedings, IEEE*, Volume 3, pages 1976–1984. 2005.

[127] S. Kumar, T.H. Lai, and A. Arora. Barrier coverage with wireless sensors. In *Proceedings of the 11th Annual International Conference on Mobile Computing and Networking*, pages 284–298. ACM, 2005.

[128] C. Gui and P. Mohapatra. Virtual patrol: a new power conservation design for surveillance using sensor networks. In *Proceedings of the 4th International Symposium on Information Processing in Sensor Networks*, page 33. IEEE Press, 2005.

[129] S.Y. Wang, K.P. Shih, Y.D. Chen, and H.H. Ku. Preserving target area coverage in wireless sensor networks by using computational geometry. In *Wireless Communications and Networking Conference (WCNC), 2010 IEEE*, pages 1–6. IEEE, 2010.

[130] K.P. Shih, Y.D. Chen, C.W. Chiang, and B.J. Liu. A distributed active sensor selection scheme for wireless sensor networks. In *Computers and Communications, Proceedings of 11th IEEE Symposium on*, pages 923–928. IEEE, 2006.

[131] J. Chen, J. Jia, Y. Wen, D. Zhao, and J. Liu. Modeling and extending lifetime of wireless sensor networks using genetic algorithm. In *Proceedings of the First ACM/SIGEVO Summit on Genetic and Evolutionary Computation*, pages 47–54. ACM, 2009.

[132] H. Zhang, H. Wang, and H. Feng. A distributed optimum algorithm for target coverage in wireless sensor networks. In *Information Processing, Asia-Pacific Conference on*, Volume 2, pages 144–147. IEEE, 2009.

[133] H. Zhang. Energy-balance heuristic distributed algorithm for target coverage in wireless sensor networks with adjustable sensing ranges. In *Information Processing, Asia-Pacific Conference on*, Volume 2, pages 452–455. IEEE, 2009.

[134] A. Dhawan and S.K. Prasad. A distributed algorithmic framework for coverage problems in wireless sensor networks. *International Journal of Parallel, Emergent and Distributed Systems*, 24(4):331–348, 2009.

[135] Z. Hongwu, W. Hongyuan, F. Hongcai, L. Bing, and G. Bingxiang. A heuristic greedy optimum algorithm for target coverage in wireless sensor networks. In *Circuits, Communications and Systems, Pacific-Asia Conference on*, pages 39–42. IEEE, 2009.

[136] M. Naderan, M. Dehghan, and H. Pedram. A distributed dual-based algorithm for multi-target coverage in wireless sensor networks. In *Computer Networks and Distributed Systems, International Symposium on*, pages 204–209. IEEE, 2011.

[137] X. Xing, J. Li, and G. Wang. Integer programming scheme for target coverage in heterogeneous wireless sensor networks. In *Mobile Ad-hoc and Sensor Networks (MSN), Sixth International Conference on*, pages 79–84. IEEE, 2010.

[138] H. Yang, D. Li, and H. Chen. Coverage quality based target-oriented scheduling in directional sensor networks. In *Communications (ICC), 2010 IEEE International Conference on*, pages 1–5. 2010.

[139] H. Liu, W. Chen, H. Ma, and D. Li. Energy-efficient algorithm for the target q-coverage problem in wireless sensor networks. *Wireless Algorithms, Systems, and Applications*, pages 21–25, 2010.

[140] S. Meguerdichian, F. Koushanfar, G. Qu, and M. Potkonjak. Exposure in wireless ad-hoc sensor networks. In *Proceedings of the 7th Annual International Conference on Mobile Computing and Networking*, pages 139–150. ACM, 2001.

[141] X. Chu and H. Sethu. A new distributed algorithm for even coverage and improved lifetime in a sensor network. In *INFOCOM 2009, IEEE*, pages 361–369. 2009.

[142] J. Carle and D. Simplot-Ryl. Energy-efficient area monitoring for sensor networks. *Computer*, 37(2):40–46, 2004.

[143] M. Cardei and J. Wu. Energy-efficient coverage problems in wireless ad-hoc sensor networks. *Computer Communications*, 29(4):413–420, 2006.

[144] H. Zhang and J.C. Hou. Maintaining sensing coverage and connectivity in large sensor networks. *Ad Hoc & Sensor Wireless Networks*, 1(1-2):89–124, 2005.

[145] M. Ye, E. Chan, G. Chen, and J. Wu. Energy efficient fractional coverage schemes for low cost wireless sensor networks. In *Distributed Computing Systems Workshops, 26th IEEE International Conference on*, pages 79–79. 2006.

[146] J.R. Jiang and T.M. Sung. Maintaining connected coverage for wireless sensor networks. In *Distributed Computing Systems Workshops, 28th International Conference on*, pages 297–302. IEEE, 2008.

[147] L. Pan, J. Luo, and J. Li. Probing queries in wireless sensor networks. In *Distributed Computing Systems, 28th International Conference on*, pages 546–553. IEEE, 2008.

[148] J. Li, S. Ji, and J. Zhu. Data caching based queries in multi-sink sensor networks. In *Mobile Ad-hoc and Sensor Networks, 5th International Conference on*, pages 9–16. IEEE, 2009.

[149] N. Trigoni, Y. Yao, A. Demers, J. Gehrke, and R. Rajaraman. Multi-query optimization for sensor networks. *Distributed Computing in Sensor Systems*, pages 467–467, 2005.

[150] L. Xie, L. Chen, S. Lu, L. Xie, and D. Chen. Energy-efficient multi-query optimization over large-scale sensor networks. *Wireless Algorithms, Systems, and Applications*, pages 127–139, 2006.

[151] O. Chipara, C. Wu, C. Lu, and W. Griswold. Interference-aware real-time flow scheduling for wireless sensor networks. In *Real-Time Systems, 23rd Euromicro Conference on*, pages 67–77. IEEE, 2011.

[152] NA Vasanthi and S. Annadurai. Energy efficient sleep schedule for achieving minimum latency in query based sensor networks. In *Sensor Networks, Ubiquitous, and Trustworthy Computing, IEEE International Conference on*, Volume 2, pages 214–219. 2006.

[153] H. Wu, Q. Luo, and W. Xue. Distributed cross-layer scheduling for in-network sensor query processing. In *Pervasive Computing and Communications, Fourth Annual IEEE International Conference on*, pages 10–pp. IEEE, 2006.

[154] Y. Zheng, M. Liu, and J. Cao. Rsqs: resource-saving multi-query scheduling in wireless sensor networks. In *Asia-Pacific Services Computing Conference, IEEE*, pages 801–806. 2008.

[155] B. Yu, J. Li, and Y. Li. Distributed data aggregation scheduling in wireless sensor networks. In *INFOCOM 2009, IEEE*, pages 2159–2167. 2009.

[156] C. Lu, B.M. Blum, T.F. Abdelzaher, J.A. Stankovic, and T. He. Rap: A real-time communication architecture for large-scale wireless sensor networks. In *Real-Time and Embedded Technology and Applications Symposium*, pages 55–66. IEEE, 2002.

[157] O. Chipara, C. Lu, J.A. Stankovic, and G. Roman. Dynamic conflict-free transmission scheduling for sensor network queries. *Mobile Computing, IEEE Transactions on*, 10(5):734–748, 2011.

[158] H. Wu and Q. Luo. Adaptive holistic scheduling for query processing in sensor networks. *Journal of Parallel and Distributed Computing*, 70(6):657–670, 2010.

[159] S. Gopalakrishnan. Optimal schedules for sensor network queries. In *Real-Time Systems Symposium (RTSS)*, pages 140–149. IEEE, 2010.

[160] H. Wu, Q. Luo, J. Li, and A. Labrinidis. Quality aware query scheduling in wireless sensor networks. In *Proceedings of the Sixth International Workshop on Data Management for Sensor Networks*, page 7. ACM, 2009.

[161] Y. Yao and J. Gehrke. Query processing in sensor networks. CIDR, 2003.

[162] W. Wang, Y. Wang, X.Y. Li, W.Z. Song, and O. Frieder. Efficient interference-aware TDMA link scheduling for static wireless networks. In *Proceedings of the 12th Annual International Conference on Mobile Computing and Networking*, Volume 23, pages 262–273, 2006.

[163] Y. Yu, B. Krishnamachari, and V.K. Prasanna. Energy-latency tradeoffs for data gathering in wireless sensor networks. In *Annual Joint Conference of the IEEE Computer and Communications Societies*, Volume 1. 2004.

[164] X. Xu, X. Li, P. Wan, and S. Tang. Efficient scheduling for periodic aggregation queries in multiple sensor networks. *IEEE Transaction on Networking*, 2011.

[165] Z. Zhang and O. Berger. Cluster based data query analysis and optimization for wireless sensor networks. In *Advanced Communication Technology, 10th International Conference on*, Volume 2, pages 953–957. IEEE, 2008.

[166] C. Yanrong and C. Jiaheng. Power efficient data query processing protocol for wireless sensor networks. In *Wireless Communications, Networking and Mobile Computing, International Conference on*, pages 2372–2375. IEEE, 2007.

[167] K. Park, B. Lee, and R. Elmasri. Energy efficient spatial query processing in wireless sensor networks. In *Advanced Information Networking and Applications Workshops, 21st International Conference on*, Volume 2, pages 719–724. IEEE, 2007.

[168] C. Ai, L. Guo, Z. Cai, and Y. Li. Processing area queries in wireless sensor networks. In *Mobile Ad-hoc and Sensor Networks, Fifth International Conference on*, pages 1–8. IEEE, 2009.

[169] J. He, Z. Cai, S. Ji, R. Beyah, and Y. Pan. A genetic algorithm for constructing a reliable MCDS in probabilistic wireless networks. *Wireless Algorithms, Systems, and Applications*, pages 96–107, 2011.

[170] B. Yu and J. Li. Minimum-time aggregation scheduling in multi-sink sensor networks. In *Sensor, Mesh and Ad Hoc Communications and Networks, Eighth Annual IEEE Communications Society Conference on*, pages 422–430. 2011.

[171] S. Chen, M. Huang, S. Tang, and Y. Wang. Capacity of data collection in arbitrary wireless sensor networks. *Parallel and Distributed Systems, IEEE Transactions on*, 23(1):52–60, 2012.

[172] P. Gupta and P.R. Kumar. The capacity of wireless networks. *Information Theory, IEEE Transactions on*, 46(2):388–404, 2000.

[173] X.Y. Li, S.J. Tang, and O. Frieder. Multicast capacity for large scale wireless ad hoc networks. In *Proceedings of the 13th annual ACM International Conference on Mobile Computing and Networking*, pages 266–277. 2007.

[174] X.F. Mao, X.Y. Li, and S.J. Tang. Multicast capacity for hybrid wireless networks. In *Proceedings of the 9th ACM International Symposium on Mobile Ad Hoc Networking and Computing*, pages 189–198. 2008.

[175] U. Niesen, P. Gupta, and D. Shah. On capacity scaling in arbitrary wireless networks. *Information Theory, IEEE Transactions on*, 55(9):3959–3982, 2009.

[176] U. Niesen, P. Gupta, and D. Shah. The balanced unicast and multicast capacity regions of large wireless networks. *Information Theory, IEEE Transactions on*, 56(5):2249–2271, 2010.

[177] X.Y. Li, J. Zhao, Y.W. Wu, S.J. Tang, X.H. Xu, and X.F. Mao. Broadcast capacity for wireless ad hoc networks. In *Mobile Ad Hoc and Sensor Systems, 5th IEEE International Conference on*, pages 114–123. 2008.

[178] S. Chen, Y. Wang, X.Y. Li, and X. Shi. Data collection capacity of random-deployed wireless sensor networks. In *Global Telecommunications Conference, IEEE*, pages 1–6. 2009.

[179] G. Sharma, R. Mazumdar, and N.B. Shroff. Delay and capacity trade-offs in mobile ad hoc networks: A global perspective. *IEEE/ACM Transactions on Networking (TON)*, 15(5):981–992, 2007.

[180] D. Lymberopoulos, N.B. Priyantha, M. Goraczko, and F. Zhao. Towards energy efficient design of multi-radio platforms for wireless sensor networks. In *Information Processing in Sensor Networks, International Conference on*, pages 257–268. IEEE, 2008.

[181] P. Kyasanur and N.H. Vaidya. Capacity of multi-channel wireless networks: impact of number of channels and interfaces. In *Proceedings of the 11th Annual International Conference on Mobile Computing and Networking*, pages 43–57. ACM, 2005.

[182] V. Bhandari and N.H. Vaidya. Connectivity and capacity of multi-channel wireless networks with channel switching constraints. In *26th IEEE International Conference on Computer Communications*, pages 785–793. IEEE, 2007.

[183] H.N. Dai, K.W. Ng, R.C.W. Wong, and M.Y. Wu+. On the capacity of multi-channel wireless networks using directional antennas. In *27th Conference on Computer Communications*, pages 628–636. IEEE, 2008.

[184] W. Cheng, X. Cheng, T. Znati, X. Lu, and Z. Lu. The complexity of channel scheduling in multi-radio multi-channel wireless networks. In *INFOCOM 2009*, pages 1512–1520. IEEE, 2009.

[185] P. Bahl, R. Chandra, and J. Dunagan. Ssch: slotted seeded channel hopping for capacity improvement in IEEE 802.11 ad-hoc wireless networks. In *Proceedings of the 10th Annual International Conference on Mobile Computing and Networking*, pages 216–230. ACM, 2004.

[186] D. Chafekar, D. Levin, V.S.A. Kumar, M.V. Marathe, S. Parthasarathy, and A. Srinivasan. Capacity of asynchronous random-access scheduling in wireless networks. In *27th Conference on Computer Communications. IEEE*, pages 1148–1156. 2008.

[187] M. Andrews and M. Dinitz. Maximizing capacity in arbitrary wireless networks in the sinr model: Complexity and game theory. In *INFOCOM 2009, IEEE*, pages 1332–1340. 2009.

[188] C.K. Chau, M. Chen, and S.C. Liew. Capacity of large-scale CSMA wireless networks. In *Proceedings of the 15th Annual International Conference on Mobile Computing and Networking*, pages 97–108. ACM, 2009.

[189] O. Goussevskaia, R. Wattenhofer, M.M. Halldórsson, and E. Welzl. Capacity of arbitrary wireless networks. In *INFOCOM 2009, IEEE*, pages 1872–1880. 2009.

[190] S. Li, Y. Liu, and X.Y. Li. Capacity of large scale wireless networks under gaussian channel model. In *Proceedings of the 14th ACM International Conference on Mobile Computing and Networking*, pages 140–151. 2008.

[191] Z. Wang, H.R. Sadjadpour, and J.J. Garcia-Luna-Aceves. A unifying perspective on the capacity of wireless ad hoc networks. In *27th Conference on Computer Communications. IEEE*, pages 211–215. 2008.

[192] W. Huang, X. Wang, and Q. Zhang. Capacity scaling in mobile wireless ad hoc network with infrastructure support. In *Distributed Computing Systems (ICDCS), 30th International Conference on*, pages 848–857. IEEE, 2010.

[193] G. Zhang, Y. Xu, X. Wang, and M. Guizani. Capacity of hybrid wireless networks with directional antenna and delay constraint. *Communications, IEEE Transactions on*, 58(7):2097–2106, 2010.

[194] VS Kumar, M.V. Marathe, S. Parthasarathy, and A. Srinivasan. Algorithmic aspects of capacity in wireless networks. In *ACM SIGMETRICS Performance Evaluation Review*, Volume 33, pages 133–144. 2005.

[195] A. Keshavarz-Haddad, V. Ribeiro, and R. Riedi. Broadcast capacity in multi-hop wireless networks. In *Proceedings of the 12th Annual International Conference on Mobile Computing and Networking*, pages 239–250. ACM, 2006.

[196] B. Liu, D. Towsley, and A. Swami. Data gathering capacity of large scale multihop wireless networks. In *Mobile Ad Hoc and Sensor Systems, Fifth IEEE International Conference on*, pages 124–132. 2008.

[197] E.J. Duarte-Melo and M. Liu. Data-gathering wireless sensor networks: organization and capacity. *Computer Networks*, 43(4):519–537, 2003.

[198] D. Marco, E. Duarte-Melo, M. Liu, and D. Neuhoff. On the many-to-one transport capacity of a dense wireless sensor network and the compressibility of its data. In *Information Processing in Sensor Networks*, pages 556–556. Springer, 2003.

[199] H. El Gamal. On the scaling laws of dense wireless sensor networks. In *Proceedings of the Annual Allerton Conference on Communication Control and Computing*, volume 41, pages 1393–1401. 2003.

[200] X. Wang, Y. Bei, Q. Peng, and L. Fu. Speed improves delay-capacity trade-off in motioncast. *Parallel and Distributed Systems, IEEE Transactions on*, 22(5):729–742, 2011.

[201] M. Franceschetti, O. Dousse, NC David, and P. Thiran. Closing the gap in the capacity of wireless networks via percolation theory. *Information Theory, IEEE Transactions on*, 53(3):1009–1018, 2007.

[202] X. Zhu, B. Tang, and H. Gupta. Delay efficient data gathering in sensor networks. *Mobile Ad-hoc and Sensor Networks*, pages 380–389, 2005.

[203] S. Chen, Y. Wang, X.Y. Li, and X. Shi. Order-optimal data collection in wireless sensor networks: delay and capacity. In *Sensor, Mesh and Ad Hoc Communications and Networks, 2009. SECON'09. 6th Annual IEEE Communications Society Conference on*, pages 1–9. 2009.

[204] T. Moscibroda. The worst-case capacity of wireless sensor networks. In *Information Processing in Sensor Networks, 6th International Symposium on*, pages 1–10. IEEE, 2007.

[205] Z. Wang, H. Sadjadpour, and J.J. Garcia-Luna-Aceves. The capacity and energy efficiency of wireless ad hoc networks with multi-packet reception. In *Proceedings of the 9th ACM International Symposium on Mobile Ad Hoc Networking and Computing*, pages 179–188. 2008.

[206] Y. Xu and W. Wang. Scheduling partition for order optimal capacity in large-scale wireless networks. In *Proceedings of the 15th Annual International Conference on Mobile Computing and Networking*, pages 109–120. ACM, 2009.

[207] M. Garetto, P. Giaccone, and E. Leonardi. On the capacity of ad hoc wireless networks under general node mobility. In *26th IEEE International Conference on Computer Communications*, pages 357–365. 2007.

[208] A. Ghosh, O.D. Incel, V.S.A. Kumar, and B. Krishnamachari. Multi-channel scheduling algorithms for fast aggregated convergecast in sensor networks. In *Mobile Adhoc and Sensor Systems, IEEE 6th International Conference on*, pages 363–372. 2009.

[209] M. Alicherry, R. Bhatia, and L.E. Li. Joint channel assignment and routing for throughput optimization in multi-radio wireless mesh networks. In *Proceedings of the 11th Annual International Conference on Mobile Computing and Networking*, pages 58–72. ACM, 2005.

[210] B. Han, V.S.A. Kumar, M.V. Marathe, S. Parthasarathy, and A. Srinivasan. Distributed strategies for channel allocation and scheduling in software-defined radio networks. In *INFOCOM 2009, IEEE*, pages 1521–1529. 2009.

[211] X. Lin and S.B. Rasool. Distributed and provably efficient algorithms for joint channel-assignment, scheduling, and routing in multichannel ad hoc wireless networks. *Networking, IEEE/ACM Transactions on*, 17(6):1874–1887, 2009.

[212] V. Bhandari and N.H. Vaidya. Capacity of multi-channel wireless networks with random (c, f) assignment. In *Proceedings of the 8th ACM International Symposium on Mobile Ad Hoc Networking and Computing*, pages 229–238. 2007.

[213] V. Ramamurthi, S.K.C. Vadrevu, A. Chaudhry, and M.R. Bhatnagar. Multicast capacity of multi-channel multihop wireless networks. In *Wireless Communications and Networking Conference, IEEE*, pages 1–6. 2009.

[214] Y. Li, M.T. Thai, F. Wang, and D.Z. Du. On the construction of a strongly connected broadcast arborescence with bounded transmission delay. *Mobile Computing, IEEE Transactions on*, 5(10):1460–1470, 2006.

[215] D.W. Matula and L.L. Beck. Smallest-last ordering and clustering and graph coloring algorithms. *Journal of the ACM (JACM)*, 30(3):417–427, 1983.

[216] S. Ji, Z. Cai, Y. Li, and X. Jia. Continuous data collection capacity of dual-radio multichannel wireless sensor networks. *Parallel and Distributed Systems, IEEE Transactions on*, 23(10):1844–1855, 2012.

[217] S. Ji, J. (S.) He, A. S. Uluagac, R. Beyah, and Y. Li. Cell-based snapshot and continuous data collection in wireless sensor networks. *ACM Transactions on Sensor Networks*, 2012.

[218] C. Wang, C. Jiang, Y. Liu, X.Y. Li, S. Tang, and H. Ma. Aggregation capacity of wireless sensor networks: Extended network case. In *INFOCOM 2011 Proceedings*, pages 1701–1709. IEEE, 2011.

[219] C. Wang, C. Jiang, X.Y. Li, S. Tang, and P. Yang. General capacity scaling of wireless networks. In *INFOCOM 2011 Proceedings, IEEE*, pages 712–720. 2011.

[220] Y. Wang, X. Chu, X. Wang, and Y. Cheng. Optimal multicast capacity and delay tradeoffs in manets: A global perspective. In *INFOCOM 2011 Proceedings, IEEE*, pages 640–648. 2011.

[221] P. Li, X. Huang, and Y. Fang. Capacity scaling of multihop cellular networks. In *INFOCOM 2011 Proceedings, IEEE*, pages 2831–2839. 2011.

[222] H. Li, Y. Cheng, P.J. Wan, and J. Cao. Local sufficient rate constraints for guaranteed capacity region in multi-radio multi-channel wireless networks. In *INFOCOM 2011 Proceedings, IEEE*, pages 990–998. 2011.

[223] P. Li, M. Pan, and Y. Fang. The capacity of three-dimensional wireless ad hoc networks. In *INFOCOM 2011 Proceedings, IEEE*, pages 1485–1493. 2011.

[224] E.I. Asgeirsson and P. Mitra. On a game theoretic approach to capacity maximization in wireless networks. In *INFOCOM 2011 Proceedings, IEEE*, pages 3029–3037. 2011.

[225] K. Lee, Y. Kim, S. Chong, I. Rhee, and Y. Yi. Delay-capacity tradeoffs for mobile networks with lévy walks and lévy flights. In *INFOCOM 2011 Proceedings, IEEE*, pages 3128–3136. 2011.

[226] L. Fu, S.C. Liew, and J. Huang. Effective carrier sensing in csma networks under cumulative interference. In *INFOCOM 2010 Proceedings, IEEE*, pages 1–9. 2010.

[227] Y. Li, L. Guo, and S.K. Prasad. An energy-efficient distributed algorithm for minimum-latency aggregation scheduling in wireless sensor networks. In *Distributed Computing Systems (ICDCS), 2010 IEEE 30th International Conference on*, pages 827–836. 2010.

[228] X.H. Xu, S.G. Wang, X.F. Mao, S.J. Tang, and X.Y. Li. An improved approximation algorithm for data aggregation in multi-hop wireless sensor networks. In *Proceedings of the 2nd ACM International Workshop on Foundations of Wireless Ad Hoc and Sensor Networking and Computing*, pages 47–56. 2009.

[229] Federal Communications Commission. ET Docket 03-237. Notice of Proposed Rule Making and Order. 2003.

[230] I.F. Akyildiz, W.Y. Lee, M.C. Vuran, and S. Mohanty. Next generation/dynamic spectrum access/cognitive radio wireless networks: a survey. *Computer Networks*, 50(13):2127–2159, 2006.

[231] R. Gandhi, S. Parthasarathy, and A. Mishra. Minimizing broadcast latency and redundancy in ad hoc networks. In *Proceedings of the 4th ACM International Symposium on Mobile Ad Hoc Networking and Computing*, pages 222–232. 2003.

[232] S.C.H. Huang, P.J. Wan, J. Deng, and Y.S. Han. Broadcast scheduling in interference environment. *Mobile Computing, IEEE Transactions on*, 7(11):1338–1348, 2008.

[233] Y. Song and J. Xie. A distributed broadcast protocol in multi-hop cognitive radio ad hoc networks without a common control channel. In *INFOCOM 2012 Proceedings, IEEE*, pages 2273–2281. 2012.

[234] Y.R. Kondareddy and P. Agrawal. Selective broadcasting in multi-hop cognitive radio networks. In *Sarnoff Symposium, IEEE*, pages 1–5. 2008.

[235] O. Ertuğrul and F. Buzluca. An efficient broadcasting scheme for cognitive radio ad hoc networks. In *Proceedings of the 4th International Conference on Cognitive Radio and Advanced Spectrum Management*, page 5. ACM, 2011.

[236] C. Liyana Arachchige, S. Venkatesan, R. Chandrasekaran, and N. Mittal. Minimal time broadcasting in cognitive radio networks. *Distributed Computing and Networking*, pages 364–375, 2011.

[237] R. Gandhi, A. Mishra, and S. Parthasarathy. Minimizing broadcast latency and redundancy in ad hoc networks. *IEEE/ACM Transactions on Networking (TON)*, 16(4):840–851, 2008.

[238] S. Huang, P.J. Wan, X. Jia, and H. Du. Low-latency broadcast scheduling in ad hoc networks. *Wireless Algorithms, Systems, and Applications*, pages 527–538, 2006.

[239] S.C.H. Huang, P.J. Wan, X. Jia, H. Du, and W. Shang. Minimum-latency broadcast scheduling in wireless ad hoc networks. In *INFOCOM 2007. 26th IEEE International Conference on Computer Communications*, pages 733–739. 2007.

[240] R. Gandhi, Y.A. Kim, S. Lee, J. Ryu, and P.J. Wan. Approximation algorithms for data broadcast in wireless networks. *Mobile Computing, IEEE Transactions on*, 11(7):1237–1248, 2012.

[241] Z. Chen, C. Qiao, J. Xu, and T. Taekkyeun Lee. A constant approximation algorithm for interference-aware broadcast in wireless networks. In *INFOCOM 2007. 26th IEEE International Conference on Computer Communications*, pages 740–748. 2007.

[242] R. Mahjourian, F. Chen, R. Tiwari, H. Zhai, and Y. Fang. An approximation algorithm for conflict-aware broadcast scheduling in wireless ad hoc networks. In *Proceedings of the 9th ACM International Symposium on Mobile Ad Hoc Networking and Computing*, pages 331–340. 2008.

[243] S.C.H. Huang, S.Y. Chang, H.C. Wu, and P.J. Wan. Analysis and design of a novel randomized broadcast algorithm for scalable wireless networks in the interference channels. *Wireless Communications, IEEE Transactions on*, 9(7):2206–2215, 2010.

[244] P. Wang, I.F. Akyildiz, and A.M. Al-Dhelaan. Percolation theory based connectivity and latency analysis of cognitive radio ad hoc networks. *Wireless Networks*, 17(3):659–669, 2011.

[245] W. Ren, Q. Zhao, and A. Swami. On the connectivity and multihop delay of ad hoc cognitive radio networks. *Selected Areas in Communications, IEEE Journal on*, 29(4):805–818, 2011.

[246] J.F.C. Kingman. *Poisson Processes*, Volume 3. Oxford University Press, 1993.

[247] G. Wegner. Uber endliche kreispackungen in der ebene. *Studia Sci. Math. Hung*, 21:1–28, 1986.

[248] R.A. Leese. A unified approach to the assignment of radio channels on a regular hexagonal grid. *Vehicular Technology, IEEE Transactions on*, 46(4):968–980, 1997.

Index

Milton Keynes UK
Ingram Content Group UK Ltd.
UKHW031126141024
449569UK00006B/406